Ricki Lewis

Instructor's Manual with Test Item File to accompany Life

Heather R. McKean
Eastern Washington University

James L. Hanegan
Eastern Washington University

D1450728

WCB Wm. C. Brown Publishers

Contents

Preface

Using the Text as a Road Map

Life is an instructional aid, not a scenario of the course. The text lends itself to higher levels of understanding by providing interesting examples and problems that illustrate biological principles. Scientific method is integrated throughout the text to support you in demonstrating how science works. Though factual knowledge is well covered, it is deemphasized by Ricki Lewis's narrative style, which makes reading more enjoyable and increases understanding. Objectives and key concepts are clearly identified by setting them off in the front of each chapter and after each section. The questions at the end of each chapter entitled "To Think About" prod the student to integrate what he or she has learned and to think in a scientific way. A list of suggested readings at the end of each chapter is provided for students seeking further information for term papers or just for interest.

Use of the Instructor's Manual

This *Instructor's Manual* provides an annotated version of the key concepts for each chapter that have been prepared in outline form to give you a quick overview of the chapter. It is also available as software in WordPerfect ^{T} format entitled *Extended Lecture Outline,* which gives you the flexibility of a prepared outline with the editorial capabilities of a software program. The terms bold faced in the text are listed for convenience. Answers are provided for the end-of-chapter "Questions" and the "To Think About" questions. These questions are particularly helpful in getting students to understand the meaning of biological principles.

The testbank questions are drawn primarily from the body of each chapter, excluding, for the most part, the "Readings." They include simple factual as well as application questions. Approximately one question was taken from each major paragraph, allowing for full coverage of the material. Page numbers have been provided to assist in selection of questions for writing examinations. It is not necessary that students complete the questions at the end of the chapter to do well on the testbank questions, but they are helpful study guides. These questions, along with those in the *Student Study Guide*, provide a good review of the material covered.

WCB TestPak*

WCB TestPak, a computerized testing service, provides instructions for either a mail-in/call-in testing program or the complete test item file on diskette for use with the IBM PC, Apple, or Macintosh computer. WCB TestPak requires no programming experience.

WCB QuizPak, a part of TestPak, provides an additional 40–50 multiple choice questions for each chapter that can be used for student tutorials. Using this portion of the program will help students to prepare for examinations. Also included with the WCB QuizPak is an on-line testing option to allow professors to prepare tests for students to take on the computer. The computer will automatically grade the test and update a gradebook file.

WCB GradePak, also a part of TestPak, is a computerized grade management system for instructors. This program tracks student performance on examinations and assignments. It will compute each student's percentage and corresponding letter grade, as well as the class average. Printouts can be made utilizing both text and graphics.

*ISBN *Extended Lecture Outline*		*TestPak	
Mac 14294		Mac 10176	
IBM 5.25	14291	IBM 5.25	10174
Apple 3.5	14292	IBM 3.5	10173
Apple 5.25	14293	Apple 3.5	13958
IBM 3.5	10173	Apple 5.25	10175

List of Transparencies/Slides

1 Thinking Scientifically

KEY CONCEPTS

Biology is the study of the unity and diversity
  of life.
Science is a cycle of inquiry using the scientific
  method.
The scientific method includes making
  observations, formulating a hypothesis,
  designing an experiment, collecting and
  interpreting data, and reaching conclusions.

KEY TERMS

double-blind   hypothesis
epidemiology   placebo
experimental control   scientific method

CHAPTER OUTLINE

I. Biology and You
   A. Biology is the study of the unity and
      diversity of life.
   B. Understanding the similarities and
      differences between types of organisms
      can provide a broader perspective on
      human experience.
II. A Biologist's View of the Living World
    A. The scientist follows a systematic
       approach to cataloging and
       interpreting observations of life called
       the scientific method.
    B. The scientific method involves
       observing, reasoning, predicting,
       testing, concluding, and interpreting.
    C. The scientific method (fig. 1.3).
       1. Observations.
          a) The first step of the scientific
             method.
          b) Jim Schlatter in 1965 made
             the observation that a
             chemical he was testing had
             a sweet taste.
       2. Hypothesis.
          a) An educated guess made
             after observation.
          b) Jim Schlatter hypothesized
             that if the small protein
             was low in calories and
             harmless, it could be used
             for a low-calorie sweetener.
       3. Experiment.
          a) Sample size considerations.
             (1) A large sample size
                 ensures that results
                 are statistically
                 significant.
             (2) Conclusions cannot
                 accurately be based on
                 fewer than 30 samples.
          b) Experimental control.
             (1) Efforts are made to
                 eliminate the effects of
                 more than one variable.
             (2) A placebo is a substance
                 that is similar to the
                 experimental substance
                 being tested but lacks
                 the effect.
             (3) In a double-blind
                 experiment neither the
                 researchers nor the
                 subjects know who has
                 been given the
                 substance being
                 evaluated and who has
                 been given the placebo.

c) Jim Schlatter's experiment involved giving the substance to one group of mice while giving a placebo to another group and then measuring for any differences in development (fig. 1.2).

D. Science is a cycle of inquiry.
   1. A conclusion must be placed into perspective with existing knowledge.
   2. An underlying characteristic of scientific thinking is that the sequence of observing, reasoning, predicting, testing, concluding, and interpreting is a cycle, with new ideas spawned at every step.

E. Designing experiments.
   1. The science of biology is a collection of facts that describes and explains the workings of the world.
   2. Types of experiments.
      a) Animal model experiments.
         (1) Nonhumans are exposed to treatments and situations that might provoke ethical objections if they were conducted on humans, and the results are extrapolated to humans (figs. 1.4-1.6).
         (2) They are sometimes used to examine biological phenomena simply for the sake of expanding our knowledge and with no immediate intention of extending the results to humans.
         (3) Some are used to assess the value of continuing a particular line of investigation on humans.

      b) Epidemiological studies.
         (1) Epidemiology is the analysis of data derived from real-life situations ranging from one-person accounts to large-scale studies involving thousands of individuals.
         (2) One type of study pairs individuals of like characteristics such as sex, race, age, diet, and occupation, and then compares differences by one factor.
         (3) In some situations, epidemiological information can only demonstrate a correlation and not a cause-and-effect relationship between two observations.

F. Limitations of the scientific method.
   1. Interpreting results.
      a) There may not be a sufficient number of experimental subjects available.
      b) Evidence may lead to an unexpected or multiple conclusion, failing to provide a definitive answer.
      c) Observations or experimental results can be misinterpreted.
      d) It may be difficult to maintain objectivity in drawing conclusions.
   2. When the scientific method is unethical.
      a) Testing vaccines may be unethical as in the case of Edward Jenner, who took a great risk in testing his smallpox vaccine on children.

2

b) Withholding the vaccine treatment for known AIDS patients who are given a placebo instead may be unethical since it was believed that the vaccine would work.

3. Expect the unexpected.
   a) The investigator needs to keep an open mind toward observations and not allow biases or expectations to cloud the interpretations of results.
   b) Sometimes the results of experiments are ignored as in the case of Barbara McClintock, whose idea that genes were capable of jumping from one chromosome was ignored by the scientific community for decades.

4. Scientific discovery by "accident" (fig. 1.7).
   a) Alexander Fleming realized that agents in the air kept bacteria from growing, which lead to the discovery of penicillin (fig. 1.8).
   b) The cause of phenylketonuria disease was discovered by a chemist by analyzing the urine of his retarded relatives after their mother noticed it smelled unusual.

III. We All Think Like Scientists at Times
   A. The cycle of questioning and answering is second nature to a practicing biologist.
   B. The steps of the scientific method can be used everyday to answer questions about ourselves and the environment.

LEARNING OBJECTIVES

After reading this chapter, the student should be able to answer these questions:

1. What are some of the ways in which biology affects you?
2. How do scientists use the scientific method to explain biological phenomena?
3. How are scientific experiments designed?
4. What are the limitations of the scientific method?
5. How do you use the scientific method in your everyday life?

ANSWERS

"Questions"

1. *Observations:* complex chemicals can be formed under specific laboratory conditions in the presence of electrical sparks and heat; hot vents in the ocean floor contain simple chemicals like those used in Miller's experiment.
   Hypothesis: Deep sea vents are too hot to form and maintain stable complex chemicals.
   Experiment: Measure the formation and stability of complex chemicals formed in temperatures like those of sea vents.
   Collection of data: Complex chemicals required for life did form under these conditions of temperature but disintegrated immediately.
   Conclusion: Life could not have arisen in oceanic hot spots.

2. a. The use of a placebo allows the experimental and control groups to be treated in the same way by providing a substance to the tester that is undistinguishable from the test substance.
   b. Withholding information allows both the experimenter and the subjects to remain unbiased in their judgments as to the effectiveness of the test substance since neither knows which group is which.
   c. Two large populations that are identified to be similar in profile (age, sex, nationality, etc.) are given a sample of what they believe is the test sample. One group actually receives the test substance while the other is given the placebo. The experimenter records the results reported by each individual by a code number without knowing what substance was taken.
   d. Because of the unpleasant side effect, the individual taking DMSO will likely know that he or she is not taking the placebo. This will be obvious to the experimenter, who may be biased when collecting the data.
3. Individuals may be biased in their judgement or actually lie, looking for a correlation between exposure to the chemical and symptoms they may or may not have. At best this study might serve to indicate whether serious studies should be done in this area.
4. A compromise to using live or sacrifice animals is to use parts of organisms grown in the laboratory.
5. You should have identified the following components: observation, hypothesis, experimental design, collection of data, and interpretation.

"To Think About"

1. The study is invalid because there were no controls and the sample size of one is inadequate.
2. The extrapolation to the normal population from data collected on a group of individuals, with a specific problem related to the study, is inappropriate. The hypercholesterolemic individuals may respond differently than the average person due to the disease.
3. No background information was given on the individuals tested. The results of the genetic examination may be a reflection of hereditary or genetic damage caused by agents unrelated to Love Canal. No controls were used and the sample size was inadequate. All that can be concluded from this study is that a more serious study should be done.
4. Though this study meets all criteria of the scientific method, the conclusion should be qualified by adding that vasectomy does not increase the risk of heart disease for the average person.
5. Assuming that the control patients are also young, strong, and healthy, the study itself as a preliminary study is appropriate in design. It is the interpretation by the news media that is in error. They inappropriately extrapolated the experimental group data to the general population.
6. Since the research is individualized, the sample size is inadequate. Also, since the selection of candidates is by those who can pay, the sampling may be biased. It is conceivable that wealthy individuals may have led a different life-style than the average person. Extrapolation of the results of these studies to the general public should be done cautiously.

AUDIOVISUAL MATERIALS

Aristotle and the Scientific Method, Coronet
    Instructional Films, 14 min.
Biology: Exploring the Living World,
    Encyclopaedia Britannica Educational Corp.,
    19 min.
Nature of Science—Forming Hypotheses, Coronet
    Instructional Films, 16 min.

Nature of Science—How Ideas Change, Coronet
    Instructional Films, 14 min.
Nature of Science—Obtaining Facts, Coronet
    Instructional Films, 11 min.
Nature of Science—Testing Hypotheses, Coronet
    Instructional Films, 14 min.
Scientific Methods and Values, Human Relations
    Media, filmstrip.

2 The Diversity of Life

KEY CONCEPTS

Organisms interact with each other in complex chains and webs.

Evidence indicates that organic molecules gave rise to prokaryotes, which gave rise to simple eukaryotes, which gave rise to the complex eukaryotes.

Diversity of life reveals adaptations to the biotic and abiotic world.

Observations of similarities and differences among organisms give insight into relatedness.

Taxonomic systems organize living things into categories of relatedness.

Organisms are classified into kingdoms according to the complexity of their cells and their strategies for meeting life's challenges.

KEY TERMS

algae	metazoa
alteration of	Monera
generations	mutualism
amino acids	mycelium
angiosperms	notochord
Animalia	nucleus
ascomycetes	organelles
autotrophic	parasitism
bacteria	phloem
basidiomycetes	photosynthesize
binary fission	pilus
bryophytes	placenta
chlorophyll	Plantae
coelom	prokaryotic cells
commensalism	proteins
cyanobacteria	Protista
eukaryotic cells	protozoans
Fungi	species
gametes	sporophyte
gametophyte	symbiosis
gymnosperms	taxonomy
heterotrophic	tracheophytes
hypha	xylem
lichen	zygomycetes

CHAPTER OUTLINE

I. Journey in the Present
   A. A living world of contrasts.
      1. Living organisms are found everywhere, including unlikely places.
      2. Microorganisms are found on hot desert rocks and in cracks in the ocean floor known as thermal vents.
   B. Life within life.
      1. Life often exists on or even in other organisms, forming a relationship called symbiosis.
      2. Types of symbiotic relationships.
         a) Mutualism.
            (1) Both partners benefit, and, in some cases, one might not be able to survive without the other.
            (2) An example is the acacia tree that grows in Mexico and is covered with ants.
         b) Commensalism.
            (1) One member of the symbiotic pair benefits without affecting the other member.
            (2) An example is the human body being a host to many microorganisms that do not threaten our health.
         c) Parasitism.
            (1) The parasite (usually small by comparison) derives benefit from the host while harming but not usually killing it.
            (2) Examples include acne and athlete's foot.

C. Life affects life.
  1. All life is connected by complex chains and webs of "who eats whom."
  2. The importance of relationships between organisms becomes apparent when disaster strikes, upsetting the balance of ecosystems.
II. Journey Through the Past (table 2.1)
  A. 4 billion years ago (fig. 2.4):
    1. The land was dotted with volcanoes.
    2. The thin atmosphere allowed high levels of ultraviolet light to reach the surface of the planet.
  B. 3 billion years ago:
    1. Nearly every part of the planet was inhabited by microscopic organisms.
    2. This was the age of bacteria.
  C. 1 billion years ago:
    1. Some microorganisms were complex in organization, containing specific structures called organelles that carry out specific functions.
    2. These first eukaryotic cells aggregated to form colonies.
  D. 500 million years ago: Distant ancestors of modern-day clams, squids, trilobites, and the ancestors of insects arrived.
  E. 300 million years ago:
    1. Life appeared on land.
    2. There were forests of ferns and primitive trees, dragonflies, grasshoppers, and crickets, all giant versions of their present-day relatives.
  F. 240 million years ago:
    1. Development of watertight skin and eggs made living and breeding solely on land possible.
    2. Gradually, the age of the dinosaurs, or "terrible lizards," dawned.
    3. Insects, the earliest mammals, spent 135 million years underfoot of the reptilian monsters, and the age of mammals was just around the corner.
  G. 180 million years ago:
    1. The giant reptiles vanished.
    2. The temperature plummeted.
    3. The small, furry, insect-eating mammals were abundant.
  H. 30 million years ago:
    1. The great plains and forests were full of life.
    2. The treetops were home to the ancestors of modern monkeys.
III. The Science of Biological Classification—Taxonomy
  A. Characteristics to consider (fig. 2.5).
    1. The branch of biology dealing with classification, called taxonomy, attempts to classify life forms by assigning a different name to each kind of organism.
    2. A taxonomic classification is based upon a set of distinguishing characteristics that reflects an organism's evolutionary descent from an ancestor.
    3. A taxonomic "name" provides information about an organism's evolutionary background.
    4. Criteria for classification.
      a) Structural, biochemical, and behavioral similarities as well as habitat preferences.
      b) Each kind of organism described by a taxonomic name is a species—a group of similar organisms that interbreed in nature and are reproductively isolated from all other such groups.
      c) It can be difficult to tell whether similar characteristics among different species represent a common ancestor, or if they developed the trait independently in response to a shared environmental challenge.
      d) Molecular approach.

(1) Changes in the sequence of amino acids in a protein reflect changes in the organism's genetic material.

(2) The comparison of sequences of amino acids between two species may reveal divergence from a common ancestor.

(3) The degrees of difference in protein sequence are translated into "tree" diagrams, which make it possible to estimate the times at which those species diverged.

(4) The relationships and time estimates that result from comparing such sequences are called a "molecular clock."

B. Problems in taxonomy.

1. Taxonomists classify differently.

a) Investigators who perceive many differences among related kinds of organisms and thus assign them to many taxonomic groups are termed "splitters."

b) Those who see few important differences and assign them to fewer groups are called "lumpers."

2. Assignment of relative value of a characteristic for determining relatedness between species is questionable.

3. Classification according to structural characteristics can also be confounded by organisms that look different at different stages of their life cycles.

C. The Linnaean system of biological classification.

1. The modern system of classifying life was introduced in the eighteenth century by the Swedish botanist Carolus Linnaeus.

2. The taxonomic scheme.

a) A series of organizational levels is used to describe the ancestry and characteristics of organisms that are increasingly restrictive.

b) The levels proceed from the most general, kingdom, to division (in plants) or phylum (in all other organisms), to class, order, family, genus, and finally species (table 2.3).

3. The prefix "super" extends the limits of the designation, and the prefix "sub" restricts the limits.

4. Scientists usually abbreviate an organism's taxonomic position to just the genus and species.

IV. A Look at the Kingdoms

A. Kingdom classifications.

1. Taxonomists disagree on whether there should be three-, four-, five-, or even six-kingdom classifications.

2. The members of each kingdom demonstrate a slightly different strategy in satisfying the same life requirements.

3. A basis for modern classification schemes is on whether an organism is a prokaryotic or eukaryotic type.

a) Prokaryotes lack organelles, and their genetic material is not enclosed in a membrane.

b) Eukaryotic cells have membrane-bound organelles, which are the structures that specialize in certain subcellular functions, such as energy production and secretion, with the genetic material enclosed in the nucleus.

4. The most common five-kingdom system provides detailed descriptions: Monera, Protista, Fungi, Plantae, and Animalia (fig 2.7).

8

5. Organisms are assigned to their kingdoms on the basis of the complexity of their cells, whether they are unicellular or multicellular, and the ways in which they obtain nourishment.

B. Monera.
  1. Bacteria.
    a) These are single-celled prokaryotic organisms.
    b) They convert themselves into structures called spores that are resistant to extremes of temperature and moisture.
    c) Criteria for classification (fig. 2.8).
      (1) Shape is defined by rods or cocci.
      (2) Energy source may be obtained directly from inorganic molecules for autotrophic or from other organisms in the environment.
        (*a*) Autotrophic bacteria convert inorganic compounds obtained from the environment into energy-rich organic compounds.
        (*b*) Heterotrophic bacteria obtain organic compounds from dead organisms or from the living and function as decomposers.
        (*c*) When they use a living organism as a food source, heterotrophic bacteria are parasites.
    d) Reproduction may be asexual in the form of binary fission or sexual by means of passing genetic material through a pilus.

  2. Cyanobacteria.
    a) One-celled organisms were once called blue-green algae because of their pigmentation, which allows them to photosynthesize.
    b) Different species of cyanobacteria are distinguished by the particular patterns in which the individual cells aggregate.
    c) They release oxygen gas, which is an energy-obtaining strategy for the atmosphere.

C. Protista.
  1. This kingdom is believed to have given rise to the other three kingdoms of eukaryotes.
  2. Most members are unicellular, but some form many-celled colonies and some are truly multicellular.
  3. The protozoans.
    a) These "first animals" are single-celled.
    b) Most live in fresh or salt water or in very moist terrestrial habitats.
    c) They are sometimes classified by their method of movement.
      (1) The euglena waves a taillike flagellum.
      (2) The amoeba molds itself into footlike projections called pseudopods (fig. 2.10).
    d) Algae photosynthesize.
      (1) The members of three phyla, the euglenoids, the dinoflagellates and the diatoms, are single-celled.
      (2) The red algae and brown algae consist entirely of multicellular species, and the green algae have both single-celled and multicellular members (fig. 2.11).

e) Water molds.
  (1) Most notorious is *Phytophthora infestans,* which caused the potato famine in Ireland (fig. 2.12).
  (2) Chytrids are distinguished by their single flagellum and by cell walls built of chitin.
  (3) The major difference between the two phyla of slime molds is reflected in the popular names—acellular and cellular.
  (4) The slime molds are fascinating because they can switch body form in response to environmental changes (fig. 2.13).

D. Fungi.
  1. Most are land-dwelling but a few forms are marine.
  2. They are decomposers, recycling the chemicals in once-living matter to the environment, where they can be used by other organisms.
  3. Fungal cells are eukaryotic, with cell walls built of chitin.
  4. They are generally multicellular, with the exception of the yeasts, which are unicellular.
  5. An individual fungus consists of an assemblage of threadlike filaments called hyphae.
  6. The three phyla of fungi are distinguished by their reproductive structures (fig. 2.15).
    a) Zygomycetes.
      (1) They have sexual spores.
      (2) Most feed on decaying plant and animal matter, while others are parasitic.

b) Ascomycota.
  (1) They have saclike sexual reproductive structures called asci.
  (2) Some are colorful growths on spoiled food and the fungus causing athlete's foot.
c) Basidiomycetes (phylum Basidiomycota).
  (1) These are the mushrooms, toadstools, puffballs, stinkhorns, shelf fungi, rusts, and smuts.
  (2) They have characteristic spore-containing structures called basidia.

E. Plantae
  1. Land-dwelling, multicellular organisms that extract energy from sunlight by using the pigment chlorophyll contained in their cells.
  2. The cells of a plant are eukaryotic, and they are enclosed in cell walls built of cellulose.
  3. The life cycle of a plant is divided into two reproductive phases.
    a) Gametophyte stage.
      (1) The plant produces sex cells known as gametes, which are either eggs or sperm.
      (2) A gamete must combine with a gamete of the opposite type to form a new individual, known as the sporophyte.
    b) Sporophyte phase: A plant produces spores that can develop into a new individual, known as the gametophyte, without fusing with another cell.
    c) This dual reproduction strategy of gamete-producing phases is termed alternation of generations.

4. The 10 divisions of the kingdom Plantae can be separated into two categories, the bryophytes (or nonvascular plants) and the tracheophytes (or vascular plants).
   a) Bryophytes.
      (1) The most primitive plants, all belonging to the single division Bryophyta; "nonvascular."
      (2) A bryophyte spends most of its existence in the gametophyte stage.
   b) Tracheophytes.
      (1) Xylem tissue transports water and mineral nutrients from roots to leaves, and the phloem tissue conducts sugars from "sources" (such as photosynthesizing leaves) to "sinks" (such as growing shoot tips and fruits).
      (2) Sports are produced that are larger, longer-lived, and more obvious to the human observer.
      (3) The most primitive species of vascular plants include the ferns, the whisk ferns, the lycophytes or club mosses, and the horsetails.
      (4) Five divisions of seed plants are recognized in which four of these are gymnosperms, or "naked-seed plants," including the conifers.
      (5) The angiosperms (seeds in a vessel), or flowering plants (division Anthophyta), have flowers built of concentric arrangements of highly specialized leaflike appendages and house either male or female reproductive organs, or both.
      (6) Extracts are used in insecticides, aspirin, gums, waxes, tannins, oils, resins, dyes, rubbers, flavorings, drugs, and pesticides (table 2.6).
F. Animalia (32 phyla) (table 2.7).
   1. Characteristics.
      a) Built of eukaryotic cells, derive energy from food, and have nervous systems.
      b) Animal phyla are distinguished from one another by basic differences in body form, which can be considered in order of increasing complexity.
   2. The subgroup mesozoans (phylum Mesozoa) includes organisms with 20 to 30 cells organized into two layers.
   3. The subgroup Parazoa includes the phylum Placozoa, whose organisms are built of two cell layers with fluid in between, and the sponges (phylum Porifera).

4. The remaining 29 animal phyla are known as Eumetazoa and are divided into two subgroups. Radiata and Bilateria.

   a) Radiata.
      (1) The phylum Cnidaria (hydroids, sea anemones, jellyfish, horny corals, and hard corals) and the phylum Ctemophora (sea walnuts and comb jellies).
      (2) These organisms all have radial symmetry, in which the body parts are arranged around a central axis, like the spokes of a wheel.

   b) Bilateria.
      (1) The other 27 animal phyla have a bilateral (two-sided) body plan and are further subdivided into the groups based upon certain characteristics of the embryo (such as the origin of the mouth and anus and the pattern with which the first cells of the embryo divide).
      (2) Protostomia is the more primitive group, and it is subdivided into three groups that are defined by the type of central body cavity (or coelom) that is present.
         (a) The most primitive protostomes are the acoelomates, which lack a coelom and include three phyla—the flatworms (phylum Platyhelminthes), ribbon worms (phylum Nemertina), and jaw worms (phylum Gnathostomidula).
         (b) More complex are the pseudocoelomates, which have a coelom that lacks a lining seen in more advanced animals, and include the Nematoda.
         (c) The remaining protostomes are the eucoelomates, which possess a lined coelom that forms complex organ systems and includes the three major ones, Mollusca, Annelida, and Arthropoda.
      (3) The phyla in the group Deuterostomia consist of bilateral animals with more complex embryos and include the most advanced animal phylum, Chordata, which includes the subphylum Vertebrata, characterized by bone or cartilage around the spinal cord.
         (a) The superclass Agnatha includes the jawless fish.
         (b) The superclass Gnathostomata includes six classes of the sharks and rays (class Chondrichthyes), the bony fishes (Osteichthyes), amphibians (Amphibia), reptiles (Reptilia), birds (Aves), and mammals

(Mammalia) (fig. 2.17).

(c) Humans are members of the placental Mammalia, the order Primate, suborder Anthropoidea, and genus species *Homo sapiens* (fig. 2.18).

LEARNING OBJECTIVES

After reading this chapter, the student should be able to answer these questions:

1. Where on earth are living organisms found?
2. What types of living organisms populated the earth in the distant past?
3. What criteria do biologists use to classify organisms?
4. What are the major groups of organisms that are living today?
5. How does biological classification reflect evolutionary relationships?
6. What is the full biological classification of humans?

ANSWERS

"Questions"

1. The conditions on earth today are believed to be different than at the time life first appeared. It is believed that the early atmosphere of earth provided a reducing environment, where the polymerization of simple chemicals required for life could have taken place. As life evolved, the contribution of oxygen to the atmosphere made it an oxidizing environment that broke down large molecules into small, thus eliminating the possibility for life to arise a second time.

2. The Monera are separated from the other kingdoms by being prokaryotic cells. The other kingdoms are separated according to whether they are multicelled or singular, the means of getting energy, and the mode of reproduction. The three phyla of Fungi are distinguished by their reproductive structures. Plantae divisions are determined by the mode of reproduction as well as by the means by which the plant gets water. The subphylum Vertebrata is divided into classes according to reproductive structures and adaptations to environments. The members of the various orders of placental mammals are distinguished from each other by their dental patterns, specializations of their limbs, toes, claws, and hooves, and the complexity of their nervous systems. Superfamilies within the suborder Anthropoidea are distinguished by nose structure, thumb position, tail use, and other characteristics.

3. By definition if the human and the space creature produced viable, fertile offspring, then they are of the same species.

4. As life evolved, new species had to become more specialized and complex to compete for limited niches.

5. There is no right answer as demonstrated by disputes among taxonomists, but the student should consider the significance of biochemical similarities as opposed to anatomical and physiological differences when proposing an answer. That is, do biochemical similarities reflect closer genetic ties than physical characteristics?

6. Comparing similar characteristics among different species can be misleading because it can be difficult to tell whether those species inherited the characteristic from a common ancestor or if they developed the trait independently in response to a shared environmental challenge. Another problem is in determining which characteristics should receive the most weight when determining relatedness. The molecular approach may be a more accurate way of determining relatedness since the DNA among a species does not vary and reflects what is seen physically.

7. Certainly a taxonomist will want to consider the lineage of the species involved when determining the classification of a newly engineered organism. However, since the parents may only be remotely related, the new organism may be placed in a new classification by itself.

8. The following represents just one example of human disease caused by different agents: the bacteria *Diplococcus pneumoniae* cause bacterial pneumonia; the influenza virus causes the flu; members of the genus *Plasmodium,* a protista, cause malaria; members of the division of Fungi called Deuteromycota may cause athlete's foot or ringworm; members of the class Cestoda are tapeworms; plant pollen may cause allergies.

9. The phylum Cnidaria includes beautiful aquatic organisms whose cells are organized into distinct tissues and have activities coordinated by a nervous system. The basic body plan is a hollow container made up of two tissue layers. The tube may be either vase-shaped and called a polyp or bowl-shaped and referred to as the medusa. They capture and eat prey by means of tentacles that form a circle around the mouth. The phylum is divided into only three classes by body form, nervous system, behavior, and mode of reproduction. The Hydrozoa has a polyp as the dominant body form and a nervous system that integrates the body into a functional whole. They are social organisms. The Scyphozoa predominantly consists of the medusa body form, which contains nervous tissue in a ring around the opening of the medusa. Contractions of the bell allow for a rhythmic propelling through the water. Reproduction is by the budding of the medusa as well as asexual reproduction of the lesser polyp form. The Anthozoans have no medusa stage. The polyp reproduces both asexually by budding, division, or fragmentation and sexually by the production of gametes. Offspring remain in colonies.

"To Think About"

1. Humans, for what might turn out to be trivial reasons, have greatly altered the diversity found on earth by inadvertently destroying some species through habitat destruction while producing others through genetic technology. Since species are interrelated in ecosystems, the removal and/or introduction of a new species can affect many other species. It is possible that through human error, noxious weeds and plant diseases could greatly affect world food supplies. The availability of rare species that might provide insight or offer solutions to human problems is gradually declining.

2. The Linnaean classification assumes that species evolved from other species, which is in contrast to divine creation of individual species that have not changed over time.

3. To answer this question the student should consider factors such as which organisms have existed the longest, which cover the most diverse environments, and which can most readily adapt to new environments.

4. The accurate assessment of intelligence even among the human species has been questioned. It is possible that the scientist is unable to determine the intelligence of other organisms by the limitations of our technology and knowledge. Consequently to use it as a taxonomical tool may be presumptuous.

5. The revival of extinct species would bring new genetic material into existing gene pools if crossing ancestral members with current members occurred.

6. Probably biologists will always disagree on classification schemes, however as technological advances are made, the degree of relatedness may be better determined.

AUDIOVISUAL MATERIALS

Algae, Indiana University, 16 min.

Angiosperms: The Flowering Plants, Encyclopaedia Britannica Educational Corp., 21 min.

Arthropods: Insects and Their Relatives, Coronet Instructional Films, 10.5 min.

Bacteria, Encyclopaedia Britannica Educational Corp., 19 min.

Classified Plants and Animals, Coronet Instructional Films, 10.5 min.

Five Kingdoms: An Illustrated Guide to the Phyla of Life on Earth, Carolina Biological Supply, slides.

Gymnosperms, Encyclopaedia Britannica Educational Corp., 17 min.

Invertebrates, Coronet Instructional Films, 13 min.

Microorganisms: Beneficial Activities, Indiana University, 15 min.

Microorganisms: Harmful Activities, Indiana University, 15 min.

Origin of Land Plants: Liverworts and Mosses, Encyclopaedia Britannica Educational Corp., 14 min.

Plant Life: Fungi, Encyclopaedia Britannica Educational Corp., 16 min.

Protists: Threshold of Life, National Geographic Educational Services, 28 min.

Protozoa: The Single-Celled Animals, Encyclopaedia Britannica Educational Corp., 17 min.

World Within Worlds, National Geographic Educational Services, 23 min.

3 The Chemistry and Origin of Life

KEY CONCEPTS

Living things are organized into cells, tissues, organs, and organ systems.

Biological function depends upon structure.

Characteristics exhibited by living organisms are metabolism, irritability, growth, and reproduction.

Viruses, viroids, and prions, which have some but not all characteristics of life, may represent a continuum between living and nonliving.

Species are perpetuated by sexual or asexual reproduction.

Atoms are composed of a nucleus containing protons, neutrons, and orbital shells that contain electrons.

The properties of an element are determined by the number of subatomic particles.

The formation of molecular bonds is dependent upon the interactions of the subparticles of the atoms.

Special properties of water make life possible.

Living things are built almost entirely of organic compounds containing carbon, hydrogen, oxygen, nitrogen, phosphorus, and sulfur.

Carbohydrates are the major energy source; fats are a storage form of energy and a key component of cell membranes; nucleic acids serve as the genetic code and energy carriers; proteins provide structure and catalytic activity in the cell.

Vitamins and minerals are essential for enzymatic activities and structural components of cells and tissues.

Life could have evolved from simple organic molecules in a reducing environment supplied by energy from lightning or volcanic activity.

KEY TERMS

acid
adipose cells
adaptation
adenine
amino acids
asexual reproduction
atoms
base

active site
carbohydrates
hypothesis
complex carbohydrates
compound
conformation
covalent bonds
cytosine
dehydration synthesis
deoxyribonucleic acid
    (DNA)
disaccharides
electrons
element
enzyme
fats
gene
guanine
hydrocarbons
hydrogen bond
hydrolysis
ion
ionic bond
irritability
isotope
lipids
metabolism
molecules
monosaccharides

base
naked-gene

neutrons
nucleotides
organs
oxidation reduction
peptide bond
pH scale
polymer
primary structure
prions
proteinoid theory
proteins
protons
quaternary structure
saturated
secondary structure
sexual reproduction
simple carbohydrates
spontaneous
    generation
tertiary structure
thymine
tissues
unsaturated
viroids
virus
vitamins

CHAPTER OUTLINE

I. The Characteristics of Life
   A. Organization (fig. 3.2).
      1. Organ parts are organized into tissues, which are made up of cells.
      2. All life consists of chemicals and has organization.
      3. Structural organization is closely tied to function.
   B. Metabolism.
      1. A living system acquires energy to build new structures, to repair or break down old ones, and to reproduce.

2. The sum total of the chemical reactions that direct this energy of life is termed metabolism.
C. Irritability and adaptation.
1. Irritability is an immediate response to stimulus.
2. Adaptation is an inherited characteristic or behavior that enables an organism to survive a specific environmental challenge and differs from one species to another (figs. 3.4 and 3.5).
3. Over many generations, those characteristics that provide adaptive advantages in the particular environment become more common in the population.
D. Reproduction.
1. Reproduction is the passing on of biochemical instructions (genes) to carry the characteristics of the organism from one generation to the next.
2. Reproduction can be either asexual or sexual.
   a) In asexual reproduction in single-celled organisms, cellular contents are doubled, and the cell then splits in two.
   b) In sexual reproduction, genetic material from two individuals combines to begin the life of a third individual.
E. What is the simplest form of life?
1. A virus is s tiny piece of genetic material wrapped in another chemical, a protein that commandeers the components of the cells it invades and uses them to mass produce copies of itself.
2. Viroids consist only of highly wound genetic material (fig. 3.6).
3. Prions are composed only of protein and invade cells and take over their chemical machinery.

II. Chemistry Basics
A. The atom.
1. Atoms react with one another, but they cannot be further broken down by chemical means.
2. Subatomic particles.
   a) The nucleus contains a positively charged particle with a mass of one unit.
   b) It also contains a neutron with a mass of one unit but has no charge.
   c) Electron shell.
      (1) Refers to the cloudlike distribution of constantly moving electrons about the nucleus.
      (2) An electron has a negligible mass and a negative charge.
      (3) The Bohr model illustrates electrons as dots in concentric circles, much like planets (fig. 3.7).
      (4) The shell closest to the nucleus consists of a single orbital containing up to two electrons, the second and third shells each consist of four orbitals, and each of these orbitals can contain two electrons (fig. 3.8).
      (5) Electron arrangement determines the chemical properties of an atom.
   d) The net charge and mass of an atom are the sum of the charges and masses of its constituent.
   e) Atomic weight, a measure of the mass of an atom, is calculated by adding the number of protons and neutrons.
   f) Atomic number equals the number of protons in an atom.

3. Elements.
   a) Atoms of a single type, each with the same number of protons, constitute an element.
   b) The 106 elements are arranged in a chart called the periodic table.
      (1) The table provides information about element characteristics because they are ordered from left to right by atomic number and vertically have the same number of electrons in the outermost shell.
      (2) Atomic weights in the periodic table are actually averages.
   c) The atoms of elements.
      (1) Atoms of different weights of an element are called isotopes.
      (2) The atoms of a particular element can vary in the number of neutrons they contain and thus have slightly different masses.
B. Atoms meeting atoms.
   1. Molecules.
      a) A compound.
         (1) A molecule containing different kinds of atoms.
         (2) It takes on characteristics that are different from those of the elements it contains.
         (3) It may be a solid, liquid, or gas at particular temperatures and pressures.
      b) Molecular formulas.
         (1) Molecules are described by writing the symbols of the constituent elements and indicating the numbers of atoms of each element in the molecule as subscripts.
         (2) An example of this is $C_6H_{12}O_6$, the sugar molecule glucose.
      c) Chemical reaction.
         (1) Atoms and molecules react with one another by gaining, losing, or sharing electrons to produce new types of molecules.
         (2) A chemical reaction entails the making and breaking of attractive forces, called bonds, between atoms.
         (3) The type of chemical bond that forms between atoms depends upon the number of electrons the atoms have in their outermost shells.
   2. Ions and ionic bonds—opposites attract (table 3.1).
      a) Octet rule.
         (1) There is a chemical tendency to fill the valence shell (outer shell).
         (2) Atoms of some of the elements most prevalent in living things (carbon, nitrogen, oxygen, phosphorus, and sulfur) are more chemically stable when they have eight electrons in their valence shells.

b) Ionic bond (fig. 3.10).
   (1) An ion is an atom that has lost or gained electrons, which gives it an electrical charge.
   (2) Atoms that lose electrons carry a positive charge, whereas atoms that gain electrons are negative.
   (3) It is the attraction between oppositely charged ions that results in an ionic bond.
3. Acids, bases, and salts.
   a) Acid.
      (1) Ionically bonded molecules tend to break up to form ions in the presence of water.
      (2) A molecule that releases hydrogen (H^+) ions into water is an acid, such as hydrochloric acid (HCl).
   b) Base.
      (1) A molecule that releases hydroxide (OH^-) ions into the water is a base.
      (2) An example of a base is sodium hydroxide (NaOH).
   c) Electrolyte solution refers to water containing ions that may carry an electrical charge.
   d) pH scale.
      (1) A solution's acidity or basicity influences its interactions with other molecules.
      (2) The measurement of how acidic or basic a solution is in terms of its concentration of hydrogen ions.
      (3) The pH scale ranges from 0 to 14, with 0 representing strong acidity and 14 representing strong basicity. A neutral solution has a pH of 7.
      (4) The pH of blood ranges from 7.35 to 7.45 (fig. 3.11).
4. Covalent bonds—sharing electron pairs (fig. 3.12).
   a) Atoms that have three, four, or five electrons in their valence shells are more likely to share electrons in a covalent bond than to form ionic bonds.
   b) Two or three electron pairs can also be shared in covalent bonds, which are termed, respectively, double and triple bonds.
   c) Carbon atoms can form all three types of covalent bonds with other carbon atoms, building the frameworks of a biologically important class of compounds called hydrocarbons.
5. Hydrogen bonds.
   a) Sometimes the sharing of electrons in a covalent bond is not equal, as in the case of water (H_2O) (fig 3.14).
   b) The positive charge is attracted to other, negatively charged atoms nearby.
   c) This attraction, which is weak compared with ionic and covalent bonds, is a hydrogen bond.

6. Chemical bonding in biology.
   a) Organic molecules are "carbon-containing" molecules, which in living systems are quite large and are termed macromolecules.
   b) Molecular weight, which is a measure of a molecule's size, is calculated by adding the atomic weights of the constituent atoms.
   c) Two common metabolic reactions are oxidation, which entails the loss of electrons, and reduction, which entails the gain of electrons.
   d) In organisms, sequences of redox reactions form electron-transport chains, in which an electron and its associated hydrogen atom are lost by one molecule and gained by another.

III. Life's Chemical Components
   A. About 99% of any living thing is composed of organic molecules, which contain the six elements carbon (C), hydrogen (H), nitrogen (N), oxygen (O), phosphorus (P), and sulfur (S).
   B. Characteristics of water (table 3.2).
      1. Molecules of water tend to stick to each other due to hydrogen bonding, a property called cohesiveness.
      2. Water bonds to many other compounds, a property called adhesion.
      3. Imbibition is the tendency of water to be absorbed by certain substances, causing them to swell.
      4. Water maintains temperature control because it has a high heat capacity, which is the resistance to temperature change.
      5. Water has a high heat of vaporization, which means that a lot of heat is required to make water evaporate.

6. Water is contained in blood, sweat, tears, saliva, intestinal juice, cerebrospinal fluid, lymph, and amniotic fluid.
7. Too much or too little water can affect health.

C. Water in the human body: accounts for more than 50% of all living matter and more than 90% of the living matter of most plants.
D. Organic compounds of life.
   1. Carbohydrates (fig. 3.15).
      a) Carbohydrates include the sugars and starches and contain the elements carbon, hydrogen, and oxygen, with twice as many hydrogens as oxygens.
      b) Monosaccharides.
         (1) Contain five or six carbons.
         (2) Examples are glucose (blood sugar), galactose, and fructose (fruit sugar).
      c) Disaccharides.
         (1) Form when two monosaccharides join and, in the process, lose a molecule of water (H_2) in a process called dehydration synthesis.
         (2) In hydrolysis (breaking with water), a disaccharide molecule and a molecule of water react to form two monosaccharide molecules.
      d) Monosaccharides and disaccharides (simple carbohydrates) provide energy and are all broken down to, or converted into, glucose before energy is extracted from their chemical bonds.

e) Complex carbohydrates (fig. 3.16).
  (1) Chains of simple sugars (monomers) are linked by dehydration synthesis into long molecules called polymers.
  (2) The long molecule (polymer) has many monomers (one unit).
  (3) Starch, which is found in plants, and glycogen, its animal equivalent, are energy-storing complex carbohydrates.
  (4) Glycogen is broken down to release energy-rich sugar molecules.
  (5) Cellulose is a complex carbohydrate that provides support in plants and is the major component of dietary fiber, which speeds the movement of feces through the intestines.
  (6) Chitin, another complex carbohydrate, forms the outer coverings of many organisms, including insects, crabs, and lobsters.

2. Fats (fig. 3.17).
  a) They contain the same elements as carbohydrates but with proportionately less oxygen.
  b) They are insoluble in water.
  c) They are built of a small molecule of glycerol, which serves as a backbone from which extends three fatty acids.
  d) Adipose cells specialize in storing fat.
    (1) White adipose tissue accounts for most of the fat in human adults and is packed with energy-rich lipids.
    (2) Brown adipose tissue is rare in adults, but it is found in layers around the neck and shoulders and along the spine of newborns.
  e) Fats are excellent energy sources, providing more than twice as much energy as equal amounts (by weight) of carbohydrate or protein, and can convert their energy directly into heat.
  f) Lipids are major components of membranes and the basis of the steroid hormones.
  g) A fat is saturated when its fatty acids contain all the hydrogens that they possible can.
  h) A fat is unsaturated if it has one double bond between carbons.
  i) The blood-thinning effects of omega-3 fatty acids are thought to be caused by still unknown influences on body chemicals called prostaglandins.

3. Proteins (table 3.4).
  a) Their functions are intimately tied to their structures (fig. 3.20).
    (1) Hemoglobin shuttles oxygen from the lungs to the rest of the body.

(2) Collagen and elastin are part of the connective tissue that literally holds the human body together.

(3) Proteins turn genetic material on.

b) Proteins are polymers built of one or more chains of amino acids.

    (1) All amino acids have the same basic formula $H_2N-C-R-COON$ but differ by R group.

    (2) The R group can be any of several other chemical groups that distinguish the 20 different amino acids commonly found in living things (fig. 3.19).

    (3) A peptide bond is formed when two adjacent amino acids are joined by dehydration synthesis.

    (4) A long chain of peptides is termed a polypeptide, which breaks down into its constituent amino acids by hydrolysis.

c) Attractive forces between amino acids in a polypeptide chain cause it to take on a three-dimensional shape.

d) Some of the amino acids needed in the human body can be manufactured from other body chemicals.

e) The proteins in food are broken down into amino acids by digestion, travel in the circulatory system to cells throughout the body, and are then built up again into human proteins.

E. Enzymes are important proteins in living things.

1. They alter the rates of chemical reactions without being used up in the process, which is called catalysis.

2. They are essential to biological growth, repair, and waste disposal.

3. They function under specific conditions of pH and temperature, and they are usually also specific to particular chemical reactions.

4. A missing or defective enzyme can be devastating, as in Tay-Sachs disease.

5. The key to an enzyme's specific action lies in a region of its surface called the active site (fig. 3.22).

6. Enzyme-substrate complex is when the substrates fit into the active site to form a short-lived partnership (fig. 3.23).

7. An enzyme can hold two substrate molecules that react to form one product, or it can hold a single substrate molecule that splits to yield two product molecules.

8. Once the enzyme and the substrate (or substrates) have formed the complex, the reaction takes place very rapidly.

9. Genetic production of an enzyme is when a particular enzyme is inserted into bacterial cells, which then churn out the enzyme in large quantities.

F. Nucleic acids.

1. The chemical building blocks of nucleic acids are called nucleotides.

a) Each nucleotide contains a five-carbon sugar, a phosphate group (PO_4), and one of four types of nitrogen-containing ring compounds, or nitrogenous bases (fig. 3.24).

b) The nitrogenous bases are adenine (A), guanine (G), thymine (T), and cytosine (C).

2. The nucleic acid deoxyribonucleic acid (DNA) is the genetic material of all species and many viruses, and it stores information that specifies the construction of proteins.
3. A sequence of DNA that specifies a particular polypeptide is a gene.
4. DNA structure.
   a) The DNA molecule resembles a spiral staircase, with the sugar and phosphate groups of the nucleotides forming the sides of the staircase and the nitrogenous bases pairing to form the "steps."
   b) Pairs form only between adenine and thymine and between guanine and cytosine, forming a symmetrical double helix.
5. DNA replication.
   a) The spiral staircase splits down the center and unwinds, separating each nucleotide pair.
   b) Each half of the pair now chemically attracts a free-floating nucleotide of the proper complementary type, and little by little the spaces created by the splitting of the original molecule are filled in.
G. Vitamins.
   1. Vitamins are organic chemicals that are essential in small amounts for the normal growth and function of an organism, but that cannot be synthesized by it.
   2. C and B-complex vitamins are soluble in water and are excreted in the urine.
   3. Fat-soluble vitamins are A, D, E, and K, which tend to accumulate in the body so too much of them can result in illness.
H. Inorganic compounds in life—minerals.
   1. Without minerals, our muscles would not contract, our bones would not support our weight, our nerves would not relay messages, and many enzymes would be stopped.
   2. Minerals are master regulators, controlling blood clotting, heartbeat, oxygen transport, and the pressures of body fluids.
   3. Bulk minerals include calcium, phosphorous, potassium, sulfur, sodium, chloride, and magnesium.
   4. The trace elements, including zinc, iron, manganese, copper, iodine, cobalt, fluoride, chromium, and selenium, are needed only in very small amounts.
   5. Ultratrace elements, like cadmium, lead, and arsenic, are needed in even smaller amounts.
IV. The Origin of Life on Earth
   A. Spontaneous generation.
      1. This refers to the hypothesis that the nonliving gives rise to living organisms.
      2. Experiments that disproved the hypothesis:
         a) Francesco Redi filled two jars with meat, leaving one open and covering the other one lightly with cloth, and then soon observed that only the uncovered jar produced maggots and then flies.
         b) Lazzaro Spallanzani found that boiled soup in sealed flasks produced no microorganisms, while flasks of boiled soup with only cork seals reveal the presence of microorganisms.

B. Life from space (fig. 3.25).
   1. One scenario of extraterrestrial chemicals "seeding" life on earth envisions simple organic molecules in interstellar dust clouds forming complex organic compounds in comets.
   2. The very fact that these chemicals have been found in many unearthly rocks and that they have even been synthesized in laboratories by simulating conditions in outer space suggests that they may be simply the result of common chemical phenomena.
C. Common ancestry.
   1. All life forms use nucleic acids as their genetic material, they use the same genetic "code" to translate the nucleic acids into proteins, and they use the same energy-generating molecules.
   2. All organisms use the same 20 amino acids to build proteins.
D. Chemical evolution (fig. 3.26).
   1. Before life (or even molecules suggestive of life) appeared, certain chemical changes had to occur.
   2. The earliest molecules believed to be in the atmosphere were the gases ammonia (NH_3), hydrogen (H_2), methane (CH_4), carbon dioxide (CO_2), and water vapor (H_2O).
   3. "Protein first" view, or proteinoid theory.
      a) Some scientists believe protein may have been the first organic polymer to form.
      b) This theory points to the observation that, in the laboratory, certain combinations of amino acids, other molecules, and heat produce tiny spheres that have some characteristics of living cells.

   4. Naked-gene hypothesis.
      a) DNA is the only molecule that can replicate, and it controls protein synthesis.
      b) A "protocell" developed a way to manufacture and put together its own amino acids so that it was no longer dependent upon random chemical reactions among the prebiotic molecules in its environment; it developed a way to generate energy.

LEARNING OBJECTIVES

After reading this chapter, the student should be able to answer these questions:

1. What characteristics distinguish living things from nonliving things?
2. What are the simplest forms of life?
3. What chemical components constitute living things?
4. What chemical compounds are important to human health?
5. How might living matter have evolved from nonliving chemicals?

ANSWERS

"Questions"

1. Structural organization is closely tied to function. Disrupt the structural plan and function ceases, thus life ceases.
2. An automobile cannot respond to stimuli or adapt to a new environment. It is incapable of reproduction.
3. Silicon requires four electrons to complete its outer energy level like carbon.
4. The chemical formula for biotin is $C_{10}H_{16}O_3N_2S$.
5. The substance found in fish that may avert coronary heart disease is omega-3 fatty acids, which has blood-thinning effects.

6. The enzyme derived from *Aspergillus oryzae* increases the rate at which water breaks down lactose into galactose and glucose.

7. The sources of carbon, hydrogen, oxygen, and nitrogen in the early atmosphere of earth were likely ammonia (NH_3), methane (CH_4), hydrogen (H_2), carbon dioxide (CO_2), and water vapor (H_2O).

"To Think About"

1. There is no right answer but the student should consider both biological and philosophical characteristics of what entails life and his or her own feelings.

2. The student may wish to think about the function of individual parts and how they interact in the living organism to keep it alive. Also consider Reading 3.1, "The Definition of Death."

3. Before answering this question the student should consider the theories regarding how life may have formed on the prebiotic earth and experiments designed to test the possibilities.

4. At the very least large studies should be conducted on populations of healthy, average individuals. Experimenters should take into account other factors such as sex, race, age, and life-style before making broad generalizations.

5. Since the moon of Jupiter would have formed during the same time period as the Earth, it supports the idea that a similar atmosphere was possible here. A reducing atmosphere is predicted as having been necessary for life to have begun on Earth. It also suggests that similar conditions may have or presently exist on other planets, making it plausible that life exists elsewhere.

6. The viroids, viruses, and prions are noncellular organisms composed of polymers of nucleic acids or proteins.

7. The "spontaneous generation" theory accounted for complex organisms arising out of dust, which was not supported by known laws of science. The origin of life from the "organic soup" theory is consistent with chemical, physical, and biological properties as we know them and allows the scientist to predict that under specific conditions, it would be possible. The theory of spontaneous generation has been eliminated due to experimentation. Religious ideas offer predictions based upon phenomena that are inconsistent with natural laws as we understand them and so cannot be tested. Scientific researchers can at best try to predict conditions and simulate them in the laboratory, but even if life was created in a test tube it would not necessarily reenact the origin of life on early earth. With increased technology and knowledge, it is likely that we will come closer to knowing how life began.A

AUDIOVISUAL MATERIALS

Basic Chemistry for the Biologist, Carolina Biological Supply, filmstrip.

Chemical Bond and Atomic Structure, Coronet Instructional Films, 16 min.

Chemistry of the Cell—Structure of Proteins and Nucleic Acid, CRM McGraw-Hill, 21 min.

The Chemistry of Life, Human Relations Media, filmstrip.

Evolution and the Origin of Life, CRM McGraw-Hill, 36 min.

Introducing Chemistry: How Atoms Combine, Coronet Instructional Films, 11 min.

Proteins: Structures and Function, John Wiley & Sons, 15 min.

4 Cells and Tissues

KEY CONCEPTS

The cell is the basic unit of life.

Each kind of cell meets the requirements for life.

Viruses require a host cell's machinery to replicate.

Through the development of microscopic technology, our knowledge of cell structure and function has expanded.

Prokaryotic cells are the simplest form of cells, characterized by lack of organelles, and they include the bacteria and blue-green algae.

Organelles are specialized compartments within eukaryotic cells that function in secretion, metabolism, structure, waste disposal, and replication.

Eukaryotic cells may have arisen from incorporation of smaller prokaryotes into larger functional units.

Cells in multicellular organisms specialize in function due to a process of selective expression of genetic material called differentiation.

Tissues are specialized for protection, absorption, secretion, movement, structure, and communication.

Epithelial tissue protects, absorbs, and transports nutrients and is involved in secretion.

Connective tissue protects and gives support to other tissues.

Nervous tissue conveys and integrates information.

Muscle tissue moves bones, food, blood, a fetus, sperm, milk, saliva, and sweat.

KEY TERMS

bone	Golgi apparatus
cardiac muscle	grana
cartilage	lysosomes
cells	messenger RNA
cell theory	microtubules
cell wall	mitochondria
centrioles	muscular tissue
chlorophyll	nervous tissue
chloroplast	neurons
chromosomes	neurotransmitters
cilia	nucleoid
compound microscope	nucleolus
connective tissue	peroxisome
cristae	photosynthesis
endocytosis	plasma
endoplasmic reticulum	platelets
endosymbiont theory	red blood cells
epithelial tissue	ribosomes
epithelium	skeletal muscle
exocytosis	smooth muscle
fibroblast	thylakoids
flagella	white blood cells

CHAPTER OUTLINE

I.   Cells—Biological Efficiency and Organization
    A.   Unicellular organisms consist of a single cell, such as bacteria and protists.

B. Multicellular organisms.
  1. They are built of many cells.
  2. Structures within the cells of multicellular organisms and within some more complex unicellular organisms, called organelles (little organs), carry out specific functions.
C. All cells have some structures in common that allow them to perform the basic life functions of reproduction, growth, response to stimuli, and energy conversion.

II. Viruses—Simpler Than Cells (fig. 4.2)
A. A virus consists of a nucleic acid (DNA or RNA) surrounded by protein.
B. A virus must be within a cell to reproduce and is thus called an obligate parasite.
C. Many viruses, such as HIV, cannot survive outside of a living cell.
D. Some other viruses are afforded protection from the physical environment by their protein coverings.
E. A virus reproduces by injecting its DNA and RNA into the host cell, where it situates itself within the host's DNA.
F. An RNA virus, such as HIV, is called a retrovirus and must first make a replica of its RNA in DNA form.
G. Once viral DNA integrates into the host's DNA, it can either remain there and be replicated along with the host's DNA whenever the cell divides, but not cause harm, or the viral DNA can actively take over the cell, leading eventually to the cell's death.
  1. To do this, some of the virus's genes direct the host cell to replicate viral DNA rather than the host DNA.
  2. As viral DNA accumulates in the cell, some of it is used to manufacture proteins.
  3. Some of the proteins wrap around the DNA to form new viral particles that cut through the host cell's outer membrane, causing the cell to burst and release new viruses.

4. A particular type of virus, however, infects only certain species, which constitute its host range (fig. 4.3).

III. Viewing Cells—The Development of the Microscope
A. By the thirteenth century, the value of such "lenses" in aiding people with poor vision was widely recognized.
B. Johann and Zacharius Janssen (two Dutch spectacle makers).
  1. The origin of a double-lens compound microscope is traced to them.
  2. Similar double-lens systems were constructed to focus on objects too small to be seen by the naked human eye.
C. Robert Hooke.
  1. In 1660, he melted together strands of spun glass to create lenses that were optically superior to any that had been available before.
  2. He was the first human to see cells, the fundamental structural units of life.
D. Anton van Leeuwenhoek.
  1. He used only a single lens, but it was more effective at magnifying and produced a clearer image.
  2. He discovered bacteria and protozoa (figs. 4.4 and 4.5).

IV. The Cell Theory
A. The first part of the theory was contributed by German biologists Matthias J. Schleiden and Theodor Schwann, who stated that all living matter is composed of cells, and that cells are the basic structural and functional units of life.
B. Rudolph Virchow added that all cells come from preexisting cells and suggested that human disease results from changes taking place on the cellular level.

V. Characteristics of Cells (table 4.3)
- A. A cell requires energy, genetic information to direct biochemical activities, and structures to carry out these activities, such as movement that occurs within living cells and cells moving about in the environment.
- B. The prokaryotic cell.
    1. Prokaryotes flourish today, comprising the majority of living cells on earth.
    2. The cell wall.
        a) It is built of peptidoglycans (peptide sugars).
        b) Species whose cell walls turn purple in the presence of the Gram stain are termed gram positive; those whose cell walls turn pink are called gram negative.
    3. Beneath the prokaryote's cell wall is a cell membrane, or plasmalemma.
    4. In some prokaryotes, taillike appendages called flagella, which enable the cell to move, are anchored in the cell wall and underlying cell membrane.
    5. The genetic material of a prokaryote is a single circle of DNA (fig. 4.7).
        a) It is described as "naked" because it is not complexed with protein.
        b) The part of a prokaryotic cell in which the DNA is located is called the nucleoid.
    6. Ribosomes.
        a) These are spherical structures built of RNA and protein.
        b) They enable the cell to utilize DNA sequence information to direct the manufacture of proteins, which is a rapid process in prokaryotes.
    7. The cyanobacterium's membranes are studded with pigment molecules that absorb and extract energy from sunlight.
- C. The cell's problem.
    1. A large cell lacking the means to bring in required chemicals from the environment or to eliminate wastes might die.
    2. As cells grew larger, they could survive only if they could somehow increase their surface areas relative to their increasing volumes, so the large cell needed to divide into two.
- D. The eukaryotic cell (figs. 4.9, 4.10, and 4.11).
    1. Another cellular solution to the problem of increasing size is to divide into compartments, or organelles, much like a growing store is subdivided into departments.
    2. Organelles are established by biological membranes, which are barriers composed of lipids and proteins (table 4.4).
    3. Organelles have access to the environment outside the cell and to each other by networks of bubble like structures called vesicles (or vacuoles) that bud off from the membranes.
    4. Cells that have organelles are termed eukaryotic.
    5. Plants and animals are eukaryotic.
    6. Membranes.
        a) They surround organelles, keeping within them chemical reactions whose products might harm other parts of the cell.
        b) Some organelles are constructed of membranes that are studded with enzymes, allowing certain chemical reactions to occur on their surfaces, and some have different enzymes that are organized according to the sequences of biochemical reactions.

7. Nucleus.
   a) This is the most prominent organelle and contains the genetic material (DNA).
   b) The remainder of the cell consists of other organelles and a jellylike fluid called cytoplasm.
8. The protoplasm (cytoplasm) and the organelles are considered the living parts of the cell.
9. Arrays of protein rods within a animal cell form a framework called the cytoskeleton, which helps to give the cell it shape.
10. Organelles in action—secretion (fig. 4.12).
    a) The ability of individual cells to manufacture the remarkably complex milk is made possible by the interaction of organelles, which function together to form a secretory network.
    b) Nucleus.
       (1) In humans, 23 pairs of rod-shaped chromosomes contain information that other parts of the cell use to construct proteins, with each chromosome consisting of millions of DNA building blocks (nucleotides), long sequences of which comprise genes.
       (2) The nucleotide sequence of a gene is transcribed, or rewritten, into another type of nucleic acid, messenger RNA.
       (3) RNA building blocks are stored in a structure in the nucleus called the nucleolus.
       (4) Messenger RNA exits the nucleus by passing through holes, called nuclear pores.
    c) The endoplasmic reticulum is a maze of interconnected membranous tubules and sacs that winds from the nuclear envelope to the cell membrane.
       (1) The portion of this membranous system near the nucleus is flattened and studded with ribosomes, and this region is called rough ER because of its appearance in the electron microscope.
       (2) Smooth ER is where lipids are synthesized and added to the proteins that are transported from the rough ER.
       (3) The next stop in the eukaryotic production line is the Golgi apparatus, which is a system of flat, stacked, membrane-enclosed sacs where sugars bond to one another to form starches, or they bond to proteins to form glycoproteins or to lipids to form glycolipids.
       (4) Exocytosis is the process where the protein-carrying vesicles fleetingly become part of the plasmalemma and then open out facing the exterior of the cell, which releases free proteins outside the cell.

29

11. Other organelles and structures.
  a) Mitochondria.
    (1) They provide cellular energy through the energy-generating reactions of cellular respiration.
    (2) They have outer membranes similar to those of the ER and Golgi apparatus and intricately folded inner membranes.
    (3) The folds of the inner membrane are called cristae, and they contain many of the enzymes that take part in cellular respiration.
    (4) They contain their own genetic material.
  b) Lysosomes.
    (1) These are sacs that bud off of the Er or Golgi apparatus.
    (2) They chemically dismantle captured bacteria, worn out organelles, and other debris.
    (3) It is a highly acidic region for the enzymes, without harming other cellular constituents.
    (4) In humans, lysosomes are abundant in liver cells.
  c) Peroxisome.
    (1) They are buds from the smooth ER.
    (2) They house enzymes important in oxygen utilization.
  d) Centrioles.
    (1) These are oblong structures built of protein rods called microtubules, and they are found in pairs in animal cells, oriented at right angles to one another near the nucleus.
    (2) Centrioles appear to play a role in organizing other microtubules to pull replicated chromosomes into two groups during cell division.
  e) Chloroplast.
    (1) It gives organisms their green color.
    (2) Chloroplasts house the chemical reactions of photosynthesis.
    (3) Photosynthesis enables the cells to capture solar energy and use it to manufacture organic molecules.
    (4) The inner membrane of a chloroplast is studded with enzymes necessary for photosynthesis and is organized into stacks, called grana, of flattened membranous disks, called thylakoids.
    (5) Chloroplasts are the most abundant of a general class of pigment-containing organelles, called plastids, found in plant cells.
  f) Plant cells are surrounded by a rigid cell wall built of the carbohydrate cellulose, which supports the cell and protects its contents.

VI. Specialized Cells Form Tissues
  A. Levels of organization.
    1. Although all cells contain the structures and organelles necessary for survival, the combinations of these components give many cells specialized or differentiated characteristics.
    2. The differentiated cells of humans and other multicellular organisms are grouped into tissues.
    3. Tissues are grouped to form organs, which lead to organ systems.
    4. The cells constituting different tissues are specialized in structure, in function, and in the kinds of molecules they manufacture in large quantities (fig. 4.13).
  B. Epithelium—the "covering" tissue
    1. The human body has many surfaces, and the lining tissue is called epithelium, which consists of closely aggregated cells with very little extracellular material between them.
    2. Different types (table 4.5, fig. 4.14).
      a) Simple epithelium is one cell thick.
      b) Stratified epithelium is two or more cells thick.
      c) Pseudostratified epithelium is a single layer of cells whose nuclei are at different levels and gives the illusion of stratification.
      d) Flat epithelial cells are called squamous and accumulate a hard protein, called keratin, and become so thin that the cells flake off.
      e) Epithelium can also be cube-shaped (cuboidal) or tall (columnar).
    3. Functions.
      a) It protects the inner and outer surfaces of organs, the epithelial linings of blood vessels, and the digestive cavities, where they participate in the absorption and transport of nutrients.
      b) Epithelial cells lining parts of the respiratory system are fringed with waving protein projections called cilia, which move dust particles up and out of the body.
  C. Connective Tissue
    1. This fills in spaces, attaches epithelium to other tissues, protects and cushions organs, and provides mechanical support.
    2. It also follows an anatomical plan of cells embedded in a nonliving substance called a matrix.
    3. Types of connective tissue (table 4.6).
      a) A very abundant type is the fibroblast, which manufactures two types of protein fibers that are part of the matrix (fig. 4.15).
        (1) Collagen is a flexible white protein that resists stretching, and elastin is a yellowish protein that stretches readily.
        (2) The matrix also consists of a thin gel made of proteoglycans, which are complex carbohydrates linked to proteins.
      b) Loose connective tissue is the "glue" of the body, consisting of widely spaced fibroblasts and a few adipose (fat) cells surrounded by a meshwork of collagen and elastin fibers.
      c) Fibrous connective tissue is built of dense tracts of collagen and forms ligaments.

d) Blood.
  (1) A complex mixture of different cell types suspended in a matrix called plasma.
  (2) Red blood cells transport oxygen and constitute the bulk of the cells.
  (3) White blood cells protect against infection and help to clear the body of its own cells that have worn out or become abnormal.
  (4) Blood also contains cell fragments called platelets, which release chemicals that promote blood clotting.
  (5) The blood plasma is about 92% water, and it carries cells, dissolved salts and gases, proteins, nutrients, and waste products.
e) Cartilage.
  (1) A connective tissue that cushions organs and forms a structural framework to keep tubular organs from collapsing.
  (2) In joints, cartilage can sustain weight while allowing bones to move against one another.
  (3) It forms the skeleton in the embryo, and it is gradually replaced with bone, which is a much harder tissue.
  (4) It has a single cell type, the chondrocyte, lodged within oblong spaces called lacunae embedded in a collagen matrix.
  (5) The strong networks of collagen and elastin fibers give cartilage great flexibility.
  (6) It contains proteoglycans consisting of protein chains bonded to long chains of the disaccharide hyaluronic acid, which attracts tremendous amounts of water and provides support as well as resiliency.
  (7) It is covered with a thick and tough shell of collagen and lacks nerves and blood vessels.
f) Bone.
  (1) This is supportive and provides maximum strength with minimum weight and protects other tissues and organs.
  (2) It follows the connective tissue organization of widely separated cells.
  (3) The mineral hydroxyapatite contains calcium and phosphate and constitutes most of the mineral phase of bone.
  (4) The organic phase of bone consists almost entirely of collagen.
  (5) Cells called osteocytes occupy spaces called lacunae, and long narrow passageways called canaliculi connect the lacunae.
  (6) These are arranged around larger passageways called Haversian canals, which surround blood vessels.
  (7) Other canals connect the inner ones to the marrow cavity within, where blood cells are manufactured.

32

(8) Osteoblasts secrete bone matrix, while large cells with many nuclei, called osteoclasts, degrade bone matrix.

(9) Osteoprogenitor cells line the passageways of bone and serve as a reserve supply of cells that can transform into osteoblasts or osteoclasts in the event of growth or injury.

D. Nervous tissue
1. Structure.
   a) Neurons are nerve cells structurally supported by neuroglia, which constitute the nervous tissue.
   b) A neuron consists of a cell body, a thick branch called the axon, and several thinner branches called dendrites (fig. 4.16).
2. Function.
   a) Neurotransmitters are chemicals that receive information and are released from the axon of another neuron or from direct energy stimulation such as light, heat, or pressure.
   b) The arrival of this neurotransmitter or sensory stimulation alters the plasmalemma of the receiving cell's dendrite so that different types of ions can enter and leave the cell.
   c) This membrane change alters the electrical potential of the receiving cell, and the electrochemical change is sent along the neuron's cell membrane.

   d) When it reaches the end of the axon, the electrochemical "wave" triggers the release of a neurotransmitter.
   e) The information is thereby passed on to another cell, usually a neuron but sometimes a muscle.
3. Neuroglia are supportive cells.
   a) One abundant type of neuroglia, Schwann cells, has very fatty membranes that wrap around axons, forming an insulating sheath called myelin.
   b) Other neuroglia provide a structural scaffolding.
E. Muscle tissue.
1. Muscles contract when two types of protein filaments (actin and myosin) slide past one another, shortening their total length.
2. Muscle cells have many mitochondria that provide the energy for contraction.
3. There are four types of contractile cells.
   a) Skeletal muscle.
      (1) This consists of one huge cell with many nuclei.
      (2) Striations are caused by the arrangement of proteins.
      (3) This muscle tissue makes possible voluntary movements.
   b) Cardiac muscle.
      (1) This is found in the heart and is striated with cells of single nuclei.
      (2) They are joined together by disclike structures.

c) Smooth muscle.
    (1) This is not striated, and its involuntary contractions are slow when compared with those of other contractile cells.
    (2) Smooth muscle cells are responsible for the pulsations along the digestive tract that help to move food along and for erecting hairs at the back of the neck.
d) Myoepithelial cells are not striated and are found in the epithelium, where they contract to expel secretory products, such as milk, saliva, and sweat, from the glands that produce them.

VII. The Origin of Eukaryotic Cells
  A. Endosymbiont theory states that eukaryotic cells formed from large prokaryotic cells that incorporated smaller and simpler prokaryotic cells.
  B. Evidence in support of the endosymbiont theory.
    1. The mitochondria and chloroplasts found in eukaryotic cells bear striking resemblances to prokaryotic cells.
    2. These also resemble bacteria in size, shape, and membrane structure, and each reproduces by splitting in two and contains its own DNA.
    3. They function in close association, which is how prokaryotes use their genes to make proteins.
    4. Pigments in the chloroplasts are similar to pigments used by cyanobacteria to carry out photosynthesis.
  C. The endosymbiont theory specifically proposes that mitochondria descended from aerobic (oxygen-using) bacteria, that the chloroplasts of red algae descended from cyanobacteria, and that the chloroplasts of green plants descended from yet another type of photosynthetic microorganism.

LEARNING OBJECTIVES

After reading this chapter, the student should be able to answer these questions:

1. How are viruses different from cells?
2. How are the two basic types of cells (prokaryotic and eukaryotic) alike and how are they different?
3. How did compartmentalization and division of labor within the cell become ways to maintain functions as cells grew larger?
4. Which organelles carry out the processes of secretion, waste removal, inheritance, and obtaining energy? How are secretion and waste removal carried out in eukaryotic cells?
5. What are the four tissue types in humans? What are their functions?
6. How might eukaryotic cells have evolved from prokaryotic cells?

ANSWERS

"Questions"

1. a. A virus is not a cell but rather a single or double strand of DNA or RNA that may be encapsulated in a protein sheath. Because it is dependent upon a host cell's machinery to reproduce, some biologists question whether it should be classified as living.
   b. A bacterium is one of the simplest forms of cells called prokaryotes. It has only a nuclear region and no distinct organelles. Its genetic material is a loop of DNA that is not associated with proteins.
   c. A human cell is eukaryotic, which means it has a true nucleus and organelles. Its genetic material is associated with proteins to form chromosomes. It is differentiated for a specific function.

2. The shape of a cell, and in part, its function, is determined by its cytoarchitecture. Protein tubules and rods extend the cell membrane in characteristic ways. They are also involved in cell movement as in the case of the white blood cell.

3. As the dimensions of a cube increase, the volume increases at a faster rate than the surface area. A 3-inch box has a surface area equal to 54 inches^2 and a volume equal to 27 inches^3, giving it a surface area/volume ratio of 2:1. A 5-inch box has a surface area equal to 150 inches^2 and a volume equal to 125 inches^3, giving it a surface area/volume ratio of 1.2:1. A 7-inch box has a surface area equal to 294 inches^2 and a volume equal to 343 inches^3, giving it a surface area/volume ratio of .85:1. This demonstrates that the larger the object, the smaller the surface area/volume ratio.

4. Hepatocyte = 110,000 μm^2/5,000 μm^3 = 20:1
    Pancreas cell = 13,000 μm^2/1,000 μm^3 = 13:1
    The hepatocyte has the higher surface area/volume ratio. It is likely that it is able to do more activities requiring membrane surfaces.

5. Both prokaryotic and eukaryotic cells have cell membranes, cytoplasm genetic material, and ribosomes. They are both involved in protein synthesis, cell division, and metabolism. Eukaryotic cells are about 10 times larger than prokaryotes and always require oxygen for metabolism. They have membrane-bound organelles, whereas prokaryotes have none. The DNA of eukaryotes is long, coiled, and associated with proteins, unlike the short, circular, naked form of prokaryotes. The genetic material of the prokaryote is organized in a nuclear region, whereas the eukaryote has a true nucleus. RNA and protein synthesis is spatially separated in the eukaryote as opposed to the prokaryotes. Most eukaryotes are multicellular and differentiated, unlike the single-celled prokaryotes.

6. The experiment using radioactive hydrogen is following the process of protein synthesis. Amino acids are first assembled in the ribosomes of the rough ER, modified in structure in the smooth ER, prepared for transport in the Golgi apparatus, and then secreted from the cell in the process of exocytosis. Lysosome activity could be monitored by first labeling bacteria and then inoculating a culture of white blood cells with them. One would then expect to see grains outside the cell membrane brought into the cell by endocytosis. Eventually the grains would appear inside the lysosome, indicating that the bacteria had been engulfed by that organelle for destruction.

7. a. The flagella on the sperm of this individual are nonfunctional, providing no means of transport to the oviduct of the female for fertilization of the egg.
    b. The child with Fabry's disease has a malfunction of lysosomes, which prevents the buildup of excess chemicals such as glycolipids.
    c. In cystic fibrosis there is overactivity of the smooth ER where lipids, the main component of mucus, is synthesized.
    d. In cyanide poisoning, the mitochondria are shut down so ATP is no longer produced, resulting in death.
    e. Centrioles involved in organizing replicated chromosomes for cell division are overactive in cancer cells.

8. a. Keratin—epithelial tissue
    b. Collagen—connective tissue
    c. Elastin—connective tissue
    d. Proteoglycans—connective tissue
    e. Hyaluronic acid—connective tissue
    f. Hydroxyapatite—bone
    g. Neurotransmitters—nerve cells
    h. Myelin—muscle cells

9. Epithelium tissue contains simple squamous, simple cuboidal, simple columnar, stratified squamous, stratified cuboidal, stratified columnar, pseudostratified, and secretory cells. Connective tissue contains fibroblasts, adipose cells, white blood cells, red blood cells, platelets, chondrocytes, osteoclasts, osteoblasts, and osteocytes. Nervous tissue is composed of neurons and neuroglia cells (Schwann cells). Muscle tissue is composed of skeletal, cardiac, and smooth muscle cells.

10. The cell theory states that all living matter is composed of cells; that cells are the basic structural and functional units of life; and that all cells come from preexisting cells. Evidence for this theory comes from observations of cells through microscopy and from observations that cells isolated from multicelled organisms can be kept alive in the laboratory. The endosymbiont theory states that eukaryotic cells formed from large prokaryotic cells that incorporated smaller and simpler prokaryotic cells. Evidence for this theory comes from the observation that mitochondria and chloroplasts have their own genetic material and replicate separate from the cell's genetic material.

"To Think About"

1. Though there is no right answer as demonstrated by the debate among biologists, the student should consider the following when making an argument: viruses are not cells but only DNA or RNA that is either naked or encapsulated in a protein sheath; they are unable to replicate without a host cell's machinery.

2. Bacteria's simple organization and rapid growth rate (short life span) allow the population to adapt quickly to changing environments. Consequently they have been highly competitive over time and in all environments.

3. Compartmentalization, made possible by inner membranes, increases the surface areas where enzymatic reactions can take place. It isolates chemical reactions so that they do not interfere with each other.

4. All of these tissues meet the criterion of being cells embedded in a nonliving matrix.

5. The *Pelomyxa palustris* clearly supports the endosymbiont theory by providing an existing model of one single-celled organism living within another.

6. It is possible that the genetic material found in mitochondria and chloroplasts may be that of the host cell in the form of jumping genes. If this was the case, they were likely moved there by microtubules, which are involved in moving genetic material in the cell.

AUDIOVISUAL MATERIALS

Cell Biology, Coronet Instructional Films, 17 min.

Cell Biology: Life Functions, Coronet Instructional Films, 19 min.

Cell Biology: Structure and Composition, Coronet Instructional Films, 13 min.

The Cell: A Functioning Structure (Parts 1 and 2), CRM McGraw-Hill, 30 min.

The Cell: Its Structure, Carolina Biological Supply, filmstrip.

Cell Structure and Function, educational images ltd., slides.

Inside the Cell: Microstructures, Mechanisms, and Molecules, Guidance Associates, filmstrip, video, slides.

The Living Cell, Harper & Row College Media, 27 min.

The New Cell, Carolina Biological Supply, filmstrip.

5 Cellular Architecture

KEY CONCEPTS

Cells within and between organisms can be distinguished by the pattern and types of molecules (proteins and sugars) on their surfaces.

Human leukocyte antigens predict the likelihood of development of certain diseases.

Membranes control which substances enter and leave a cell and form compartments within cells.

A membrane is built of a lipid bilayer embedded with movable proteins.

In passive diffusion, a substance moves across the membrane from high to low concentration.

Osmosis, the passive diffusion of water, influences a cell's shape.

In facilitated diffusion, a substance moves down its concentration gradient with the aid of a carrier protein.

In active transport, a substance moves against its concentration gradient using energy from split ATP.

Endocytosis involves the membrane surrounding a substance, bringing it into the cell.

In exocytosis, a bit of membrane surrounds a substance and by joining a larger membrane, the content is transported out of the cell.

A cell's shape is largely determined by its cytoskeleton, which is a network of protein rods and tubes. Microtubules are involved in cell division and the formation of cilia. Microfilaments form part of muscle tissue.

KEY TERMS

active transport	exocytosis
adenosine triphosphate	facilitated diffusion
cell membrane	flagellum
cilia	fluid mosaic
contractile vacuole	glycoproteins
cytoskeleton	hydrophilic
endocytosis	hydrophobic

lipid bilayer	passive diffusion
microfilament	phospholipid
microtubules	solution
osmosis	turgor pressure

CHAPTER OUTLINE

I. Together, the cell surface, cell membrane, and cytoskeleton form a structural framework that helps to distinguish cells from one another (fig. 5.1).
   A. The surface molecules of a cell are anchored in the cell membrane, the outer covering of a cell.
   B. Just beneath the cell membrane are protein fibers that are part of the cell's interior scaffolding of cytoskeleton.

II. The Cell Surface—Cellular Name Tags
   A. Some surface molecules distinguish cells of one species from cells of another.
   B. Other surface structures distinguish individuals within a species from one another.
   C. Within the body, a huge collection of white blood cells and the biochemicals they produce form the immune system, which recognizes the surfaces of an individual's cells as "self" and all other cell surfaces as "nonself."
   D. Surface structures also distinctively mark cells of different tissues within an individual, which is important during the development of the embryo, when different cells sort themselves out to grow into distinctive tissues and organs.
   E. The closer the match between two persons' cell surfaces, the more likely that the immune system of one person will recognize the cells of the other person as "self" (fig. 5.3).
   F. Cell surfaces and health predictions—the HLA system.

1. These are cell surface molecules that appear in different patterns in different people.
2. An HLA profile taken at birth may be used to predict disease susceptibilities.

III. The Cell Membrane—Cellular Gates
  A. The movement of molecules into and out of a cell is monitored by the cell membrane, which is a selective barrier that completely surrounds the cell.
  B. In eukaryotes, membranes are also found within cells, where they compartmentalize and protect structures.
  C. The protein-lipid bilayer.
    1. The chemical characteristics and the arrangement of the molecules that build a membrane determine the activity of the membrane.
    2. Phospholipid molecules.
      a) These make the structure of a biological membrane possible.
      b) The phosphate end of the molecule, which seeks water, is said to be hydrophilic (water-loving), while the other end, consisting of two fatty acid chains, moves away from water and is said to be hydrophobic (water-hating).
    3. Lipid bilayers are being used in the pharmaceutical industry to construct microscopic bubbles, called liposomes, which are used to encapsulate drugs for passage across membranes (figs. 5.4 and 5.5).
    4. Passageways for water-soluble molecules and ions are formed by proteins that are embedded throughout the lipid bilayer.
    5. The membranes of living cells, then, consist of lipid bilayers and the proteins within and extending out of them.
    6. The protein-lipid bilayer is called a fluid mosaic because the proteins can move and are not regularly arranged, as are lipid molecules (fig. 5.6).
    7. The number or distribution of cell membrane proteins distinguishes different cell types.
    8. The cell membrane of a nerve cell is about 80% protein.
    9. The fatty Schwann cell is about 80% lipid.
  D. Movement across membranes
    1. Solution
      a) This is a homogeneous mixture of a substance (the solute) dissolved in water (the solvent).
      b) Concentration refers to the relative number of one kind of molecule compared to the total number of molecules present, usually given in terms of the solute.
    2. Passive diffusion
      a) Diffusion.
        (1) The movement of a substance from a region where it is very concentrated to a region where it is not very concentrated.
        (2) This is called "moving down" or "following" its concentration gradient.
      b) Passive diffusion.
        (1) This requires no input of energy and reaches a point at which the concentration of the substance is the same on both sides of the membrane.
        (2) Molecules of oxygen, carbon dioxide, and water are among those that freely cross through biological membranes.

3. A special case of diffusion—the movement of water (table 5.1)
   a) Osmosis is influenced by the concentration of dissolved substances inside and outside of the cell (fig. 5.8).
   b) Most cells are isotonic with reference to their surrounding fluid—that is, solute concentration is the same within and outside the cell, so that there is no net flow of water.
   c) If a cell is placed in a solution in which the concentration of solute is lower than inside the cell (a hypotonic solution), water enters the cell and the cell swells.
   d) If a cell is placed in a solution in which the solute concentration is higher than inside the cell (a hypertonic solution), water leaves the cell to dilute the higher solute concentration outside and the cell shrinks.
   e) The paramecium has a special organelle, a contractile vacuole, which enables it to pump the extra water out (fig. 5.10).
   f) The resulting rigidity, caused by the force of water against the cell wall, is called turgor pressure (fig. 5.11).
   g) Kidney tubules have very active membranes that return valuable substances to the blood and excrete water.
   h) One chemical that is recycled in the kidney is water, and the amount returned to the blood is influenced by the intake of water as well as by other chemicals, such as caffeine and alcohol.

4. Facilitated diffusion (fig. 5.12): Some molecules that are too big to slip through a custom-fit channel in the membrane can nevertheless cross it with the aid of a carrier protein.

5. Active transport.
   a) Sometimes a cell accumulates a particular substance at higher concentrations than are present outside the cell.
   b) This is possible with the aid of both a carrier protein and energy provided by a molecule called adenosine triphosphate (ATP).
   c) Movement of a molecule through a membrane against its concentration gradient using a carrier protein and energy is called active transport.
   d) The energy-driven "sodium/potassium pump" helps to control a cell's volume by setting up solute concentrations on either side of the cell membrane.

6. Endocytosis.
   a) Large particles enter cells by endocytosis, in which the cell membrane in a localized region moves outward to surround and enclose the particles (fig. 5.13).
   b) The pocket of membrane then pinches off from the interior of the membrane, producing a vesicle (a bubblelike structure) containing the particle, which is released into the cytoplasm.

7. Exocytosis. (fig. 5.14)
   a) Inside the cell, a vesicle made of a lipid bilayer surrounds a structure that is to be transported out of the cell.
   b) An example of this is a drop of secretion such as mucus.

IV. The Cytoskeleton—Cellular Support
  A. Functions of the cytoskeleton.
    1. The cytoskeleton gives a cell its shape by supporting its outer membrane and defining spaces inside where particular organelles lodge.
    2. Another function is the control of cell movements.
    3. Within a cell in the process of division, duplicated chromosomes are distributed into two cells by the cytoskeleton, and the cytoskeleton builds the cell wall in plant cells.
  B. Microtubules (fig. 5.15).
    1. These are largely responsible for cellular movements.
    2. Each microtubule is a long chain of a protein called tubulin.
    3. Cells are in a perpetual state of flux, building up and breaking down microtubules to carry out particular functions.
    4. A sperm's "tail," or flagellum, contains microtubules that can slide past one another, generating movement.
    5. Many types of eukaryotic cells are fringed with cilia, hairlike structures built of microtubules that move in a coordinated fashion.
    6. Cilia move using energy obtained from ATP (fig. 5.16).
    7. In animals, beating cilia move liquids over cell surfaces, and in the upper respiratory tract in humans they sweep mucus and inhaled dust up to the throat.
    8. In the human female, the cilia wave, causing an egg cell to move down the tract; they are also found on the sperm cells of primitive plants.
  C. Microfilaments (fig. 5.17).
    1. This is a tiny rod made of the protein actin.
    2. They cause muscle contraction to occur and are important in endocytosis by white blood cells.

V. Coordination of Cellular Architecture—The Meeting of Sperm and Egg
  A. The key to a correct match-up lies in cell surfaces, and a pattern of proteins and microfilaments that dot an egg's surface attract sperm only of the same species.
  B. When the successful sperm makes contact with the egg's surface, it rapidly strings together actin molecules in its front tip, forming a projection that pokes at the egg.
  C. In an instantaneous response, microfilaments that protrude from the egg's surface organize to form a "fertilization cone," which rises upward to engulf the sperm head; the head is packed with genetic material from a male sea urchin and is drawn into the egg by endocytosis, leaving its microtubule tail behind.
  D. Inside the egg cell, the single engulfed sperm nucleus is pulled by microfilaments towards the egg cell's nucleus, and the microfilament guides disassemble as the two cells unite and the nuclear membranes fuse to form a single nucleus.

LEARNING OBJECTIVES

After reading this chapter, the student should be able to answer these questions:

1. What structures comprise the cellular architecture?
2. How are a cell's surface molecules and their arrangements important in biological functioning?
3. What are the components of a cell membrane and how are they organized?
4. How do substances cross a cell membrane?
5. What are the functions of the cytoskeleton, and what are its components?

ANSWERS

"Questions"

1. a. The body's immune system uses cell surface proteins to distinguish tissue that belongs to itself from that of members of its own species.
    b. The body's immune system uses cell surface proteins to distinguish tissue that belongs to itself from that of members of a different species.
    c. Surface differences between cell types are particularly important during the development of the embryo, when different cells sort themselves to grow into distinctive tissues and organs.
2. Compare the HLA profiles of diseased individuals with individuals who show no symptoms of the disease.
3. They both are composed of lipid bilayers.
4. After protein synthesis, buds carrying the surface markers break off the Golgi apparatus, forming vesicles that travel to the cell membrane. The vesicle fuses with the cell membrane in the process of exocytosis, leaving the marker positioned on the cell surface.
5. ATP is used in the process of active transport to provide the energy required to move molecules from low to high concentration. ATP is expended for the movement of cilia and flagella and during muscle contraction, when microfilaments slide past each other.
6. Cancer cells lose the characteristic surface markers of their cell line and become like embryonic cells. Cancer cells are in a perpetual flux of building up and breaking down microtubules.
7. Microfilaments align beneath the cell membranes and propel portions of the cell membrane outward to entrap particles.

1. A chemical could be designed with an active site similar to the sperm that would bind to the egg, eliminating sites for sperm binding, or a chemical with an active site that matched the receptor of the egg could be used to bind up the sperm before it attached to the egg, preventing fertilization.
2. It is possible that the proteins block the surface markers so that the mother rabbit's immune system does not recognize the fetus as a foreign substance. If it were not for this mechanism, the mother's immune system might cause the fetus to abort.
3. There is no right answer for this question but the student should consider questions such as which candidate has the most immediate need and which candidate has the best chance of surviving with the donated organ.
4. The original membrane model assumed that lipids and proteins were not integrated. Improved microscopic techniques reveal that proteins are embedded in the lipid bilayer and capable of movement, giving rise to the fluid mosaic model.
5. It is likely that the vesicles merge with the cell membrane, releasing the content into the vicinity of the egg through the process of exocytosis.
6. The amount of solvent in the 5% solution is a lower concentration than that of the cell. Consequently, solvent, moving from high concentration to low, exits the cell. The solute is delayed by the semipermeable membrane. The cell wall, which is permeable, remains intact, while the cell membrane shrinks due to lower volume. This results in the cell membrane now being visible.

AUDIOVISUAL MATERIALS

Cell Biology, Coronet Instructional Films, 17 min.

Cell Biology: Life Functions, Coronet Instructional Films, 19 min.

Cell Biology: Structure and Composition, Coronet Instructional Films, 13 min.

The Cell: A Functioning Structure (Parts 1 and 2), CRM McGraw-Hill, 30 min.

The Cell: Its Structure, Carolina Biological Supply, filmstrip.

Cell Structure and Function, educational images ltd., slides.

Inside the Cell: Microstructures, Mechanisms, and Molecules, Guidance Associates, filmstrip, video, slides.

Diffusion and Osmosis, Coronet Films, 10.5 min.

Diffusion and Osmosis, Encyclopaedia Britannica Educational Corp., 14 min.

The New Cell, Carolina Biological Supply, filmstrip.

The Living Cell, Harper & Row College Media, 27 min.

6 Biological Energy

KEY CONCEPTS

The reactions of metabolism convert energy stored in nutrient molecules to ATP.

Metabolic reactions are organized into pathways regulated by enzymes, the presence of reactants or products, and/or hormones.

Enzyme evolution may have occurred in response to substrate availability.

Basal metabolic rate (BMR) is influenced by age, weight, sex, body proportions, and activity level and regulated by the thyroid gland.

Autotrophs use light or inorganic chemicals to form organic molecules, whereas heterotrophs eat others to gain energy.

Photosynthesis transforms light energy to chemical energy stored in the bonds of glucose. It involves the light reaction where photolysis releases free oxygen, hydrogen ions, and electrons from water.

Excited electrons from plant pigments bond hydrogen ions with a carrier NADP for transport to the dark reaction. In the Calvin cycle of the dark reaction, carbon dioxide is combined with hydrogen and high-energy electrons from NADPH, and uses energy from ATP to form glucose.

Glycolysis and cellular respiration convert energy stored as glucose to ATP. In glycolysis, one glucose molecule yields two molecules each of pyruvic acid, NADH, and ATP. In anaerobic respiration, NADH formed from the breakdown of pyruvic acid is reduced to NAD^+. In aerobic respiration, pyruvic acid is converted to CO_2, NADH, and $FADH_2$. Along a respiratory chain, NADH and $FADH_2$ is used to convert ADP to ATP.

Glycolysis is probably the most ancient form of metabolism to form as a reverse reaction to photosynthesis.

KEY TERMS

acetyl CoA formation
adenosine triphosphate (ATP)
alcoholic fermentation
anabolism
autotrophs
basal metabolic rate
catabolism
cellular respiration
chloroplast
cristae
dark reactions
electron
glycolysis
grana
heterotrophs
hormones
lactic acid fermentation
light reactions
metabolism
negative feedback
photophosphorylation
photosynthesis
photosystems
respiratory chain
stroma
stroma lamellae
thylakoids
thyroid gland
transport chains

CHAPTER OUTLINE

I. Energy in Living Systems
    A. Energy.
        1. Energy is the ability to do work, and it is evidenced by movement.
        2. It cannot be created or destroyed, but it can change form.
        3. The energy of motion is called kinetic energy.
        4. Energy contained in the structure or position of matter is called potential energy.

B. ATP—biological energy currency (fig. 6.1).
1. It takes energy to make bonds, and energy is released when bonds of molecules are broken.
2. Much of this released energy of life is stored temporarily in the covalent bonds of the molecule adenosine triphosphate (ATP).
   a) ATP is a nucleotide, composed of the nitrogen-containing base adenine, a sugar group (ribose), and three phosphate groups (a phosphorus atom bonded to four oxygen atoms).
   b) When the endmost phosphate group of an ATP molecule detaches, the disruption of its bond releases energy and the molecule become adenosine diphosphate (ADP).
   c) Another phosphate bond can be broken to yield adenosine monophosphate (AMP) and another release of energy.
   d) The energy contained in the phosphate bonds of ATP can be used by another chemical reaction.
   e) These reactions use the energy in ATP's bonds to synthesize molecules, to break down large molecules into smaller ones, to power the active transport of molecules across membranes, and to move such structures as cilia.
   f) On a larger scale, ATP is responsible for the muscular motion and the transmission of nerve impulses.

C. Metabolic pathways—energy on a cellular level (fig. 6.2).
1. The reactions that take place within living cells are collectively termed metabolism.
2. The general functions of metabolism are to build structures and to convert stored energy into forms that can be directly used to power such biological functions as synthesis, motion, and transport.
3. In general, the reactions of metabolism extract chemical energy from the bonds of nutrient molecules to form phosphate bonds in ATP.
4. Metabolic reactions are organized into pathways that consist of several chemical reactions linked sequentially so that a product of one reaction becomes a reactant (starting material) of another.
5. Anabolism.
   a) These energy-requiring pathways that build large molecules from small ones are also known as biosynthesis.
   b) It is said to diverge, because a few types of precursor molecules combine to yield many different types of products.
6. Catabolism.
   a) Metabolic pathways that break down large molecules and release energy; also known as degradation.
   b) Catabolic pathways converge, in that many different types of large molecules degrade to yield fewer types of small molecules.

D. Control of metabolism.
1. In a metabolic pathway, the enzyme whose reaction proceeds the slowest controls the pathway's productivity, because each subsequent reaction requires the product of the preceding reaction to continue.
2. The reaction catalyzed by this enzyme is called the rate-limiting step, and the enzyme is a regulatory enzyme.
3. When an enzyme is turned off by the accumulation of a product, it is responding to negative feedback, for example, in the biosynthesis of amino acids (fig. 6.3).
4. Sometimes the accumulation of a particular intermediate in a metabolic pathway under the regulatory enzyme's control signals that the pathway is not active enough, and the regulatory enzyme steps up its activity, which is called positive feedback.
5. Metabolic balance is also maintained by hormones.

E. The evolution of metabolic pathways.
1. The reactions of metabolism in diverse species are remarkably similar, with many pathways virtually identical.
2. There could have been more than one genetic variety of protocell in which one evolved that made an enzyme that could convert some other nutrient (C) into the original one (D).
3. Then, a variety of organisms with an additional enzyme that converted a nutrient B to C (and then using the first enzyme, C to D), such that A–B–C–D evolved (fig. 6.4).

F. Whole-body metabolism—energy on an organismal level.
1. The moving muscles, flowing blood, and other bodily functions constitute whole-body metabolism.
2. Nutritionists measure energy in units called Calories, with a single Calorie equaling the amount of energy needed to raise the temperature of 1 kilogram (slightly more than a quart) of water by 1° C.
3. The nutritional Calorie is 1,000 times as large as a chemist's calorie (with a lowercase c), so it is preferable to refer to the nutritional Calorie as a kilocalorie.
4. The energy required by an organism simply to stay alive is described as the basal metabolic rate (BMR), which measures the kilocalories needed for heartbeat, breathing, the functioning of nerves, kidneys, and glands, and the maintenance of body temperature when the subject is awake, physically and mentally relaxed, and has not eaten anything for 12 hours.
5. Several factors can influence BMR, including age, sex, weight, and body proportions.
6. The basal metabolic rate rises from birth to about age five, then declines until the teen years, when it peaks again, and then drops in parallel to declining energy needs.
7. The thyroid gland, located in the neck, is one of several glands that affects basal metabolism and manufactures the hormone thyroxine, which increases energy expenditure.

45

8. In general, the smaller the organism, the higher its metabolic rate.
9. Smaller organisms have higher surface-to-volume ratios and therefore lose more heat to the environment than larger organisms.
10. Shivering raises the metabolic rate.
11. Plants detect the diminishing hours of daylight that herald the arrival of cold weather and drop their leaves and slow metabolism in their protected inner tissues and roots.
12. Many animals hibernate (fig. 6.5).

G. A global view of biological energy.
1. Solar energy is harnessed by green plants, algae, and certain bacteria in a series of metabolic pathways called photosynthesis, where carbon dioxide and water react, in the presence of sunlight and certain pigment molecules, to produce oxygen and energy-rich molecules.
2. Food webs of "who eats whom" are actually routes of energy transfer in the living world with each energy transfer only about 10% efficient.
3. Organisms that obtain energy from nonliving sources, such as the sun, are called autotrophs (self-feeding).
4. Chemoautotrophs are organisms that extract energy from hydrogen sulfide and then combine this energy to synthesize organic compounds.
5. The organic compounds that the bacteria synthesize with the aid of the energy derived from hydrogen sulfide are used to power their own life functions.

6. The diverse living communities of thermal vents are made possible by the chemoautotrophic bacteria.
7. The long-term storage of energy in chemical bonds of nutrient molecules is converted to short-term storage in the bonds of ATP to power cellular activities.

II. Photosynthesis
A. Overview.
1. In the reactions of photosynthesis, plants capture energy in sunlight and convert it to stable chemical energy in a series of steps.
2. The excited electrons become stabilized by losing energy in increments.
3. This energy is used by the cell to first synthesize molecules that contain the energy in their bonds for short periods and ultimately to synthesize organic molecules, primarily glucose, which store the energy indefinitely.
4. Reactions of photosynthesis convert 6 molecules of carbon dioxide and 6 molecules of water to 1 molecule of glucose and 6 molecules of oxygen: $6\ CO_2 + 6H_2O \text{------}> C_6H_{12}O_6 + 6O_2$.
5. Glucose ($C_6H_{12}O_6$) is an excellent energy source and is dismantled to carbon dioxide and water-releasing energy from its bonds to make 36 "high-energy" ATP bonds.

B. Light (fig. 6.6).
1. Electromagnetic energy consists of tiny packets of energy called photons that travel in waves.

2. The visible portion of the electromagnetic spectrum is vital to life because of its role in photosynthesis.
3. Visible light excites molecules, and when a photon impinges upon a molecule, it is absorbed, causing electrons close to its atom's nucleus to jump to a higher energy level, where they are in an excited state.
C. Chlorophyll and chloroplasts (figs. 6.7 and 6.8)
    1. The plant molecules that typically capture light energy in their excitable electrons are pigments.
        a) Chlorophyll *a* is a large pigment molecule that absorbs wavelengths corresponding to red, orange, blue, and violet and reflects green wavelengths in plants.
        b) Other pigment molecules absorb light energy.
    2. Several different pigment molecules cluster together to form photosystems.
    3. A chloroplast is constructed of two outer membranes surrounding a highly folded third membrane.
        a) The nonmembranous inner region is called the stroma, and it contains several types of proteins.
        b) Inner chloroplast membranes that are loosely packed are called stroma lamellae, and these contain the pigment molecules.
        c) Stacks of thylakoids are called grana.
D. The chemical reactions of photosynthesis.
    1. Those reactions that occur early in the process are linked to the electron-transport chain.

    a) An electron from a molecule of chlorophyll *a* that has been excited by light energy is passed along a series of electron-carrier molecules, each of which holds the electron at a slightly lower energy level than the one before.
    b) Some of the energy that is released is packaged in the form of ATP.
    c) An electron-transport chain ends with a final electron-accepting molecule.
    3. Photophosphorylation is when energy released by the electron-transport chain linking the two photosystems is stored in the phosphate bonds of ATP.
    4. Photolysis is when electrons are stripped from a water molecule (to replace electrons lost by molecules of chlorophyll *a*), which split to yield oxygen gas and protons H^+.
    5. Two sets of reactions constitute photosynthesis: light and dark reactions.
        a) Light reaction.
            (1) Light is required and water is split.
            (2) ATP is produced and a molecule called nicotinamide adenine dinucleotide phosphate ($NADP^+$) is reduced to form NADPH.
        b) Dark reaction.
            (1) It does not require light for reactions to take place.
            (2) Products of light reactions are used to produce glucose from carbon dioxide.

6. The light reactions.
   a) Photosynthesis begins in the cluster of pigment molecules that constitute photosystem II.
   b) The energy from incoming photons is absorbed by the pigment molecules in photosystem II.
   c) The energy is transferred from one pigment molecule to another in the photosystem until it reaches a particularly reactive molecule of chlorophyll *a*, one of whose electrons is excited to a higher energy level.
   d) The excited electron leaves the reactive chlorophyll *a* molecule and is accepted by the first electron-carrier molecule of the electron-transport chain that links the two photosystems (fig. 6.9).
   e) The reactive chlorophyll *a* molecule replaces its lost electron with an electron released from the splitting (photolysis) of a molecule of water into oxygen gas and protons (H^+).
   f) When the electrons that have been boosted from the reactive chlorophyll *a* molecule pass through the electron-transport chain, they lose energy, which is captured and used to add a phosphate group to an ADP molecule so that it becomes ATP.
   g) Energy is stored in the phosphate bonds of ATP.
   h) The energy of photons propels electrons from the reactive chlorophyll *a* of photosystem I to the first electron-carrier molecule in a second electron-transport chain.
   i) The boosted electrons from the reactive chlorophyll *a* molecule of photosystem I are replaced by the electrons passed down the first electron-transport chain from photosystem II.
   j) The transported electrons of photosystem I reduce a molecule of $NADP^+$ to NADPH.
   k) This NADPH, plus the ATP generated in photosystem II, are the sources of energy used for the dark reactions that follow.
   l) The overall products of the light reactions are oxygen, ATP, and NADPH.
7. The dark reactions.
   a) The end result of photosynthesis is the incorporation of carbon from carbon dioxide into glucose and other organic compounds that a living cell can use to store energy.
   b) The dark reactions "fix" carbon into biologically useful organic compounds and are powered by the products ATP and NADPH.
   c) Some of the dark reactions form a metabolic cycle known as the Calvin cycle (fig. 6.10).
      (1) This begins with the reaction of carbon dioxide and a five-carbon molecule to form a six-carbon molecule, which splits to yield two three-carbon products.
      (2) After several more reactions, the three-carbon molecule glyceraldehyde phosphate is produced.

(3) Other dark reactions convert molecules of glyceraldehyde phosphate to glucose and other organic molecules.

(4) The Calvin cycle makes two complete turns, converting 18 molecules of ATP to ADP and 12 molecules of NADPH to $NADPH^+$.

III. Energy Extraction—From Glucose to ATP

A. In most organisms, energy is retrieved from glucose in a catabolic process that occurs in two stages.

B. The first, glycolysis, takes place in the cytoplasm, and the second stage, cellular respiration, is a series of reactions that occur in the mitochondrion.

C. Plants carry out both photosynthesis and cellular respiration (fig. 6.11).

D. In the presence of oxygen, glucose is oxidized to yield carbon dioxide, water, and energy with one molecule ultimately yielding 36 ATP molecules.

E. Glycolysis (fig. 6.12).

1. In nine enzyme-catalyzed steps, one glucose molecule is rearranged and split to yield two three-carbon molecules of a compound called pyruvic acid.

2. Two molecules of one of the intermediate organic compounds formed along the pathway lose hydrogens to molecules of nicotinamide adenine dinucleotide (NAD^+), which are thereby reduced to NADH and stored temporarily.

3. A gain of hydrogens is a reduction reaction because electrons are gained; losing hydrogens is an oxidation reaction because electrons are lost.

4. Each round of glycolysis produces two molecules of pyruvic acid, two molecules of NADH, and two molecules of ATP.

5. This process occurs in all organisms.

F. Fermentation—in the absence of oxygen.

1. Species that live in environments lacking oxygen are called anaerobes, and after glycolysis they utilize a short metabolic pathway called fermentation or anaerobic respiration.

2. The reactions that follow glycolysis, whether anaerobic or aerobic, remove hydrogens from NADH to produce NAD^+.

3. Two types of fermentation pathways accomplish this, and both occur in the cytoplasm.

4. Yeast cells convert pyruvic acid to ethanol and carbon dioxide in alcoholic fermentation, converting NADH to NAD^+ in the process to manufacture baked goods and alcoholic beverages (fig. 6.13).

5. Lactic acid fermentation converts the pyruvic acid of glycolysis to the three-carbon compound lactic acid in a single step, converting NADH and NAD^+; this occurs in human muscle cells when they are exercising.

G. Aerobic respiration—in the presence of oxygen.

1. Aerobic respiration is a far more efficient pathway than glycolysis or anaerobic respiration, because it allows the dismantling of more of the bonds in glucose.

2. It occurs in the mitochondrion, an organelle constructed of an outer membrane and a highly folded inner membrane called cristae, which contains enzymes and electron-carrier molecules that participate in aerobic respiration (fig. 6.15).

3. The role of oxygen is to accept electrons that have been passed through a series of electron-accepting molecules.

4. The energy that is released along the way is harnessed to convert ADP to ATP.

5. The process starts where glycolysis ends with a reaction called acetyl CoA formation

6. Once inside the mitochondrion, each molecule of pyruvic acid loses a carbon dioxide, converts NAD^+ to NADH, and attaches to an enzyme called coenzyme A to form a new molecule, acetyl CoA (fig. 6.16).

7. In addition to its production from pyruvic acid, acetyl CoA is part of the catabolic pathways of fats and amino acids.

8. In energy metabolism, the formation of acetyl CoA is a bridge between glycolysis and aerobic respiration (fig. 6.22, table 6.3).

9. For each acetyl CoA that enters the Krebs cycle, one ATP molecule and three NADH molecules are produced, and one molecule of another electron-carrier molecule, flavin adenine dinucleotide (FAD) is reduced to $FADH_2$ (fig. 6.17).

10. In the final stage of extracting energy from glucose, the energy contained in the NADH and $FADH_2$ molecules generated is used to convert ADP into ATP.

11. This occurs on a series of electron-accepting enzymes that are embedded in the inner mitochondrial membrane, forming a respiratory chain.

12. Electrons that enter glucose catabolism as part of hydrogen atoms in the intermediates of glycolysis and the Krebs cycle are now transferred from NADH and $FADH_2$ to electron-carrier molecules in the respiratory chain, losing energy as they pass from one carrier to the next (fig. 6.18).

13. Some of this released energy is captured to manufacture ATP.
    a) One molecule of NADH is used to convert three molecules of ADP to ATP.
    b) Energy from one molecule of $FADH_2$ is used to convert two molecules of ADP to ATP.
    c) Oxygen is the final electron acceptor, combining with hydrogen to form water.

14. The ATP output from the Krebs cycle is 24.

IV. How Did the Energy Pathways Evolve?
   A. It is likely that glycolysis was the first of the energy pathways to form, because it is common to nearly all cells and is the simplest and most common (fig. 6.21).
   B. Photosynthesis occurs only in green plants, algae, and cyanobacteria.
   C. Fermentation is restricted to certain species.
   D. Aerobic respiration occurs only in cells that utilize oxygen.
   E. Glycolysis probably evolved when the earth's atmosphere lacked oxygen, and these reactions enabled the earliest organisms to extract energy from simple organic compounds present in the nonliving environment.
   F. Photosynthesis may have evolved from glycolysis because some of the reactions of the Calvin cycle are the reverse of some of the reactions of glycolysis.
   G. The evolution of photosynthesis, over time, pumped oxygen into the primitive atmosphere, a mechanism that could produce organic compounds.

H. Splitting of water in the atmosphere by electrical storms released single oxygen atoms that joined with diatomic oxygen (O_2) to produce the ozone (O_3), which blocks harmful ultraviolet radiation.

I. An early photosynthetic organism was a prokaryotic cell, probably an anaerobic bacterium that used hydrogen sulfide (H_2S) instead of the water used by plants.

LEARNING OBJECTIVES

After reading this chapter, the student should be able to answer these questions:

1. How is energy essential to life, on a cellular, whole-body, and global level?
2. Where does the energy used by living organisms ultimately originate?
3. How do the reactions of photosynthesis capture energy from sunlight and use it to manufacture glucose?
4. How do the reactions and pathways of cellular respiration extract energy from a glucose molecule?
5. How are the pathways of energy metabolism interrelated?
6. How might the pathways of energy metabolism have evolved?

ANSWERS

"Questions"

1. Both metabolic processes are organized into pathways that consist of many chemical reactions sequentially linked together. They both involve enzymes and carriers. Both include the step of phosphorylation (production of ATP) and electron-transport chains. Photosynthesis differs from cellular respiration in that it is anabolic, converting the small molecules of water and carbon dioxide into glucose and oxygen. Cellular respiration is catabolic, reducing pyruvic acid to smaller molecules. Photosynthesis requires energy input in the form of sunlight captured with the aid of the molecule chlorophyll. Energy entering cellular respiration is in

the form of pyruvic acid. Energy exiting photosynthesis is in the storage form of glucose, whereas in cellular respiration it is in the usable form of ATP.

2. It is likely that an enzyme is missing that is required to catalyze a chemical reaction where the excess molecule is a reactant. Since it cannot be depleted, it would show up in excess. When chemical reactions are sequential in metabolism, the impedance of one step prevents subsequent reactions from taking place, resulting in the buildup of other reactants.

3. The signaling of a decrease in production of acetyl CoA by excess ATP is an example of negative feedback. The product of a chain of reactions inhibits an enzyme involved earlier in the chain, thus preventing excess accumulation of the product. The signaling of increased breakdown of glycogen by increased levels of AMP and ADP exemplifies a positive feedback loop where the buildup of intermediates in a reaction speeds up the activity of an enzyme involved in the reaction.

4. Humans, like all heterotrophs, are unable to derive energy from sunlight or inorganic molecules. They must utilize chemical energy stored in large organic molecules synthesized by autotrophs to produce ATP. Like other aerobes, the human also needs the oxygen released from water by plants.

5. If the bacteria were using only the carbon from carbon dioxide for the production of glucose, they would emit oxygen instead of sulfur, which comes from the H_2S.

6. The rate of a reaction is in part regulated by the availability of its reactants. Since carbon dioxide is required for the dark reaction, increasing its availability should increase the overall rate of photosynthesis.

7. Photosynthesis is an anabolic process since the small molecules of water and carbon dioxide are combined to yield the larger molecule of glucose. Glycolysis is catabolic since glucose is broken down into two molecules of pyruvic acid. Anaerobic respiration is unlike the reversal of photosynthesis since the breakdown of pyruvic acid yields ethanol or lactic acid,

neither of which are found in photosynthesis. When combined with glycolysis, aerobic respiration more closely resembles the reversal of photosynthesis, with glucose and oxygen being the reactants and carbon dioxide and water the products.

8. Six ATP molecules are produced from glycolysis, 6 from the acetyl CoA step, and 24 from Krebs cycle.

"To Think About"

1. The body's cells are unable to utilize fructose so it accumulates to toxic levels. The primary source of energy is simple sugars through glycolysis and as pyruvate in the Krebs cycle. Since fructose is unable to be used in glycolysis the cells must switch to fatty acids and amino acids as energy sources, which can also enter the Krebs cycle. The individual does not have an adequate energy source for normal functioning and also uses amino acids and fatty acids as an energy source rather than utilizing these molecules for normal growth and development.

2. Smoking suppresses hunger. When the individual stops smoking he or she suffers from hunger, which triggers the body's craving for food energy. This may lead to increased food intake, leading to a gain in weight or a change in the weight setpoint mechanism, slowing metabolism, which also leads to weight gain.

3. The advantage would be a source of glucose for cellular metabolism. Humans would not need to rely on agriculture for food sources. The disadvantage is that it would be extremely difficult to regulate the energy flow in the body and to obtain a balance of required nutrients for normal growth and development.

4. She exceeded her ability to provide energy aerobically for muscle contraction. Her muscles generated the energy necessary by anaerobic metabolism with the accumulation of lactic acid. She had to breath rapidly following running to convert the lactic acid back into pyruvate, which then enters the Krebs cycle. The lactic acid may also have caused the painful sensations in the muscles following the exercise period.

5. The anaerobic pathway of cellular respiration is probably used by cancer cells as shown by their ability to survive in an acid environment that is low in oxygen.

6. Humans and yeast cells likely had a common cellular ancestor. Since these pathways have been maintained in evolutionary history, it indicates that they have been successful in supplying the cells with energy or molecules essential for development and growth.

AUDIOVISUAL MATERIALS

The Green Machine, ("Nova"), 49 min. 1978.
Metabolism: The Fire of Life, Carolina Biological Supply, 36 min. 1982.
Photosynthesis, Encyclopaedia Britannica Educational Corp., 20 min. 1982.
Respiration: Energy for Life, Churchill Films, videocassette, 26 min. 1982.

7 Mitosis

KEY CONCEPTS

Mitosis is responsible for growth, development, reproduction, and repair of damaged tissue.

The cell cycle is a sequence of events that describe whether or not a particular cell is dividing or preparing to do so.

Interphase involves the replication of the genetic material followed by mitosis, where genetic and cellular material are equally divided between daughter cells.

The rate, timing, and number of mitotic divisions is regulated by the action of hormones, growth factors, cell size, and proximity to other cells.

Binary fission in bacteria is an asexual form of reproduction that differs from mitosis.

The growth of bacteria depends upon space and nutrients.

Cell populations are defined by their proportion of cells that are in different stages of the cell cycle.

Cancer cells are characterized by heritability, transplantability, dedifferentiation, abnormal growth rate, and lack of contact inhibition.

Benign tumors are localized and do not metastasize.

Oncogenes can be triggered by viruses, chemicals, radiation, and deficiencies in nutrients.

KEY TERMS

anaphase	interphase
binary fission	karyokinesis
cancer	meiosis
cell cycle	metaphase
cell populations	metastasized
centromere	mitosis
chromatids	nucleolus
contact inhibition	oncogenes
cytokinesis	population growth
G_1 phase	curve
G_2 phase	prophase
growth factors	somatic cells

hormones	spindle apparatus
stem cells	transgenic organisms
telophase	

CHAPTER OUTLINE

II. Mitosis Provides Growth, Development, Repair, and Reproduction.
  A. Mitosis (fig. 7.1).
     1. A form of cell division in which two identical cells are generated from one.
     2. It occurs in somatic (body) cells for growth and repair in multicelled organisms and reproduction for single-celled organisms.
  B. Binary fission.
     1. A form of asexual reproduction similar to the mitosis that bacteria undergo.
     2. The original cell first doubles its genetic material and then distributes it equally among the two "daughter" cells and other cellular constituents.
  C. Meiosis.
     1. Mode of reproduction for most higher organisms.
     2. A form of cell division that halves the amount of genetic material to fashion the sperm and egg, which are sex cells.
II. The Cell Cycle (fig. 7.2)
  A. The cell cycle is divided into two major stages: mitosis, when the cell is actively dividing, and interphase, when the cell is not actively dividing.

B. Interphase.
  1. It is a very active time, when the basic biochemical functions of life are occurring and also replication of genetic material and other subcellular structures in preparation for splitting into two daughter cells.
  2. Interphase is divided into phases of gap (designated "G") and synthesis ("S").
      a) During the first gap phase, the G_1 phase, proteins, lipids, and carbohydrates are synthesized.
      b) The next period of interphase is the S phase.
          (1) Replication of the genetic material is undertaken over a period of 8 to 10 hours.
          (2) Microtubules are synthesized, which will be assembled into a spindle apparatus early in the mitotic process.
      c) G_2 phase (the second gap phase).
          (1) More proteins are synthesized.
          (2) The end of G_2 is signaled by the DNA winding more tightly around its associated proteins.

C. Mitosis.
  1. During mitosis, or M phase, the replicated genetic material divides.
  2. A cell's chromosomes are replicated during the S phase, resulting in two strands of identical chromosomal material, called chromatids, which are held together by a constriction called a centromere (fig. 7.3).
  3. Phases of mitosis (fig. 7.4).
      a) Prophase (the first stage of mitosis).
          (1) The DNA is coiled very tightly around the chromosomal proteins, which shorten and thicken into the chromosomes and become visible under a microscope.
          (2) The mitotic spindle forms during prophase from microtubules.
          (3) A dark spot called the nucleolus is where a type of nucleic acid RNA is manufactured, which participates in protein synthesis.
      b) Metaphase: The spindle aligns the chromosomes along the central axis of the cell, called the equatorial plate.
      c) Anaphase.
          (1) The centromeres split, relieving the tension and sending one chromatid from each pair to opposite ends of the cell, called poles.
          (2) As the chromatids separate, some microtubules in the spindle shorten and some lengthen in a way that actually moves the poles farther apart, stretching the dividing cell.
      d) Telophase.
          (1) The spindle is disassembled, and nucleoli and nuclear membranes reform at each end of the stretched-out cell.

(2) Cytokinesis is when the other cellular contents, including organelles and macromolecules, are distributed among the two forming daughter cells.

(3) Plant cells complete an additional step when they divide, building a new cell wall.

4. Variations in mitosis among different species.

   a) Bacteria.

   (1) The replicated DNA molecules of bacteria are attached directly to the cell membrane.

   (2) The cell membrane and the cell wall elongate between the attached DNA molecules so that the two copies of the genetic material are pulled apart.

   (3) Bacterial cell division is technically not considered to be mitosis because bacteria lack nuclei, chromosomal proteins, and microtubules.

   b) Single-celled eukaryotes.

   (1) Dinoflagellate chromosomes are free of associated proteins.

   (2) Microtubules press on the nuclear membrane to guide the movements of chromosomes from outside the nucleus.

   c) In higher organisms, the nuclear membrane actually breaks down, permitting direct interaction of chromosomes and the microtubules that pull them into two daughter cells.

III. How is Mitosis Controlled?

   A. A mitosis "trigger."

   1. One theory is that an individual cell may be induced to divide by the presence of a still-undiscovered "trigger molecule," perhaps functional in cancer cells.

   2. Experimental evidence: When two cells in different stages of the cell cycle are fused in the laboratory to form a large cell with two nuclei, the "younger" nucleus quickly "catches up" to the stage of the older nucleus.

   B. A cellular clock.

   1. The Hayflick limit is the rule that mammalian cells seem to obey an internal "clock," which allows them to divide a maximum number of times.

   2. A fibroblast taken from a human fetus, for example, divides from 35 to 63 times, the average being about 50 times.

   C. Hormones.

   1. Certain cells divide frequently at some times, yet infrequently or not at all at others, due to the influence of biochemicals called hormones.

   2. A hormone is manufactured in a gland and travels in the bloodstream to another part of the body, where it exerts an effect.

   D. Growth factors.

   1. Growth factors are proteins that mediate healing.

   2. Epidermal growth factor.

      a) It stimulates epithelium (lining tissue) to undergo mitosis.

      b) An example of EGF action is the filling in of new skin underneath the scab of a skinned knee.

   3. Fibroblast growth factor.

      a) It stimulates division of endothelial cells in blood vessels.

b) FGF also provokes mitosis in fibroblasts, which secrete collagen, a protein that also helps build blood vessels.

4. Platelet-derived growth factor (PDGF) is synthesized by the large cells that give rise to blood platelets.

5. Growth factors can be produced in the laboratory using genetic engineering techniques (fig. 7.6).

E. A cell's size.

1. A cell may divide when its surface-to-volume ration becomes too small for the cell to obtain enough nutrients and to excrete sufficient wastes.

2. Experiments with amoebas demonstrate that if cytoplasm is removed just before the cell would normally divide, mitosis is delayed until the cell grows.

F. A cell and its neighbors—the effect of crowding.

1. When a cell comes in contact with others it slows or halts mitosis in a process called contact inhibition.

2. Division of bacterial cells (fig. 7.8).

a) They are limited by the proximity of other cells, which compete for space and nutrients.

b) Bacterial cell division that is not limited proceeds in an exponential or logarithmic pattern.

(1) Doubling time refers to the duration of a bacterial division.

(2) In the lag phase of population growth, the population is adjusting to their new environment, so equal numbers of cells die as arise by division, with no net gain.

(3) The log phase is when cell division reaches maximum rate.

c) The stationary phase is when the log phase ends, when nutrients and growth factors are depleted, or perhaps when antibiotic drugs or immune system biochemicals begin to kill more bacteria.

d) Finally, wastes accumulate, resources are exhausted, and as a result cell death rate overtakes cell division rate.

G. Mitosis during development is highly regulated (fig. 7.9).

1. The balance of cell death and cell reproduction ensures that the organization of the tissues that form organs is maintained in a growing individual.

2. Many tissues have cells that divide often, termed stem cells.

H. Cell populations.

1. Renewal cell populations are those where cells are actively dividing to maintain linings within animal bodies, which are constantly being shed.

2. Expanding cell populations are those where up to 3% of the cells are dividing when a tissue is injured and new cells are required to repair it.

3. Static cell populations are cells that are highly specialized and no longer divide, as in the case of nerve and muscle cells that grow only by cell enlargement.

IV. Cancer—When the Cell Cycle Goes Awry

A. Cancer is a group of disorders in which certain cells lose normal control over both mitotic rate and the number of divisions they undergo.

B. It begins with a single cell, which divides to produce others like itself, growing into a mass called a cancerous or malignant tumor or traveling in the blood.

C. Characteristics of cancer cells (fig. 7.11).
   1. Cancer cells can divide uncontrollably and eternally, given sufficient nutrients and space.
   2. They have a rate of mitosis that is faster than the rate for the normal cell type from which they arose (table 7.1).
   3. Cancer is inheritable, which means when a cancer cell divides both daughter cells are cancerous.
   4. A cancer cell is transplantable, meaning that if a cancer cell is injected into a healthy animal, the disease spreads.
   5. A cancer cell is somewhat dedifferentiated, exhibiting less specialization than the normal cell type from which it derives.
   6. Some tumors are cancerous and are called malignant, while benign tumors are not.
   7. Cancerous cells have surface structures that enable them to squeeze into any available space (invasiveness) and can spread (metastasis) (fig. 7.12).
D. The causes of cancer.
   1. Oncogenes.
      a) Genes that normally control cell division, but whose ill-timed or ill-placed activation leads to cancer, are called oncogenes.
      b) The new technology of transgenic organisms, in which a multicellular organism is engineered to contain a particular oncogene in every cell, is being used to study the development of cancer (fig. 7.14).
      c) When an oncogene is placed next to a gene that it is not normally next to, perhaps by a virus, it may boost its expression.
      d) Burkitt's lymphoma is a cancer in which a virus triggers chromosome breakage, which places an oncogene next to an antibody gene (fig. 7.15).
      e) The childhood kidney cancer called Wilms' tumor is caused by the absence of a gene that normally halts mitosis in the rapidly developing kidney tubules in the fetus.
   2. Carcinogens.
      a) These are chemicals, nutrient deficiencies, or radiation that may cause cancer (table 7.2).
      b) They may be identified both by epidemiological studies and on the Ames test, in which the ability of a chemical to cause genetic change in bacteria is taken as a strong indication that it may cause cancer in higher organisms.
      c) Carcinogens may be the direct cause of cancer when placed next to a cell, they may act only after the chemical has been metabolized into a intermediate compound (procarcinogin), or they may be chemicals that make other carcinogens more powerful (promoters) (table 7.3).

LEARNING OBJECTIVES

After reading this chapter, the student should be able to answer these questions:

1. Under what circumstances do cells undergo mitosis?
2. What happens within and to a cell during the cell cycle?
3. What happens to a cell when it is not actively dividing?
4. What are some of the ways in which mitotic rate and the number of divisions are controlled?

5. How do the growth properties of cancer cells reflect a loss of control of mitosis?
6. What are the characteristics of cancer cells?
7. How can genes that normally control the cell cycle cause cancer?
8. What are some factors that might contribute to causing cancer?

ANSWERS

"Questions"

1. During interphase, the DNA and other subcellular structures are replicated in preparation for splitting into two daughter cells. Prophase marks the advent of cell division, when stained chromosomes become visible and the nuclear membrane and nucleolus disappear. During metaphase, spindle microtubules attach to chromosomes at the centromere, aligning them at the center of the cell. In the next stage, anaphase, each sister chromatid is pulled toward opposite poles. Telophase is the terminal phase of mitosis, when cytokinesis occurs resulting in two daughter cells.
2. The G_1 phase is the normal operating time of the cell cycle. Some cells are arrested in this state since they do not go through mitosis. During this time, proteins, lipids, and carbohydrates are synthesized. S phase is the time when replicating of genetic material occurs and the spindle apparatus is manufactured. In G_2, more proteins are synthesized and membranes for the new cells are produced.
3. Cancer is a good example of mitosis out of control. The drain to systems to support these demanding cell lines is enough to kill the organism. On the other hand, the inability of nerve cells to go through cell division means that severe damage to brain or spinal tissue cannot be repaired, resulting in reduced or lack of function.
4. The cell is likely to continue to divide the remaining number of times predestined for that cell line according to the Hayflick limit. If it is a fibroblast it will divide approximately 16 to 44 times.

5. If a layer of cells is torn, mitosis sometimes fills in the missing cells. In the laboratory, experiments with torn cornea tissue treated with EGF divides to restore a complete cell layer. In the human body, new skin is produced by mitosis under the scab that has formed over a wound.
6. Cancer cells are normal cell lines that have an uncontrollable rate of mitosis due to lack of contact inhibition. This characteristic is passed on to its clones (inheritability) and can even be passed to noncancer cells (transplantability). They appear dedifferentiated like embryo cells. They are capable of moving (metastasis) and invading other tissues.
7. The results indicate that the population of cells within the tumor were not clones; given new environments, each expressed its unique qualities.
8. A virus infecting a cell may insert its genetic material next to an oncogene. When the viral DNA begins to reproduce itself, it also stimulates the expression of the oncogene. Another idea is that through chromosomal breaks caused by viruses, oncogenes can be placed in new locations, which give them new properties.
9. Since benign tumors contain all the properties of cancer except metastasis, they can still affect the cells in their environment by growing and impeding normal tissue function. It is possible that some cells within the tumor were metastatic and traveled throughout the body.
10. Oncogenes are genes involved in the regulation of mitosis that are expressed at the wrong time. One type of cancer may be caused by an oncogene that promotes expression of genes used in mitosis. Another type of cancer may suppress the activation of genes used in mitosis.

AUDIOVISUAL MATERIALS

Cell Biology: Mitosis and DNA, Cornell University Film Library, 16 min.

Cell Division: Mitosis and Meiosis—Biology Today Series, CRM McGraw-Hill, 1974

Mitosis, 2d ed., *Heredity and Adaptive Change Series,* Encyclopaedia Britannica Educational Corp., 1980

Mitosis and Meiosis, McGraw-Hill, 1974.

Mitosis in Animal Cells, Harper & Row, 1979

"To Think About"

1. Assuming that all the cells are under identical environmental stimulation to go through cell division, the stage that takes the longest period of time will be most represented.
2. Group a—G_1 phase, Group B—G_2 phase?
3. Epithelial and blood cells are fast-dividing cell lines.
4. Since chemicals used to treat cancer can cause significant side effects it is advantageous to identify an effective drug for the particular cancer in the laboratory to save the patient the agony caused by testing various chemicals.

8 Human Meiosis and Reproduction

KEY CONCEPTS

Male and female reproductive systems are composed of homologous structure for gamete production and transport.

If fertilization occurs, the embryo implants in the uterus, otherwise the endometrium and unfertilized ovum are sloughed off in the menstrual flow.

Meiosis results in haploid gametes, which allow for the maintenance of chromosome number after fertilization.

Independent assortment of homologs and crossing over during meiosis result in increased genetic variability.

Reduction division results in halving the number of chromosomes.

Equational division is mitotic and results in four genetically unique haploid daughter cells.

A diploid spermatogonium divides mitotically, yielding one stem cell and one cell that accumulates cytoplasm, becoming a primary spermatocyte.

The morphology of the mature spermatocytes is adapted for travel through the female reproductive tract and for penetration of the ovum.

The million oocytes that a female is born with are arrested in prophase 1, and meiosis is completed only when a secondary oocyte is fertilized.

KEY TERMS

acrosome
alleles
asexual reproduction
cervix
crossing over
diploid
embryo

endometrium
epididymis
equational division
fallopian tubes
follicle cells
gametes
haploid

homologous pairs
independent assortment
meiosis
oocyte
oogenesis
ovaries
polar body
prostate
reduction division

seminal vesicles
sexual reproduction
somatic cells
sperm
spermatogenesis
synapsis
testes
uterus
vas deferens

CHAPTER OUTLINE

I. The Human Male Reproductive System
   A. Sperm cells are manufactured within a 125-meter-long network of tubes called seminiferous tubules, which are packed into paired, oval organs called testes.
   B. The testes lie within a sac called the scrotum.
   C. Each testis is a tightly coiled tube, the epididymis, in which sperm cells mature and are stored.
   D. The vas deferens bend behind the bladder.
   E. This joins the urethra, the tube that also carries urine out through the penis.
   F. Three glands contribute secretions called seminal fluid.
      1. The vas deferential.
         a) These pass through the prostate gland.
         b) They produce a thin, milky, alkaline fluid that activates the sperm to swim.

2. The seminal vesicles add the sugar fructose for energy and leave hormonelike prostaglandins, which may stimulate contractions in the female reproductive tract that help the meeting of sperm and ovum.

3. Bulbourethral glands, each about the size of a pea, are attached to the urethra where it passes through the body wall.

4. All of these secretions combine to form the seminal fluid, in which the sperm cells travel.

II. The Human Female Reproductive System
A. Ovaries.
1. The female sex cells develop within paired organs in the abdomen called the ovaries
2. Within each ovary of a newborn female are about a million oocytes.
   a) An individual oocyte is surrounded by follicle cells, which nourish it.
   b) Once a month, the most mature oocyte from one ovary is released and swept by a current generated by beating cilia on the fingerlike projections of the nearest of the paired fallopian tubes.
   c) The oocyte can be fertilized only during the first 24 hours of this period.
   d) If the oocyte encounters a sperm cell in the fallopian tube and the cells combine and their nuclei fuse, the oocyte completes its development and is called a fertilized ovum.
B. Fallopian tubes carry the oocyte into a muscular saclike organ, the uterus, or womb.
C. Uterus.
1. The endometrium is a blood-rich uterine lining.
2. If the oocyte is not fertilized, it and the endometrium are expelled as the menstrual flow.

D. The lower end of the uterus narrows to form the cervix.
E. The vaginal opening is protected by two pairs of fleshy folds.
1. The major folds are called the labia major.
2. The inner flaps of tissue are called the labia minora.
F. The clitoris is anatomically similar to the penis and stimulates females to experience orgasm.

III. Meiosis
A. The sperm and ovum are often referred to by other, more general names: gametes, germ cells, or sex cells.
1. Each germ cell contains 23 different chromosomes, whereas other cells, called somatic (body) cells, contain two of each type of chromosome, for a total of 46 chromosomes, or 23 pairs.
2. Germ cells are termed haploid (or n) to indicate that they have only one of each type of chromosome.
3. Somatic cells are diploid ($2n$), signifying their double chromosomal load.
4. Germ cells are formed from certain germ-line cells by a special form of cell division called meiosis.
5. Meiosis plus maturation constitute gametogenesis (making gametes).
B. Stages of meiosis.
1. Meiosis entails two divisions of the genetic material.
   a) The first division is called reduction division (or meiosis I) because it reduces the number of chromosomes in humans from 46 to 23.
   b) The second division, called the equational division (or meiosis II), is much like a mitotic division, producing four cells from the two cells formed in the first division.

2. Interphase.
   a) This is a period during which the genetic material replicates.
   b) The germ-like cell in which meiosis begins has two of each type of chromosome.
   c) Homologous pairs look alike and carry the genes for the same traits in the same sequence.
      (1) One of each homologous pair comes from the person's mother.
      (2) One of each homologous pair comes from the father.
      (3) When meiosis begins, each homolog is replicated to form two chromatids joined by a centromere.
3. Prophase I.
   a) This is when the chromosomes condense and become visible (fig. 8.4a).
   b) The homologs line up next to one another, gene by gene, a phenomenon called synapsis.
   c) The synapsed chromosomes pull apart.
   d) These points of contact are the sites at which the two homologs exchange parts in a process called crossing over (fig. 8.5).
   e) New gene combinations arise from crossing over when the parents carry different forms of the same gene, called alleles—blue or brown alleles of an eye-color gene, for example.

4. Metaphase I.
   a) Each chromosome of a homologous pair attaches to a spindle fiber that goes to one pole of the cell (fig. 8.4c).
   b) Each member is anchored to an opposite end of the cell.
   c) When the chromosome pairs align down the center of the spindle of the metaphase I cell, whether a maternally or paternally derived chromosome goes to one pole or the other occurs at random.
   d) Using the formula 2^n, where n equals the number of homologous pairs, we can calculate that the 23 chromosome pairs of a human can line up in 2^{23}, and this arrangement of the members of homologous pairs in the metaphase cell is called independent assortment.
5. Homologs separate in anaphase I.
6. They complete their movement to opposite poles in telophase I.
7. Second interphase is when the chromosomes once again become invisible and spread into very think threads.
8. Prophase II is when the chromosomes are again condensed and visible (fig. 8.7a).
9. In metaphase II the chromosomes (each consisting of a pair of chromatids) align down the center of the spindle (fig. 8.7b).

10. In anaphase II, the centromeres split, each chromatid pair divides in two, and resulting single-chromatid chromosomes are pulled to opposite poles (fig. 8.7c).
11. In telophase II, nuclear envelopes form around the four nuclei (fig. 8.7d).
    a) These are from each of the two cells produced by meiosis I.
    b) Cytokinesis completes the process of meiosis by separating the four nuclei into individual cells (fig. 8.7e).
    c) The net result of meiosis is four haploid cells, each carrying a new assortment of genes and chromosomes.
C. Meiosis leads to genetic variability.
    1. Crossing over mixes up the genetic contributions from the previous generation.
    2. Independent assortment introduces more variability in the way that chromosomes are packaged into daughter cells.
IV. Development of the Sperm—Spermatogenesis
A. The product of male meiosis is the motile, lightweight sperm.
B. The differentiation of sperm cells is called spermatogenesis.
    1. A diploid cell destined to produce sperm cells is called a spermatogonium.
    2. The spermatogonia accumulate cytoplasm and replicate their genetic material, becoming primary spermatocytes.
    3. Each primary spermatocyte divides to form two equal-sized haploid cells called secondary spermatocytes in meiosis I.
    4. In meiosis II, each secondary spermatocyte divides to yield two equal-sized spermatids; each spermatid then specializes, developing the characteristic sperm "tail," or flagellum.

5. The tail has many mitochondria and ATP molecules, and this energy system enables the sperm to swim once it is inside the female reproductive tract.
6. After spermatid formation, some of the cytoplasm is stripped away, leaving mature, tadpole-shaped spermatozoa, or sperm cells for short (fig. 8.9).
7. Each sperm cell consists of a tail, body, and head region and has a small protrusion on its front end called the acrosome.
8. This bump contains enzymes that will help the cell penetrate the ovum's outer membrane.
9. Male meiosis begins in the seminiferous tubules.
10. When a spermatogonium divides, one daughter cell moves towards the lumen and accumulates cytoplasm, becoming a primary spermatocyte.
11. The entire process, from spermatogonium to spermatocyte, takes about 2 months, and a human male manufactures trillions of sperm in his lifetime (fig. 8.10).
V. Development of the Ovum—Oogenesis
A. Meiosis in the female is called oogenesis (egg-making), and it begins, like spermatogenesis, with a diploid cell.
    1. This cell is called an oogonium.
    2. Oogonia are not attached to each other but are surrounded by a layer of follicle cells.
    3. Each oogonium grows, accumulating cytoplasm and replicating its chromosomes, becoming a primary oocyte.
    4. In meiosis I, the primary oocyte divides to become a small cell with very little cytoplasm, called a polar body.

5. The larger cell is called a secondary oocyte (fig. 8.11).
6. In meiosis II, the tiny polar body may divide to yield two polar bodies of equal size, or it may simply decompose.

B. After puberty, meiosis I is completed in only one or a few oocytes each month, and these oocytes stop meiosis again, this time at metaphase II.
C. At specific hormonal cues each month, one such secondary oocyte is released from an ovary, or ovulated; if the oocyte membrane is penetrated by a sperm, then meiosis is completed, and a fertilized ovum forms.
D. A female will only ovulate about 400 oocytes between puberty and menopause.

LEARNING OBJECTIVES

After reading this chapter, the student should be able to answer these question:

1. What structures form the human male and female reproductive systems?
2. Why is it necessary for germ cells to have half the number of chromosomes found in other human cells?
3. What steps accomplish the halving of the chromosome number in germ cell formation?
4. How does germ cell formation increase genetic variability?
5. How do the male and female germ cells differentiate their specialized characteristics?
6. Where in the reproductive systems of each sex does each stage of germ cell formation take place?
7. How does germ cell formation differ in the two sexes?

ANSWERS

"Questions"

1. 2^{39}, or over 54 trillion, combinations are possible for each gamete. To determine the number of combinations for the puppies, this number is multiplied by itself.
2. A single drop of semen still contains adequate numbers of spermatids to make fertilization possible.
3.
   a. Crossing over—the exchange of parts between homologous chromosomes during synapsis of prophase 1.
   b. Gamete—the reproductive cells of sexually reproducing organisms, which contain one half of the genetic material of the parent.
   c. Haploid—refers to one copy of each type of chromosome that a gamete contains.
   d. Homolog—one pair of similar chromosomes that carries the genes for the same traits, one coming from each parent in a sexually reproducing organisms.
   e. Synapsis—the physical joining of homologs during prophase 1 of meiosis, which allows for crossing over.
4. Each system has paired structures in which the sperm and ova are manufactured, a network of tubes to transport these cells, and various hormones and glandular secretions that control the entire process.

5. a. An oogonium—46
   b. A primary spermatocyte—23
   c. A spermatid—23
   d. A cell during anaphase of meiosis I, from either sex—46
   e. A cell during anaphase of meiosis II, from either sex—46
   f. A secondary oocyte—23
   g. A polar body derived from a primary oocyte—23

"To Think About"

1. If the number of chromosomes passed from the parent to the offspring was not halved, the resulting zygote would receive double the number of chromosomes. This much genetic material would make normal gene function impossible.

2. Since the child only receives one-half of each parent's genetic material, the chances of putting together the exact combination of one of the parents is highly unlikely.

3. The Vietnam veteran exposed to Agent Orange may have had the spermatogonia affected by this mutagenic chemical. When spermatogonia divide, one cell remains a spermatogonium (the stem cell) while the other becomes a primary spermatocyte. Consequently, mutations caused by the agent will affect future generations of spermatids.

4. During oogenesis, nuclear divisions occur only without cytokinesis. This results in four nuclei, one of which stays with the cellular components, while the other three, known as polar bodies, migrate to the cell membrane. A polar body does not have the cellular components necessary for the building of an embryo.

5. The male testes are suspended outside of the body in a scrotal sac to lower the temperature to one that is optimal for sperm production. The testes contain coiled tubes where sperm are produced. They are then transported to the epididymis, where they are stored until ejaculation. Nutrients and pH buffers from the male accessory glands combine with the sperm to form the seminal fluid, which is ejaculated through the penis into the vagina of the female. The female ovaries are housed deep inside the abdominal cavity for protection. Approximately once a month, one egg is released from one of the ovaries and is swept into the fallopian tube. If sperm is present it will become fertilized. The zygote is moved by cilia into the uterus, where it implants in the endometrium of the uterus. Nine months later the fetus will move out of the mother's womb, by muscle contractions, through the cervix and vagina. If no pregnancy occurs, the endometrium and unfertilized egg are sloughed off as menstrual flow.

AUDIOVISUAL MATERIALS

Cell Division: Mitosis and Meiosis—Biology *Today Series,* CRM McGraw-Hill, 1974.

Genetics: Chromosomes and Genes—Meiosis, Coronet Instructional Films, 1968.

Meiosis, 2d ed., *Heredity and Adaptive Change Series,* Encyclopaedia Britannica Educational Corp., 1980.

Mitosis and Meiosis, McGraw-Hill, 1974.

9 A Human Life—Development Through Aging

KEY CONCEPTS

Various mechanisms ensure that only one egg and one sperm fuse to form a zygote.

Cell divisions follow fertilization, taking one cell to first the morula and then the blastocyst stages.

The outer portion of the trophoblast invades the endometrium, while the inner mass develops into the embryo and extraembryonic membranes.

Human chorionic gonadotrophin (HCG) secreted by the trophoblast maintains the pregnancy.

Cells in a specific germ layer become part of particular organ systems.

The embryonic period establishes the placental connections to the mother and develops the basic body plan of the child.

During the fetal period of development the child grows in size and internal organs become capable of sustaining life without the mother.

Birth is a process of physical separation of mother and child.

Aging is a natural process of development that begins prior to conception and involves passive and active changes at all levels of organization.

KEY TERMS

amniocentesis	extraembryonic
blastocyst	membranes
blastomeres	free radicals
capacitation	gerontology
chorionic villi	human chorionic
cleavage	gonadotropin
corona radiata	inner cell mass
ectoderm	mesoderm
embryonic induction	morula
endoderm	neural tube
endometrium	notochord

placenta	trophoblast
primitive streak	zon pellucida
progerias	zygote
pronuclei	

CHAPTER OUTLINE

I. The Stages of Prenatal Development (table 9.1)
   A. Morphogenesis is the series of events that take a single fertilized ovum through development into a newborn baby.
   B. Development occurs over 38 weeks in distinct stages.
II. The Preembryonic stage
   A. Fertilization.
      1. The gametes.
         a) Secondary oocyte.
            (1) It remains arrested during metaphase until after fertilization.
            (2) The oocyte is surrounded by the membranes zona pellucida and corona radiata (fig. 9.2*a*).
         b) Sperm.
            (1) Hundreds of millions of sperm are deposited in the vagina.
            (2) They are propelled to the fallopian tube via sweeping cilia lining the cervix and their own flagella.
      2. Conception (fig. 9.3).
         a) It occurs when membranes of sperm and oocyte come in contact in the fallopian tube.
         b) The enzymes of the sperm's acrosome digest the outer membranes of the oocyte (entry of more than one sperm).

c) Block to polyspermy occurs through excitation of the membrane.

d) The oocyte pronuclei completes meiosis and then fuses with the sperm.

B. Cleavage (figs. 9.4 and 9.5, table 9.2).

1. The blastomere stage is the first mitotic division about 30 hours after fertilization.

2. The morula stage forms a solid ball of 16 or more cells, which arrives in the uterus 3 to 6 days after fertilization.

3. Blastocyst stage.

a) This is when the solid ball becomes hollow.

b) The trophoblast refers to the outer layers of cells, which will form the chorion, the fetal portion of the placenta.

c) The inner cell mass will become the embryo.

d) The blastocyst cavity fills with fluid from the uterus.

C. Implantation (figs. 9.6 to 9.8).

1. Attachment of the blastocyst to the endometrium (uterine lining) occurs 5 to 7 days past fertilization when digestive enzymes from the trophoblast rupture blood vessels of the endometrium.

2. Projections grow from the trophoblast into the endometrium.

3. Release of the embryonic hormone human chorionic gonadotropin (HCG), prevents menstruation.

D. The primordial embryo.

1. Gastrulation.

a) An amniotic cavity forms between the inner cell mass and the trophoblast.

b) The inner cell mass flattens to form the embryonic disc, which develops into three layers (fig. 9.9).

(1) Ectoderm refers to the outer layer, which develops into the nervous system, the sense organs, and an outer skin layer.

(2) Mesoderm cells in the middle layer become bone, muscles, blood, the inner skin layer, reproductive organs, or connective tissue.

(3) Endoderm cells of the inner layer become digestive, respiratory, and urinary systems.

2. Methods of detection.

a) There is the modern method of HCG detection in the urine of pregnant women.

b) Previously urine from a pregnant woman was injected into a rabbit, which caused its ovaries to swell if HCG was present.

III. The Embryonic Stage

A. This includes the development of the embryo and formation of the extraembryonic membranes.

B. Supportive structures.

1. Placenta.

a) It is formed by the chorionic villi and blood pools from the mother (fig. 9.11).

b) Nutrients and oxygen pass across thin membranes that separate mother from embryo.

c) They are completely developed by the 10th week.

d) It secretes hormones that maintain pregnancy.

e) Most chemicals and viruses can pass to the embryo.

67

2. Yolk sac (fig. 9.12).
   a) It manufactures blood cells until the liver can take over.
   b) Parts of it become intestines and germ cells.
3. The allantois contributes blood cells and gives rise to fetal arteries, veins, and bladder.
4. Amniotic fluid.
   a) It is derived from the mother's blood.
   b) It maintains constant temperature and pressure and protects the embryo in the event of the mother falling.
   c) Amniocentesis is the sampling of amniotic fluid as a prenatal test.
   d) Chorionic villi sampling is a prenatal test where cells from the chorion are extracted for study.
5. The umbilical cord is attached to the placenta and contains two arteries and one vein.
C. The embryo (fig. 9.13).
1. Embryonic induction is where the specialization of one group of cells causes the specialization of adjacent cells.
2. Organogenesis is the transformation of germ layers into organs.
3. Third-week events.
   a) The primitive streak elongates to form an axis that the organs form around and becomes the notochord, which will give rise to the skeleton.
   b) A primitive heart forms and begins to beat.

4. Fourth-week events.
   a) Primitive blood vessels fill with blood cells.
   b) The neural tube forms, which will give rise to the central nervous system.
   c) Small buds form beginnings of arms and legs.
   d) A distinct head, jaw, eyes, ears, and nose shape.
   e) The gastrointestinal tube begins to form.
5. Fifth-week events.
   a) Brain cells are rapidly differentiating.
   b) Tiny ridges on hands will form fingers.
6. Seventh and eighth weeks (fig. 9.14).
   a) A cartilaginous skeleton is present.
   b) The placenta is almost fully formed.
   c) Formed eyes are sealed shut.
IV. The Fetal Period
A. Third month.
1. Ears lie low.
2. Bone tissues begins to replace cartilage.
3. The fetus begins to move.
4. Sex organs differentiate (fig. 9.15).
B. Fourth month.
1. The fetus grows hair, eyebrows, lashes, nipples, and nails.
2. The fetus is approximately 8 inches long.
C. Fifth month.
1. The fetus becomes covered with an oily substance called vernix caseosa, which is held in place with white down hair called lanugo.
2. The fetus is approximately 9 inches long.

D. Sixth month.
1. Skin appears pink and wrinkled.
2. The fetus is approximately 12 inches long (fig. 9.16).
V. Labor and Birth (figs. 9.71 and 9.18).
A. Occurs approximately 266 days after fertilization.
B. Onset.
1. It is signified by the breaking of the amniotic sac.
2. The discharge of blood and mucus from the vagina, called the bloody show, occurs 1 to 2 days prior to contractions.
C. The first stage of labor is when uterine contractions prompted by hormones push the baby against the cervix until it dilates to 10 centimeters.
D. The second stage is the actual delivery of the baby, called parturition.
E. The third stage is when the afterbirth of placenta and extraembryonic membranes leaves the woman's body.
F. Caesarean section.
1. This is the removal of the baby via an incision in the abdominal region.
2. It is the preferred method of delivery when problems arise in normal delivery.
VI. Maturation and Aging
A. Aging over a lifetime.
1. Aging begins before birth, when some cells die and are not replaced.
2. Gradual changes occur over a lifetime, but there are interesting highlights (fig. 9.19).
B. What triggers aging?
1. Tracing the primary causes of aging is complicated.
2. The field of gerontology examines the biological changes of aging.
C. Aging as a passive process.
1. DNA repair decreases, which may cause cells to die of faulty genetic instruction.
2. The breakdown of lipids decreases.
3. Mitochondria degenerate.

4. Free radicals originating from radiation or chemical attack on stable molecules may cause cellular degradation.
5. There is degeneration of the thymus.
6. There is an accumulation of fat on arteries.
D. Aging as an active process.
1. There is a buildup of lipofuscin.
2. Autoimmunity increases when the body attacks itself.
3. Programmed cell death of certain cell lines occurs.
E. Accelerated aging disorders provide clues to the normal aging process.
1. Hutchinson-Gilford syndrome is where children develop aging characteristics and die by age of 12 (fig. 9.21).
2. Werner's syndrome is the adult version of Hutchinson-Gilford syndrome.
3. Alzheimer's disease (fig. 9.22).
   a) This disease occurs around age 40 with memory loss and inability to reason.
   b) Some cases may be inherited.
   c) It can only be detected with a show of brain lesions in an autopsy.
F. Dealing with aging.
A. No fountain of youth has been discovered, though many still look for it.
B. The number of aged is increasing each year (fig. 9.23).

LEARNING OBJECTIVES

After reading this chapter, the student should be able to answer these questions:

1. What structures form during the three major stages of human prenatal development? What are their functions?
2. What structures support the embryo and fetus? How do they do so?
3. What happens during labor and birth?
4. What evidence of aging is noticeable at different stages in a human lifetime?

5. In what ways is aging a passive process? How can aging also be considered an active process?
6. What problems are faced by the elderly, and why might these problems become more pervasive in the years to come?

ANSWERS

"Questions"

1. Zygote, morula, blastocyst, gastrula, embryo, fetus.
2. The events of fertilization include capacitation of sperm; the action of acrosomal enzymes on the secondary oocyte and its surrounding zona pellucida and corona radiata; and the establishment of the block to polyspermy as the sperm enters the oocyte. The chromosomes from the male and female cells meet to form the zygote.
3. Cells from the trophoblast layer develop into a membrane called the chorion, which eventually forms the fetal portion of the placenta, bringing oxygen and nutrients to the fetus and removing wastes. The inner mass becomes the embryo plus supportive extraembryonic membranes. The endoderm tissue will give rise to the digestive, respiratory, and urinary systems. The mesoderm develops into bone, muscles, blood, dermis, reproductive organs, or connective tissue. The ectoderm becomes most of the nervous system, sense organs, epidermis, hair, nails, or skin glands.
4. The development of organs begins in the third to eighth week of prenatal development. If chemicals are present at this time, they may alter cell lines by toxic or mutagenic effects. Alterations that affect entire organ systems often result in gross deformities or death of the embryo.

5. a. HCG, human chorionic growth hormone, is released by the chorionic tissue of the fetus. It elicits and maintains the development of the placenta. Presence of HCG in the urine of a female indicates that she is pregnant.
   b. For genetic screening purposes, tissue from the chorionic villi, which is derived from the embryo, can be sampled at less risk than taking amniotic fluid. It can be taken earlier in the pregnancy in the event that abortion is an option or to reduce the fears of the parents.
   c. Amniotic fluid, which surrounds and protects the developing embryo, contains sloughed off cells from the embryo. By abstracting a small sample, embryonic cells can be grown in the laboratory and analyzed for genetic disorders.
   d. Alpha fetoprotein is a substance contained in the brain and spinal cord. If an opening of the neural tube allows a protrusion of the brain or spinal cord, the chemical leaks into the mother's circulation. It can be detected by doing a blood test on the mother.
6. Though all of the organ rudiments have formed by the end of the first trimester, they are immature and incapable of maintaining homeostasis. Without the support systems provided by the mother's body, maintenance of life is impossible.

7. a. The allantois contributes blood cells and gives rise to the fetal umbilical arteries and vein.
   b. The amnion consists of the amniotic sac and amniotic fluid. This bag of water cushions the embryo, maintains a constant temperature and pressure, protects the embryo if the mother falls, and allows development to proceed unhampered by the forces of nature.
   c. The placenta is the fusion of tissue from the endometrium of the mother with the chorionic villus of the fetus. It provides a link between the circulatory systems of the mother and child for the exchange of nutrients, wastes, and oxygen. It also secretes the hormones that maintain the pregnancy.
   d. During the fifth month, the fetus becomes covered with vernix caseosa, an oily substance that looks like cottage cheese. It protects the developing skin, which is just beginning to grow.
   e. The lanugo is a white downy hair that forms during the fifth month of development to protect the developing skin beneath it.

"To Think About"

1. The direct injection of a sperm into an egg does not result in the development of an embryo probably because activation steps have not been taken. Biochemicals in the female reproductive tract activate the sperm in a process called capacitation. Activation of the secondary oocyte requires stimulation of acrosomal enzymes released from the head of the sperm.

2. Indication of dying placental cells is a sign of immediate danger to the fetus, who cannot survive without them. If the pregnancy has gone to 41 weeks, the child is probably too large to pass into the birth canal. A Caesarean section would reduce the risk to the fetus by immediate expulsion.

3. Rats have been used for many biological experiments to make predictions about humans. The study with rats raised in an enriched environment may suggest that humans who live in a healthy and supportive community live longer. The lower temperature and caloric intake studies suggest that humans who live in a cool climate and are underweight might have increased longevity.

4. Aging begins before fertilization if the age of the oocyte is considered. At the very least, it is a lifelong process beginning before birth. Hearing begins to decline as early as age 10. The thymus begins a lifelong atrophy beginning at age 12 or 14 and is reduced to a fraction of its original size by age 70. Muscle strength and height begin declining in the 30s. Metabolism slows, resulting in increased weight, and the immune system becomes less efficient in the 40s.

5. Genetics is a key to longevity, but there is evidence that good habits such as proper eating, not smoking or drinking, regular exercise, and reduction of stress can result in a better quality of life and possibly increase longevity.

6. Human aging studies are complicated by variables such as life-styles, environmental factors, and attitudes toward growing old. Since humans live for relatively long periods of time, it is difficult to study changes that occur very slowly. Studies on animal models remedy some of these problems but introduce the problem of how well the data applies to humans.

7. Though there is no right answer the student should consider factors such as what can be done to improve the health and the financial security of aging Americans and how to change attitudes towards the elderly. An individual can increase his or her own quality of retirement years by leading a healthy life-style that includes exercise and stress management. He or she can plan for financial security in old age by contributing to retirement fund programs. Just as important is to develop a healthy attitude towards growing old and make plans for the "golden years."

AUDIOVISUAL MATERIALS

Development, AIBS McGraw-Hill, 1960.
Developmental Biology, Coronet Instructional Films, 1981.
Development of the Chick Embryo, AIBS Michigan, 1967.
Development and Differentiation, Ziff-Davis Publishing, 1974.
Development of the Embryo, Moreland-Latchford, 1969.
Development of Organs, AIBS McGraw-Hill, 1961.
Genesis: The Origin of Human Life, Esselte Bonnier Ave., Stockholm, 1974.

10 Human Reproductive and Developmental Problems

KEY CONCEPTS

The inability of a couple to produce a child may be due to reproductive problems of either partner or a combination of both.

Male infertility often involves poor quality or low quantity of sperm due to anatomical, hormonal, or immunological reasons.

Female infertility often is due to problems with ovulation or implantation, obstruction of the fallopian tube, or immunological incompatibility with the mate's sperm.

Spontaneous abortions result from chromosomal defects of the embryo, underproduction of human chorionic gonadotrophin (HCG), or anatomical problems of the uterus.

Premature infants are at risk due to immature reflexes and their inability to regulate body temperature.

The embryo is particularly sensitive to faulty genes or teratogenic agents such as viruses and chemicals because this is the critical period for most organ systems.

The use of medical intervention before and immediately after birth is possible and sometimes controversial.

Artificial insemination involves the concentration of sperm outside the body for insertion by artificial means into the female.

Surrogate motherhood involves the conception of a child under ordinary circumstances but the mother then gives up her parental rights and privileges to another.

In vitro fertilization involves the removal of the gametes from the parents, with fertilization occurring in an artificial environment and viable embryos then introduced into the womb.

Gamete intrafallopian transfer (GIFT) involves the removal and concentration of gametes from the parents, followed by reintroduction into the fallopian tube, where fertilization occurs under natural conditions.

Embryos produced under artificial conditions can be frozen and used by the biological parents for future pregnancies or adopted by women who provide a surrogate womb.

Reproductive alternatives offer both resolutions to infertility problems and ethical considerations for individuals and society.

Contraceptive methods may involve obstruction of the pathway for sperm and egg to meet, such as the condom or tubal ligation; other methods alter the hormone cycle, such as the birth-control pill, while others reduce sperm viability, such as gossypol.

Sexually transmitted diseases may be both difficult to identify and serious, as in the case of pelvic inflammatory disease or AIDS.

KEY TERMS

artificial insemination	infertility
critical period	neonatology
ectopic pregnancy	teratogens

CHAPTER OUTLINE

I. Introduction
   A. Louise Brown was the first test-tube baby, a human conceived outside of the body and then transferred to the uterus (fig. 10.1).
   B. A donkey embryo was transplanted into the uterus of a horse, who eventually gave birth to it.
II. Infertility (table 10.1).
   A. Defined as the inability to conceive after 1 year of frequent intercourse without the use of contraceptives.
   B. Forty percent is caused by male problems, 40% by female problems, and 20% by contributions of both.

C. Male infertility (fig. 10.2).
  1. Low sperm count.
     a) This may be caused by hormonal imbalance, immune infertility when the man's immune system attacks his own sperm, or varicose veins of the scrotum.
     b) Sperm can be collected and kept in a sperm bank until there is a large enough pool to artificially inseminate.
  2. Sperm quality: lack of motility is caused by misshapen flagella or hormonal problems.
D. Female infertility.
  1. Irregular ovulation.
     a) This may be caused by hormonal imbalance due to a tumor in the pituitary gland, a tumor in the ovary, an underactive thyroid gland, or overproduction of prolactin, which suppresses ovulation.
     b) Fertility drugs stimulate ovulation but may result in multiple births.
  2. Blocked fallopian tubes may be caused by birth defect or infection and can sometimes be opened surgically.
  3. Defects in the uterine lining resulting in fibroids prevent pregnancy.
  4. Hostile secretions in the vagina and cervix may attack or trap sperm.
E. The age factor in female infertility.
  1. Fertility declines after age 29.
  2. Increasing endometriosis (buildup of the uterine lining) blocks the implantation of the zygote, but it may be corrected (fig. 10.3).

III. Spontaneous Abortion
  A. Defined as a pregnancy that ends naturally before the fetus is developed.
  B. It is referred to as a miscarriage if after the first trimester.

C. Causes.
  1. An abnormal number of chromosomes in the embryo.
  2. The trophoblast cells do not produce enough HCG.
  3. Incompetent cervix opens long before labor.
  4. Fibroid tumors in the uterus crowd the fetus.
  5. Ectopic pregnancy.
     a) This is when the embryo does not move out of the fallopian tube into the uterus.
     b) It is often caused by pelvic infections that have left scar tissue.

IV. Born Too Soon—The Problem of Prematurity
  A. Premature babies are born more than 4 weeks early or weigh less than 5 pounds.
  B. High frequency of "preemies" is seen with twins or triplets, teens, or malnourished women.
  C. Problem for preemies involves immature reflexes (fig. 10.4).
     1. Not enough surfactant is produced so lungs collapse.
     2. Sucking response is absent, resulting in starvation.
     3. Loss of body heat due to low body fat, immature temperature-control mechanism, and small body size results in death.
  D. Small for gestational age.
     1. A term used to describe a fetus that went full term but is low in birthweight.
     2. Often found with mothers who are malnourished or who smoke during pregnancy, or this may be caused by abnormal development of the placenta.

V. Birth Defects
  A. The critical period (fig. 10.5).
     1. The disruption of normal development reflects alterations in structures that were developing when the damage occurred.
     2. The first trimester is most critical, since all systems are undergoing development, though brain tissue development can be altered up to two years.
     3. Subtle defects such as learning disabilities are often caused by disruptions in the fetal period.
     4. Some birth defects are caused by abnormal genes that exert their effect at a specific point in prenatal development.
  B. Teratogens (fig. 10.6).
     1. A birth defect can be caused by toxic substances ingested by the mother.
     2. Thalidomide was a mild tranquilizer taken between 1957 and 1961, which resulted in phocomelia—babies born with stumps for arms and legs.
     3. Rubella infection of the mother may result in babies born with cataracts, deafness, heart defects, learning disabilities, speech problems, or juvenile-onset diabetes.
     4. Alcohol as a teratogen (fig. 10.7).
        a) As much as one to two drinks a day may cause fetal alcohol syndrome babies.
        b) Children with FAS have small heads, misshapen eyes, a flat face and nose, and minor learning disabilities to mental retardation.
     5. Cocaine causes spontaneous abortions.
     6. Cigarette smoke robs the fetus of oxygen, prevents nutrients from crossing the placenta, and causes miscarriages, stillborns, prematurity or low birth weight.
     7. The acne medication isotretinoin causes spontaneous abortions and defects of the heart, nervous system, and face.
     8. Excess vitamin C causes the baby to bruise easily at birth and be prone to infection.
     9. Other suspected teratogens are textile dyes, lead, certain photographic chemicals, semiconductor materials, mercury, and cadmium.
  C. Baby Doe dilemma.
     1. This refers to the anguish suffered by parents who give birth to a baby with severe birth defects.
     2. Neonatology is the modern medical technology that helps infants survive.
  D. The fetus as a patient.
     1. Some fetuses can be treated by giving drugs to the mother.
     2. Some fetuses have undergone prenatal surgery.
VI. Reproductive Alternatives (table 10.2)
  A. Donated sperm—artificial insemination.
     1. It is the oldest reproductive alternative.
     2. Sperm banks provide a choice of sperm from males of preferred personal characteristics.
  B. A donated uterus—surrogate motherhood.
     1. Used in cases where the mate of a normal male has a nonfunctional uterus.
     2. A surrogate mother is artificially inseminated and then gives up the baby at birth.
  C. In vitro fertilization (figs. 10.8 and 10.9).
     1. This technique is often used when a woman's fallopian tubes are blocked.
     2. A woman takes a superovulation drug.

3. Several oocytes are removed and combined with the mate's sperm in a laboratory dish.
4. Several zygotes are transferred back into the woman, while the others are frozen to be used later.

D. Gamete intrafallopian transfer (GIFT).
1. A man contributes sperm, from which the most active are separated.
2. The collected sperm and oocytes are reintroduced into the female tract past the obstruction.
3. It is about 40% successful.

E. Embryo adoption: The embryo conceived in a woman is flushed out of the uterus early in the pregnancy and transferred to a woman's uterus who could not conceive but who could carry the fetus.

VII. Sex and Health
A. Birth control.
1. Contraceptive pill (table 10.3).
   a) Contain versions of estrogen and progesterone that suppress ovulation.
   b) They may alter the uterine lining so implantation cannot occur or thicken the mucus in the cervix so sperm cannot get through.
2. Depo-Provera.
   a) It may be given as an injection every 3 months to suppress ovulation.
   b) It may be inserted under the skin as a pellet, or inserted into the vagina in the form of rubber rings.
3. Gossypol.
   a) The only form of male birth control; acts by lowering the sperm count.
   b) The side effects are appetite loss, weakness, lowered sex drive, and sometimes permanent infertility.
4. A pregnancy vaccine produces antibodies that prevent penetration of the oocyte or attack sperm.

5. Intrauterine devices (IUDs) are placed in the uterus to obstruct implantation, but they may cause infection.
6. Condoms consist of a sheath worn over the penis and have the added benefit of reducing sexually transmitted diseases.
7. Spermicidal jellies, foams, creams, and suppositories mostly contain nonoxynol-9 and chemically destroy sperm.
8. Rhythm method: The time of ovulation can be determined by charting the menstrual cycle, monitoring temperature changes, or noting changes in consistency of vaginal mucus.
9. Sterilization.
   a) Vasectomy is the cutting of the vas deferens in males to block sperm passage.
   b) Tubal ligation is the cutting of the fallopian tubes to block the pathway of oocytes to the uterus (fig. 10.10).

B. Terminating a pregnancy.
1. First trimester abortions use a sucking or scraping device.
2. Later abortions utilize more intense scraping, injection of a salt solution into the amniotic sac, or prostaglandin suppositories to trigger uterine contractions and expel the fetus.

C. Sexually transmitted diseases.
1. Symptoms (table 10.5).
   a) Burning sensation during urination.
   b) Pain in the lower abdomen.
   c) Fever or swollen glands in the neck.
   d) Discharge from the vagina or penis.
   e) Pain, itch, or inflammation in the genital or anal area.
   f) Pain during intercourse.
   g) Sores, blisters, bumps, or a rash, particularly on the mouth or genitals.
   h) Itchy, runny eyes.

2. Pelvic inflammatory disease.
   a) It is caused by gonorrhea or chlamydia.
   b) It may result in scarring of the uterus and fallopian tubes, which increases infertility and the risk of ectopic pregnancy.
3. Acquired immune deficiency syndrome (AIDS).
   a) It is caused by a virus that suppresses the body's immune system.
   b) It may be passed in body fluids such as semen, blood, milk, and tears or by use of contaminated needles.

LEARNING OBJECTIVES

After reading this chapter, the student should be able to answer these questions:

1. What are some of the physical causes of infertility in the human male and female? How can infertility be treated?
2. What are some of the possible causes of spontaneous abortion?
3. What are some of the medical problems faced by a baby born prematurely?
4. What are the effects of some toxic chemicals that interfere with prenatal development?
5. How can the reproductive technologies of artificial insemination, in vitro fertilization, and embryo transfer help infertile couples to have children?
6. What are some commonly used and experimental birth-control devices and methods and how do they work?
7. What are some sexually transmitted diseases?

ANSWERS

"Questions"

1. a. birth-control pill; b. intrauterine device; c. tubal ligation; d. gossypol; e. birth-control vaccine
2. To overcome this joint infertility, the couple could elect to have in vitro fertilization, where eggs and sperm are collected from each partner. Fertilization takes place inside a flask and then viable embryos are transplanted into the mother. Another option is to use gamete intrafallopian transfer (GIFT), where eggs and sperm are concentrated outside of the body of their donors but reintroduced into the fallopian tube, where fertilization takes place.
3. To be "fertility aware," the woman can predict when ovulation is occurring. This would allow the couple to increase intercourse frequency during the time of ovulation if a child is desired or refrain from intercourse during this time if pregnancy is not desired.
4. One fetus may have interfered with the other under crowded conditions by obstructing passage of nutrients through the umbilical cord of its twin, causing it to die before birth.
5. A fetus, protected by the temperature-regulated environment of the womb, has no need of subcutaneous fat until birth. Premature birth will find the infant very lean and in danger if its environment cannot be controlled.
6. Fetal alcohol syndrome.
7. Technically, the intrauterine device is not a contraceptive since fertilization does take place, but the embryo is not allowed to implant.

"To Think About"

1. Though there is no right answer, the student may wish to consider the following points in each case before writing an answer.
    a. The question of quality versus quantity of life.
    b. The rights of an individual versus those of the society.
    c. The legal and moral responsibilities of each party entering a contract that involves reproductive alternatives.
    d. The legal, moral, and biological rights of the surrogate mother, father, and adoptive mother.
    e. The rights of the mother, father, and the unborn embryo or fetus.
    f. The rights and responsibilities of a child and his or her parents.
2. Advantages and disadvantages will vary with the individual. However for many, older age offers the advantage of increased emotional and financial security and the disadvantage of increased risk of infertility, abortion, and genetic defects.
3. The couple should keep in mind that the father can only pass on one-half of his genetic material to the child. The traits that are randomly passed may be recessive in nature and not observable in the father. Also cultural traits such as profession and religious preference are not inheritable.
4. A study conducted to determine the effects of marijuana smoking on sperm counts should include both a control and experimental group made up of individuals of like age, race, and life-styles who do not indulge in other drug use. Marijuana use in the experimental group should be monitored and ejaculation frequency should be considered.
5. A woman who is trying to get pregnant can reduce the risk that her baby will suffer from birth defects by eliminating consumption of known teratogens.
6. There is no right answer but the student should consider what problems could arise if the medical history of the biological parent is unknown. Could the information be obtained by an outside party such as the adoption agency without divulging the identity of the parent? A few studies indicate that some behavioral tendencies are inherited such as alcoholism or learning disabilities. Should an adopting family or the child be aware of family traits that might be passed down? In artificial sperm donation or surrogate motherhood, the identity of the biological parent may or may not be disclosed.
7. Though there is no right answer, the student should consider the rights of both the unborn fetus and the parents.

AUDIOVISUAL MATERIALS

Reproductive Hormones, AIBS McGraw-Hill, 1961.

The Reproductive System, Coronet Instructional Films, 1980.

11 Mendel's Laws

KEY CONCEPTS

Many physical characteristics and abilities are inherited.

The unit of inheritance is the gene.

The law of segregation states that alleles separate during meiosis and then combine randomly in the offspring.

The law of independent assortment predicts that the transmission of a gene on one chromosome does not affect the transmission of a gene on a different chromosome.

The homozygous condition for lethal genes results in death of the progeny.

Sex-limited genes may be passed by either sex but are only expressed in one sex or the other.

A sex-influenced allele is dominant in one sex and recessive in the other.

In incomplete dominance, the heterozygous phenotype is intermediate to the homozygous phenotype.

The heterozygote may be more vigorous than either of the homozygous phenotypes.

Genotypes may vary in their penetrance and expressivity.

Environmental factors can influence the expression of genes.

More than one gene may produce the same phenotype.

Genes may mask the expression of nonalleles.

More than one form of a gene for a trait may exist in the population.

Evidence that a characteristic is genetic and not environmental comes from increased frequency among families and differences between fraternal and identical twins.

KEY TERMS

alleles	monohybrid cross
autosomes	mutant
chromosomes	mutant phenotype
codominant	mutation
deoxyribonucleic acid	overdominant
(DNA)	penetrance
dihybrid cross	phenocopy
dominant	phenotype
epistasis	pleiotropic
expressivity	recessive
genes	segregation
genetic heterogeneity	sex chromosomes
genotype	sex-influenced
heterozygous	inheritance
homozygous	sex-limited trait
incomplete dominance	test cross
independent assortment	wild type
lethal allele	

CHAPTER OUTLINE

I.   Introduction.
    A. Genes are long chains of the molecule deoxyribonucleic acid (DNA).
    B. Genes mold our physical traits.
    C. On the microscopic level, genes direct the cell's synthesis of particular proteins and control activities of other genes.

II. Mendel's Laws of Inheritance
    A. Ideas about inheritance go back as many as 6,000 years, when Mexican farmers began to sort seeds according to the best yield.
    B. Modern genetics (fig. 11.2).
       1. This field began with the work of Mendel with garden peas in 1857.

2. Mendel recognized that there were "determinants" of physical traits.
3. He knew of the cell theory but not of chromosomes.
4. His background in agriculture and mathematics gave him insight into inheritance.

C. Segregation—following the inheritance of one gene at a time.
1. Mendel crossed true-breeding tall and short plants and found all tall offspring (fig. 11.3).
2. Crosses between the offspring resulted in a three tall to one short ratio, giving Mendel the idea that some traits could mask the presence of others.
3. Mendel reasoned that if each character was separated into different gametes and then recombined randomly to make offspring, the ratios could be explained.
4. The tendency for characters to separate and then recombine is called the law of segregation.
5. The significance of Mendel's work was not recognized until the advent of the chromosome theory of inheritance in the early 1900s, which showed that characters acted very much like chromosomes.
6. In 1908, Mendel's characters were called genes.
7. Modern interpretation of Mendel's theory.
   a) When gametes form during meiosis, the two chromosomes of each type and therefore the two copies of each gene separate into different gametes.
   b) After fertilization the diploid number is restored and the nucleus once again contains two copies of a gene.
8. Glossary of terms (table 11.1).

9. Mendel's observations reflect the events of meiosis, when the two copies of a gene separate to go into each gamete (fig. 11.4).
10. Monohybrid cross.
   a) This is the study of a single trait from the crossings of two hybrids.
   b) In modern genetic terms, Mendel crossed homozygous (*TT*) plants with homozygous (*tt*) short plants to get all tall offspring with the genotype *Tt*.
   c) Next he crossed the hybrids to get a phenotypic ration of three tall plants for every one short plant (fig. 11.6).
   d) The use of a Punnett square shows how Mendel got the genotypic ration of 1 *TT*:2 *Tt*:1 *tt* (fig. 11.5).
   e) A test cross refers to the technique of crossing an individual of unknown genotype to a homozygous recessive individual to determine its genotype.

D. Independent assortment—following the inheritance of two genes at a time (fig. 11.9).
1. Mendel's next experiments were dihybrid crosses, where he followed two traits simultaneously (fig. 11.8).
2. He crossed homozygous round, yellow-seeded plants with wrinkled, green-seeded plants resulting in all round, yellow-seeded offspring.
3. The F2 generation resulted in a ratio of 9 yellow, smooth:3 yellow, wrinkled:3 green, wrinkled:1 green, wrinkled (fig. 11.10).
4. Mendel formulated the law of independent assortment, which states that a gene for one trait does not influence the transmission of a gene for another trait.

5. The law of independent assortment only works for genes found on different chromosomes.
6. The law is explained in terms of meiosis, where different types of chromosomes derived from one parent separate independently of each other in the formation of gametes.

E. Using probability to analyze more than two genes.
1. To calculate the frequencies of traits controlled by many genes located on different chromosomes, simply predict the frequencies of inheriting one gene at a time, and multiply the frequencies.
2. For example, to determine the probability of *RrYy* plants producing *rryy* plants it is 1/4 (chances of *rr*) times 1/4 (chances of *yy*) equals 1/16 chance.

III. Disruptions of Mendelian Ratios
A. Lethal alleles.
1. Certain gene combinations are lethal to the organism, resulting in some spontaneous abortion.
2. Some dominant alleles appear as recessive lethal alleles when the homozygous condition aborts, but the heterozygous condition is normal.

B. The influence of gender—of breasts and beards.
1. Sex-linked traits.
   a) Some genes are located only on the X or Y chromosome.
   b) In the case of X-linked genes, a male will express whichever allele he receives since he will only have one copy of it.
2. Six-limited traits.
   a) They affect a part or function of the body that is present in one sex but not the other (fig. 11.11).
   b) For example, genes for facial hair are passed by both parents but only expressed in sons.

3. Sex-influenced inheritance.
   a) This is when an allele is dominant in one sex but not the other.
   b) For example, the gene for pattern baldness is dominant in males and recessive in females (fig. 11.12).

C. Different dominance relationships.
1. Incomplete dominance.
   a) This is when one allele does not completely overshadow the other, resulting in a heterozygous individual with the expression of a trait that is intermediate to either homozygous condition.
   b) For example, a cross of red snapdragons (*RR*) with white (*rr*) results in pink offspring (*Rr*) (fig. 11.13).
2. Complete dominance.
   a) This is the gene relationship where two alleles are both expressed in the heterozygote.
   b) For example, people with A and B blood type genes express the AB blood type (table 11.2).
3. Overdominance is when the heterozygous is more vigorous than either of the parent homozygotes.

D. Penetrance and expressivity.
1. Penetrance refers to the percentage of individuals carrying the gene who will express the trait.
2. Expressivity refers to the degree to which that trait is expressed (fig. 11.14).

E. Influences of the environment.
1. Some genes are influenced by the environment.
2. PKU disease is only expressed if normal phenylalanine is present in the diet.
3. Temperature alters the expression of coat color in some mammals (fig. 11.15).

F. Pleiotropy and King George III.
 1. Pleiotropy is when a gene has multiple expressions.
 2. Sickle cell disease.
    a) Two copies of this defective gene result in blood cells with an abnormal sickle shape.
    b) This results in blocked blood vessels, an overworked heart and spleen, rheumatism, and pneumonia.
 3. Porphyria.
    a) This rare genetic disease, found in King George III of England, resulted in abdominal pain, constipation, passage of dark urine, weakness, fever, and nervous disorders including mental confusion.
    b) It is caused by the absence of an enzyme required to break down porphyrin, which is part of hemoglobin.
G. Genetic heterogeneity.
 1. This is when alleles of different genes produce the same phenotype but act at different points in the same biochemical pathway.
 2. Hemophilia can be caused by 11 different genes.
H. Epistasis—gene masking at "General Hospital."
 1. This is when one gene masks the presence of a different gene.
 2. In blood typing, the gene designated H controls the expression of the ABO genes (fig. 11.16).
I. Multiple alleles.
 1. In a population, more than two alleles may exist for a gene location (locus).
 2. New alleles arise through mutation (changes in genes).
 3. In rabbits there are four alleles for coat color (table 11.3).
 4. In cystic fibrosis, there may be as many as 70 alleles in the human population.

J. Phenocopies—when it's not really in the genes.
 1. Phenocopy refers to when a trait is caused by the environment and only appears to be inheritable (fig. 11.17).
 2. Because habits are passed down as well as genes, traits like aggressiveness may be mistaken for being genetic.
 3. The teratogenic effects of thalidomide were first mistaken as the genetic condition called phocomelia, which has similar symptoms.
 4. Kuru, a neurological disorder found in the Fore people of New Guinea was mistaken for being genetic until it was discovered that it was a virus passed through the practice of cannibalism (fig. 11.18).

LEARNING OBJECTIVES

After reading this chapter, the student should be able to answer these questions:

1. How and where did the modern field of genetics originate?
2. How did Gregor Mendel follow the inheritance of a single gene, and of two genes, by observing the offspring of crosses of pea plants?
3. How can Mendel's observations be explained in terms of meiotic events?
4. How can the outcomes of genetic crosses be predicted?
5. What are the terms that are commonly used in genetics?
6. How can different forms of a gene affect one another to alter the results of crosses that are expected according to Mendel's laws?
7. How can the expression of a gene vary among individuals?

ANSWERS

1. For any child in the family, the chances are 1:4 that the individual will have sickle cell disease since the gamete contributed by each parent in every case was produced independently of the others.
2. a. TT x tt = all Tt genotypes and all tall phenotypes.
   b. TT x Tt = 1 TT:1 Tt genotypes and all tall phenotypes.
   c. tt x Tt = 1 Tt:1 tt genotypes and one tall for every short phenotype.
3. a. All of the offspring will have the phenotypes of the parents since the homozygous parent can only pass the dominant gene for each trait.
   b. One-eighth of the children will have the BBHHFF genotype and one-eighth will have the BbHhFf.

To calculate the BBHHFF genotype using the probability method, multiply the chances of the heterozygous parent passing the B gene (½) times the probability of passing the H gene (½) times the probability of passing the F gene (½) times the probability of the homozygous parent passing the B gene (1) times the probability of passing the H gene (1) times the probability of passing the F gene (1). See figure 11.1.

½ X ½ X ½ X 1 X 1 X 1 = 1/8 chance of BBHHFF

The Punnett square would look like the following:

	HBF	HbF	Hbf	hBF	hbf	hBf	HBf	hbF
HBF	HHBBFF	HHBbFF	HHBbFf	HhBBFF	HhBbFf	HhBBFf	HHBBFf	HhBbFF

Figure 11.1

4. a. Photocopies
   b. Variably expressive
   c. Genetic heterogeneity
   d. Epistasis
   e. Sex-limited trait

"To Think About"

1. Regardless of what traits were chosen, the numbers of phenotypes in the F2 generation would still reveal the law of segregation but could mask the law of independent assortment if the traits chosen were linked genes.
2. Selection of traits for the study of the law of independent assortment should include those that show full penetrance in the population, have only two alleles, and are free from epistatic and environmental effects.
3. The case of dark- and light-skinned twins demonstrates how the presence of genes can be hidden by interactions between genes, such as dominance/recessiveness.
4. Parents who carry recessive lethal genes may be unaware of their presence in either family because they have been masked by the normal genes and because homozygous recessive embryos probably spontaneously aborted without detection. No evidence for dominant lethal mutations is apparent since most would spontaneously abort.
5. Kuru first appeared to be genetic because of the 3:1 ratio among women in the population, which suggested a Mendelian mode of transmission. Birth defects caused by thalidomide may have first appeared to be genetic since mothers who took the drug with the first child would more likely take the drug with the second pregnancy, showing a higher frequency of birth defects in families.

AUDIOVISUAL MATERIALS

Genetics; Chromosomes and Genes; Meiosis, Coronet Instructional Films, 1968.

Genetics; Mendel's Laws, Coronet Instructional Films, 1962.

Mendel: Father of Genetics, Films Incorporated, 1972.

Mendel's Recombination of Traits, AIBS McGraw-Hill, 1967.

Mendel's Segregation of Traits, AIBS McGraw-Hill, 1967.

12 Linkage

KEY TERMS

Barr body	homogametic sex
crossing over	inactivation
genetic marker	linkage
hemizygous	sex-linked
heterogametic sex	sex ratio

E. The farther apart two genes are on a chromosome, the more likely crossover is to occur between them.
F. Linkage maps (maps of relative gene position on a chromosome) are constructed by determining the frequency of crossover between two genes.
G. A new diagnostic tool for genetic disease is the use of genetic markers, which are pieces of easily identifiable DNA that lie next to the disease gene and so are unlikely to be separated by crossover.

III. Matters of Sex
  A. Sex linkage.
    1. Female humans have two large X chromosomes, whereas males have one X and one small Y chromosome (fig. 12.2).
    2. The X chromosome contains about 1,000 genes, whereas the Y chromosome has only a few.
    3. The X and Y pair during meiosis, but they are not really homologs.
    4. Genes on the X chromosome are said to be sex-linked.
    5. The male will express whichever allele he receives on his X chromosome because there is not a second to mask it.
    6. A male is hemizygous for sex-linked traits because he has half the number of genes as does a female.
    7. Sex-linked traits found in males are only passed to their daughters since they receive the X chromosome.
    8. Hemophilia is a sex-linked disease that was found frequently in Queen Victoria's family (fig. 12.3).
  B. Y Linkage
    1. Genes found on the Y chromosome are only passed from father to son.
    2. Hairy ear trait is one of the few Y-linked traits (fig. 12.4).
  C. X inactivation—equaling out the sexes.
    1. Early in development one or the other X chromosome in females becomes inactivated.
    2. In some cell lines the mother's X chromosome may be the one that is randomly shut off, while in others the father's X chromosome is shut off.
    3. The mixed expression of one allele or the other is called genetic mosaicism.
    4. Manifesting heterozygote refers to a woman who expresses some of the phenotype of a sex-linked trait.
    5. The inactivated X chromosome is observable during cell division as a dark spot called a Barr body (fig. 12.5).
    6. Mary Lyon's hypothesis states that the inactivated chromosome is nonfunctional.
    7. In calico cats, the patches of orange or black color are dependent upon which X chromosome is inactivated, which means male calicos can only arise through rare meiotic errors resulting in XXY genotype (fig. 12.6).
    8. Lesch-Nyhan syndrome.
      a) A sex-linked disorder in which a child has cerebral palsy, bites his or her fingers and lips to the point of mutilation, is mentally retarded, and passes painful urinary stones.
      b) The mutant allele causes a deficiency in the enzyme HGPRT, which is normally found in hair.
      c) Carriers of Lesch-Nyhan can be detected by sampling hair from widely separated parts of the head.

D. Sex determination—male or female?
1. Various myths of how sex was determined flourished until the discovery of sex chromosomes in the 1900s.
2. Sex in humans is determined by whether the oocyte is fertilized by an X- or Y-bearing sperm.
3. In some insects sex is determined only by whether there is one copy of the X (male) or two copies of the X (female) present.
4. Homogametic sex refers to the sex with two like chromosomes as in the case of human females.
5. Heterogametic sex refers to the sex with two different chromosomes as in the case of human males.
6. Abnormal sex constitutions.
   a) Klinefelter syndrome.
      (1) This refers to males with two X chromosomes and one Y.
      (2) He has small testes, is sterile and has underdeveloped secondary sex characteristics.
   b) Turner syndrome.
      (1) A female who has only one copy of the X chromosome.
      (2) She does not mature sexually, is short in stature, and has a flap of skin on the back of her neck.
7. In fruit flies, sex is determined by the X:autosome ratio.
   a) Individuals with paired Xs are normal females.
   b) Individuals with XY are fertile males not by the presence of Y but by the balanced ratio with autosomes.
   c) Individuals with only one X are sterile males.
   d) Individuals with XXY are sterile females.
8. In birds, butterflies, moths, and some fish, the female is the heterogametic sex and the male the homogametic sex.
9. In bees, fertilized eggs become females and unfertilized become males.
10. Autosomal genes can determine sex.
11. Sometimes the environment determines sexual fate, as in the case of turtles, where the temperature of the land where the egg is laid causes development of males or females (fig. 12.7).

E. Sex ratio.
1. Primary sex ratio—males outnumber females at conception, with 120 to 150 males per 100 females due to faster swimming Y-bearing sperm.
2. Secondary sex ratio—at birth males outnumber females 106 to 100.
3. Tertiary sex ratio.
   a) It is measured each decade after birth.
   b) There are about equal numbers of males and females between the age of 20 and 40.
   c) The proportion of males gradually declines beginning at age 50 due to differences in life-style, stress, and war.
   d) By the 10th decade, females outnumber males five to one.

F. Sex preselection.
   1. Prenatal tests can now screen for the sex of the child.
   2. Various factors influence conception.
      a) Intercourse close to ovulation date increases the chances of a light Y-bearing sperm reaching the oocyte first.
      b) X-bearing sperm prefer an acidic environment, whereas Y-bearing sperm prefer an alkaline one.
      c) The X- and Y-bearing sperm can be separated in the laboratory and the preferred introduced through artificial insemination.
      d) In the laboratory, eight-celled embryos can be identified through staining processes as to their sex.
      e) Some people use prenatal knowledge of sex of the embryo to decide upon abortion, particularly in the case of known sex-linked disease.
      f) According to surveys, if given the choice of sex of children, sex ratios would remain close to natural.
IV. The Changing Science of Genetics: Knowledge of hereditary mechanisms has affected health care and choices regarding child bearing.

LEARNING OBJECTIVES

After reading this chapter, the student should be able to answer these questions:

1. How are genes that are part of the same chromosome inherited differently than genes located on different chromosomes?
2. How are genes located on the sex chromosomes inherited differently in males and females? How is sex chromosome constitution rendered equal in the two sexes?
3. What are some of the ways in which sex is determined in different organisms?
4. How does the sex ratio change throughout the human life span?
5. How do people attempt to influence whether they have a male or female child?

ANSWERS

"Questions"

1. Linked genes are more frequently passed into the gamete together because they are physically linked. The process of crossing over allows separation during meiosis, but it does not occur as frequently as independent assortment.
2. Females receive two copies of the X-linked gene because they have two Xs. Males receive only one copy of the gene since they receive only one X chromosome.
3. In many mammals, sex is determined by the presence or absence of the Y chromosomes. In birds and moths, the male is the homogenic sex. In fruit flies and species of higher plants, sex is determined by the ratio of X chromosomes to autosomes. In some species, environmental cues such as temperature or hormones influence what sex an organism will be. In bees differences in the number of sets of chromosomes determine sex.
4. There is a 50% chance that a son or daughter will be color blind. The Punnett square would look like the following:

	X	X^C
X	XX	XX^C
X^C	XX^C	$X^C X^C$

5. In the 8-cell embryo experiment, X inactivation had not yet occurred. Subsequent divisions resulted in X inactivation of different X chromosomes in different tissues, giving rise to one cell line that exhibits the orange trait and one that exhibits the brown trait. In the 64-cell embryo, only one color was exhibited because X inactivation had already occurred.

6. In humans, sex is determined by the presence or absence of the Y chromosome where females are XX and males are XY. In insects femaleness is determined by the presence of two Xs and maleness by the presence of only one X.

"To Think About"

1. The inheritance pattern of linked genes does not disprove Mendel's laws because crossing over allows for separation of linked genes.

2. Since the frequency of blond hair/blue eyes is high compared with brown hair/blue eyes, it suggests that the genes for hair and eye color are linked.

3. The husband's charge could be scientifically incorrect because mutation may have occurred, hemophilia may have been caused by autosomal genes, or the disease may have low expressivity.

4. Males with hemophilia do not reproduce as frequently as carriers since their reproductive rate would be reduced due to the disease. Since a hemophilic female must receive two copies of the gene, including one from a hemophilic father, it occurs less frequently.

5. The presence of the four colors that make up the calico color in cats requires two pairs of alleles. Since the genes are X linked, a male calico must have the XXY genotype in order to express all four genes.

6. The longevity of humans is too great to make looking at a number of generations feasible. Also, this is complicated by many environmental factors, and the investigator may bring biases about humans into the study. A desirable population for genetic study would be one with a short life cycle. It should be easily maintained in the laboratory, where environmental factors could be controlled.

7. Though there is no right answer, the student may wish to consider how sex determination could result in reduction of population rates by decreasing the number of females in the population through selection for male babies or by providing parents with their first and second choices immediately without having to produce many children to get one of the opposite sex.

8. Through preference for male babies and the availability of sex selection, the relative number of males in the population could be altered.

AUDIOVISUAL MATERIALS

Genes and Protein Synthesis, Milner-Fenwick Incorporated, 17 min.

Genetic Biology, Coronet Instructional Films, 1981.

Genetic Chance, Time Life Film and Video, 1976.

Genetic Defects: The Broken Code, National Education Television Incorporated, 1973.

Genetic Engineering, Open University Educational Enterprises Media Guild, 1977.

Genetics: Functions of DNA And RNA, Coronet Instructional Films, 1968.

Genetics: Human Heredity, Coronet Instructional Films, 1968.

Genetics: Man the Creator, Document Associations Incorporated, 1971.

Sex-Linked Inheritance, McGraw-Hill, 1961.

13 Molecular Genetics

KEY CONCEPTS

A gene is a sequence of nucleotides that specifies a sequence of amino acid in a polypeptide.

The function of a protein is dependent upon its amino acid sequence.

The function of a protein in an organism is to provide an inherited trait.

The helical structure of DNA allows it to be condensed inside the nuclear membrane.

RNA is an intermediate molecule that carries the blueprint of the gene out of the nucleus to the ribosome, which is the sight of protein synthesis.

The genetic code lies in the nitrogen bases that make up the "rungs" of the double helix.

During semiconservative replication of DNA, each parent strand serves as a template for newly synthesized DNA, which is built out of free-floating nucleotides. Each daughter molecule receives one parental strand and one newly synthesized strand of DNA.

Errors in DNA replication are repaired as they occur.

Gene amplification uses the polymerase chain reaction to make many copies of a DNA sequence.

Amino Acids are specified by a three-letter codon on the mRNA.

The ribosome physically supports the process of protein synthesis, where transfer RNA brings in amino acids in the sequence specified by the mRNA.

The genetic code is nonoverlapping, universal, and degenerate.

Not all of the mRNA transcribed is translated into amino acids.

Some portions called introns are excised prior to translation.

Some genes can move to other positions on the chromosome.

Biotechnologies provide products based on altered cells or biological molecules.

Recombinant DNA technology uses restriction enzymes to cut out genes from one organism and introduce them into other organisms.

KEY TERMS

adenine	mutation
anticodon	nucleotide
antisense strand	peptide bond
biotechnology	plasmid
codon	purines
complementary	pyrimidines
cytosine	recombinant DNA
degenerate	technology
DNA polymerases	restriction enzymes
DNA replication	ribonucleic acid (RNA)
exons	ribosomal RNA (rRNA)
gene	ribosome
gene libraries	sense strand
genetic code	thymine
genome	transcription
guanine	transfer RNA (tRNA)
introns	translation
messenger RNA (mRNA)	uracil

CHAPTER OUTLINE

I. Gene and Protein—An Important Partnership (fig. 13.1)
   A. DNA is a polymer of four different nucleotides; the order of which is different for individual genes.
   B. A gene codes for a particular polypeptide, which is a polymer of 20 different amino acids.

C. The function of proteins.
1. The function is determined by the amino acid sequence.
2. Types of proteins.
   a) Pigment.
   b) Hormones.
   c) Enzymes.
   d) Structural support.
   e) Transport of oxygen.
   f) Antibodies.
D. Amino acids are taken in the diet in the form of plant and animal protein.

II. DNA Structure Makes DNA Function Possible
A. The structure maintains the integrity of the code during its replication for the manufacture of new cells.
B. The structure allows for transmission of the genetic code from the nucleus to the site of protein synthesis by the manufacture of messenger RNA.
C. The double helix.
1. It consists of two long, twisted molecules of DNA that resemble a ladder (fig. 13.2).
2. The rails of the ladder are compared of alternating strands of deoxyribose sugar and phosphate.
3. The rungs are composed of four different nitrogenous bases (fig. 13.3).
   a) The purines.
      (1) Adenine (A).
      (2) Guanine (G).
   b) The pyrimidines.
      (1) Cytosine (C).
      (2) Thymine (T).
   c) Complementary base pairs (fig. 13.4).
      (1) Adenine normally only binds with thymine.
      (2) Guanine normally only binds with cytosine.
4. A nucleotide consists of one phosphate, one sugar, and one nitrogenous base.
5. The coiling of the DNA allows it to fit into a cell nucleus (fig. 13.5).

D. The human genome is 3 billion base pairs long.

III. DNA Replication—Maintaining Genetic Information
A. DNA replication is the mechanism behind chromosomes duplication in mitosis.
1. The two strands of the double helix twist apart.
2. Free-floating nucleotides bond with their complementary base in the exposed part of the strand.
3. After a nitrogenous base attaches to its complement on the parent strand, its phosphate and sugar component attach to the forming rail of the new molecule of DNA (fig. 13.6).
4. While one parent strand is binding a guanine nucleotide to one of its cytosine, the other parent strand is binding a cytosine nucleotide to the guanine at the same location.
B. The duplication of DNA is called semiconservative replication because half of the parent molecule is always conserved in each of the new ones formed (fig. 13.7).
C. Replication forks.
1. This refers to small sections of DNA where replication is taking place.
2. Replication proceeds in both directions at one time (fig. 13.8).
3. The two strands are kept open by unwinding proteins called gyrase and swivelase.
4. With energy provide by ATP (adenosine triphosphate), the enzymes called DNA polymerases add new nucleotides (DNA polymerases are also used in correcting errors in replication).
5. DNA ligases link the sugar and phosphate groups to form the new rails of the new strand.

D. Gene amplification.
   1. This refers to the manufacture of multiple copies of a segment of DNA of interest.
      a) DNA of interest is mixed with lab-synthesized DNA of a particular gene (single strand).
      b) The synthesized gene binds to the complement on the DNA of interest while DNA polymerase attaches free nucleotides to complete the other strand.
      c) Upon the addition of heat, the strands separate and polymerase begins making many more copies in a process called polymerase chain reaction.
   2. Uses.
      a) Crime detection.
      b) Diagnosis of infectious diseases.
      c) Preserved DNA from fossils.
      d) Prenatal diagnosis.

IV. Transcription—Transmitting Genetic Information
   A. Protein synthesis occurs in two steps
      1. Transcription is the manufacture of messenger RNA in the nucleus.
      2. The messenger RNA leaves the nucleus and enters the cytoplasm to enter into the second step.
      3. Translation is the manufacture of a polypeptide chain by translating the code held by the messenger RNA.
   B. Ribonucleic acid (RNA) (fig. 13.9).
      1. Thymine is replaced by uracil.
      2. Deoxyribose is replaced by ribose sugar.
      3. Three different RNAs are involved in protein synthesis.
   C. Types of RNA.
      1. Messenger RNA (mRNA).
      2. For a particular gene, RNA is transcribed from only one strand of the DNA called the sense strand (fig. 13.10).
      3. Steps of transcription.
         a) Enzymes unwind the DNA at the site of the gene.
         b) RNA bases bond with the exposed complementary bases on the sense strand.
         c) For example, the DNA sequences GCGTATG is transcribed into the RNA sequence CGCAUAC.
      4. Gene silencers.
         a) This refers to synthetic antisense sequences used by researchers to bond to a gene's RNA, preventing it from translating the code into a polypeptide.
         b) It is used to silence DNA from disease-causing viruses.
      5. Once released the RNA takes on a particular three-dimensional shape caused by interactions among its base pairs.
      6. Each three nucleotides (codon) in sequences in an mRNA strand will code for one amino acid or serve as a start or stop signal.
      7. A different mRNA represents each gene.
      8. The average length of mRNA is 500 to 1,000 base pairs long.
   D. Ribosomal RNA (rRNA) (fig. 13.11).
      1. It ranges from 100 to 3,000 nucleotides.
      2. It joins with certain proteins to form the ribosome, which is a structural support for protein synthesis.
      3. It is composed of two subunits that float separately in the cytoplasm and then come together for protein synthesis.
   E. Transfer RNA (tRNA) (figs. 13.12 and 13.13).
      1. They are clover-shaped molecules that attach to a specific amino acid on one end.

2. The other end contains a three-base sequence (anticodon), which joins to a complementary strand of the mRNA.
3. A tRNA with a specific code always carries a particular amino acid, for example, the tRNA bearing the anticodon AAG always brings in the amino acid phenylalanine.

V. Translation—Expressing Genetic Information (fig. 13.14).
  A. The genetic code—from a gene's message to its protein product.
    1. The genetic code refers to the correspondence between the languages of mRNA and protein.
    2. The code was first cracked by Francis Crick and co-workers in 1960 by logic and experimental evidence.
      a) How many bases code specifically for one amino acid?
        (1) They determined this by calculating that if there were 20 amino acids and 4 nitrogen bases, it would take minimally 3 bases at a time to get that many different amino acids.
        (2) Combinations of 4 bases taken 3 at a time gives 43 or 64 possible codons, which would easily take care of 20 amino acids.
      b) Is the genetic code overlapping?
        (1) Crick and his colleagues determined that any one base was only part of one codon.
        (2) They predicted that if the gene code was overlapping, then certain amino acids would always follow others.

      c) Is the genetic cod punctuated?
        (1) They determined that some codons served for the function of starting and stopping.
        (2) Chemical analysis showed that nucleotides are recognized one right after another with no additional molecules involved.
      d) Do all organisms use the same genetic code?
        (1) The genetic code is universal (fig. 13.15).
        (2) The mRNA from one species can be introduced into the cell of another and translated.
      e) Which codons specify which amino acids?
        (1) Marshal Nirenberg in 1961 began to determine which codon specified which amino acid (fig. 13.1).
        (2) He did this by synthesizing various codons and then determining which amino acid was produced.
        (3) Some codons were found to be redundant, which meant they specified more than one amino acid.
        (4) Redundancy may protect against mutation.

B. Building a protein.
1. Initiation.
    a) Messenger RNA binds with a sequence of rRNA on the small ribosomal unit.
    b) The first codon of the mRNA is always AUG, which attracts a tRNA that has a modified amino acid called formylated methionine (fmet) initiating the production of a polypeptide (fig. 13.16).
    c) Initiation complex refers to the ribosomal unit, the mRNA bonded to it, and the initiator RNA.
2. Elongation (fig. 13.16).
    a) The large ribosomal subunit attaches to the initiation complex.
    b) The tRNA that is specified by the next codon to the initiator codon attaches, and its amino acid forms a peptide bond with fmet.
    c) The first tRNA is now released to pick up another amino acid.
    d) The ribosome and its attached mRNA are now bound to a single tRNA with two amino acids dangling from it.
    e) The ribosome now moves down the mRNA by one codon.
    f) The third amino acid brought in aligns with the second and attaches in continuation of the building of the polypeptide.
    g) The second tRNA is now released.
3. Termination
    a) Elongation halts when one of the mRNA "stop" codons appears because they do not have corresponding tRNAs.
    b) As the last tRNA is released, the polypeptide is released and takes on its three-dimensional shape.
    c) It may now be further modified.

C. Translation is efficient.
1. Many mRNA transcripts can be made from a single gene.
2. An mRNA transcript can be translated by several ribosomes simultaneously each at a different point along the message (fig. 13.17).
3. Prokaryotes can begin translation on the same mRNA that is presently being transcribed because they are not separated by a nuclear membrane as in eukaryotes.
4. Some antibiotics work by disrupting the process of protein synthesis of bacteria (fig. 13.18).

VI. The Changing View of the Gene
A. Genes in pieces—introns.
1. In some eukaryotic genes there are noncoding pieces of DNA that do not appear in the mRNA.
2. The stretches of coding DNA are called extrons (fig. 13.19).
3. After the mRNA is transcribed (now called pre-mRNA), introns are removed by a complex of small nuclear RNA and a protein called "snurps."
4. The number, size, and arrangement of introns varies from gene to gene.
5. Introns may control transcription, they may be future genes, or they may be junk left over from parasitic DNA.
B. Overlapping genes: discovered in viruses where one gene lies within another (fig. 13.20).
C. The usually universal genetic code.
1. Mitochondria, organelles that do cellular respiration for the cell, have their own DNA.

2. Some of the codons in the mitochondria specify for different amino acids than those in the nucleus.
3. This supports the endosymbiont theory, which suggests that mitochondria were once free-living organisms that became engulfed by primitive eukaryotic cells.

D. Variations of the double helix.
1. The helix actually can assume five different types of double helices, which vary by direction of twist, angle between bases, and the rails (fig. 13.21 and 13.22).
2. The DNA described by Watson and Crick was *B* DNA.

E. Jumping genes.
1. Barbara McClintock discovered that genes can jump from one chromosome to another, disturbing the expression of the genes into which they insert (fig. 13.23).
2. They can stop transcription if inserted into a gene.

VII. Biotechnology
A. This refers to the alteration of cells or biological molecules with specific applications in mind.
B. Applications.
1. Monoclonal antibodies are produced in a pure form from constructed immortal immune system cells called hybridomas.
2. Large-scale cell cultures are produced that secrete valuable biochemicals.
3. Recombinant DNA technology refers to alterations of an organism's gene that might include insertion of DNA from another organism to produce desired proteins.
C. Recombinant DNA technology.
1. In 1975, 140 molecular biologists convened at Asilomar and drew up guidelines to regulate recombinant DNA technology.

2. They outlined measures of physical containment that would keep recombinants in the laboratory.
3. In 1985, they reconvened and concluded that the industry was safe.

D. Constructing recombinant DNA molecules.
1. Restriction enzymes (fig. 13.24).
   a) These are molecular scissors used to cut a gene from its normal location, insert it into a circular piece of DNA, and then transfer the circle of DNA into cells of another species.
   b) They are naturally found in bacteria, where they inactivate infecting viruses.
2. Vectors.
   a) This refers to any piece of DNA to which DNA from one type of organism can be attached and transferred into a cell of another organism.
   b) Plasmids, which are small circles of double-stranded DNA found in bacteria, are used as vectors (fig. 13.25).
3. Steps in producing a recombinant.
   a) DNA is isolated from a donor species by restriction enzymes that cut out the desired gene.
   b) The cut leaves single-stranded ends that bear a specific sequence (fig. 13.26).
   c) A plasmid is cut with the same restriction enzyme so their ends will stick together forming a recombinant.
   d) Recombinants are produced that contain genes that allow them to grow in the presence of antibiotics so they can be separated from nonrecombinants.

e) After insertion in a host bacterial cell, the recombinant is replicated along with the cell to produce large cultures of recombinant-bearing genes.

f) The bacteria then manufacture the gene products, which can be isolated and used.

E. Applications of recombinant DNA technology (table 13.2).

1. Mass production of drugs.
   a) Human insults.
   b) Clotting factor.
   c) Human growth hormone.
   d) Monoclonal antibodies.

2. Production of the enzyme renin, which is used in the manufacture of cheese.

3. Recovery of fossilized genes that are extinct, which give insight into taxonomy and evolution.

F. Transgenic organisms.

1. Transgenic refers to the resulting organism from an engineered cell that contains injected foreign genes.

2. The technology permits the rapid introduction of new traits such as the introduction of a gene that produces an insecticide into tobacco plants.

3. Transgenic animals.
   a) Transgenic sheep and mice are producing human clotting factor in their milk, which is then extracted to be used in hemophiliacs.
   b) Transgenic mice are manufacturing multiple copies of a gene that causes human breast cancer for the purpose of studying it.
   c) Gene targeting refers to a multicellular organism that has a particular gene inactivated.

G. Sequencing the human genome.

1. It has been made possible by automation of DNA sequencing techniques.

2. The process will use gene libraries, which are collections of recombinant bacterial cultures, each cell containing a piece of the human genome.

3. A gene library is constructed by cutting all of the DNA in a human cell with several restriction enzymes and inserting each resulting piece into plasmids and then bacteria.

4. Different libraries are produced by using different restriction enzymes, and they are contributed to centralized human DNA repositories.

5. The human genome is approximately 10,000 genes long.

LEARNING OBJECTIVES

After reading this chapter, the student should be able to answer these questions:

1. How are the structures of a gene and a protein similar? How are they different?
2. What is the relationship between inherited traits and proteins?
3. What is the structure of a gene (DNA), and how does it contain information?
4. Why is it necessary for the genetic material to be replicated?
5. How does the structure of DNA provide a mechanism for its replication?
6. How is RNA transcribed from DNA?
7. What are the three different types of RNA, and how does each participate in the construction of a polypeptide?
8. What is the genetic code, and how is it deciphered?
9. What are the steps of protein synthesis?
10. What new discoveries have altered our picture of the gene?
11. What is biotechnology? How will it affect our lives?

ANSWERS

"Questions"

1. a. See figure 13.1

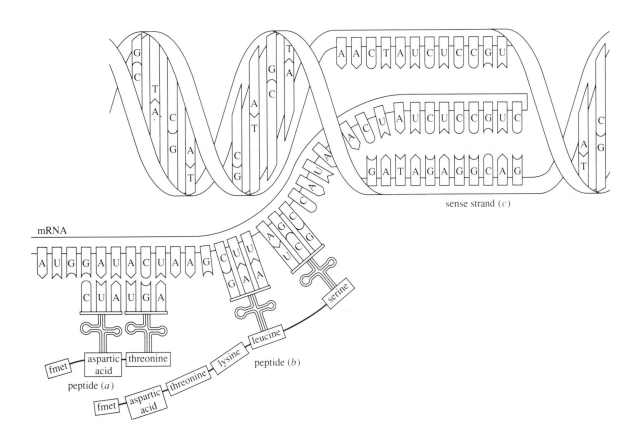

Figure 13.1

b. histidine-isoleucine-serine-valine
c. AUCUCCGUC
d. ATCTCCGTC
e. No, the code would have to be UAA, UAG, or UGA for a stop signal.
f. It would take on a three-dimensional shape and be used as an enzyme, hormone, pigment, or structure.

2. a. CGTTTTGGCGCTAATAGTACGAAG
b. CGUUUUGGCGCUAAUAGUACGAAG
c. arg-phe-gly-ala-asp-ser-thr-lys
d. arg-val-phe-phe-leu-try-gly-ala-arg-ala-leu-stop
e. Phenylalanine would be brought in, in place of asparagine.

3. a. It is a double helix where rungs are composed of complementary bases.
b. This allows for semiconservative replication where one original DNA strand is passed on to each daughter molecule along with a newly synthesized strand of DNA.
c. The mRNA that forms in transcription is complementary to the sense strand of the DNA.
d. The mRNA leader joins to a short sequence of bases on the rRNA.
e. The three-letter codon on the mRNA is complementary to the anticodon on the tRNA.

97

f. The paired nucleotides give the tRNA its three-dimensional shape.
g. The DNA probe attaches to a complementary strand of DNA in the invading agent or the abnormal gene causing the genetic disease.

4. a. Hydrogen bonds form between complementary base pairs in DNA replication and protein synthesis.
b. Peptide bonds form between adjacent amino acids in protein synthesis.
c. Covalent bonds form between alternating phosphate and sugar molecules in the formation of DNA or RNA.

5. The work of Barbara McClintock demonstrated that genes are not fixed but rather can move to different loci on the chromosome. In viruses, evidence suggests that genes may be overlapping, with one gene contained within a second gene that is read from the same bases but in a different reading sequence. Not all of the nucleotides of mRNA are translated into amino acids. Sections called introns are excised prior to translation. Four codons serve as start and stop signals and do not specify a particular amino acid. In mitochondria, some amino acids are specified by different codons than nuclear DNA.

6. Recombinant DNA technology has produced various drugs, hormones, and enzymes. It has made possible the recovery of genes from extinct organisms and the production of transgenic animals that can be used for various purposes such as the manufacture of the human gene for factor VIII.

"To Think About"

1. The genetic code has for the most part been delineated, and it has been determined that it is universal, meaning that the same molecules are found in the genetic material of all species. Though each person is unique in the sequencing of their DNA, the code is the same for all.

2. Often the last nucleotide in a codon is degenerative, which means it can be substituted by another base without altering the amino acid that it codes.

3. Prokaryotes are simple systems that do not require the separation of the processes of transcription and translation. Eukaryotic systems require separation of the two processes because the number of chemical reactions occurring simultaneously in the cytoplasm are far greater and would interfere with each other.

4. Proteins are involved all along the process of protein synthesis. Unwinding enzymes relax the DNA and open it at replication forks. Polymerase adds nucleotides while the backbone is fused by ligases.

5. Noncoding segments of DNA may be extra copies of genes or just extra nucleotide sequences that may allow for mutation.

6. The student should review the safety restrictions outlined at Asilomar and decide if these are enough, particularly in light of the clean record attained since the meeting in 1975.

7. You may need a drug or chemical such as insulin or clotting factor. Those available through biotechnology are more effective and inexpensive than those extracted from animals or cadavers.

8. Knowing the sequence of the human genome may allow for identification of genetic conditions or diseases. This could provide insight into a remedy for the problem. It would also provide insight into human genetic mechanisms and evolutionary relatedness. Funding for the human genome project has come under fire. Some scientists fear that major allocations of funds to this project may rob from other projects that offer more immediate benefits, such as cancer research.

AUDIOVISUAL MATERIALS

DNA, National Film Board of Canada, 1969.

DNA: Key to Life, BFA Educational Media, 1984.

DNA Replication, Bay State Film Production, 1982.

The DNA Story, John Wiley & Sons, 1973.

DNA: The Thread of Life, Open University Education Enterprises LTD, 1981.

Genetics Man the Creator, Inform: DOCA, 1982.

14 Human Genetics

KEY TERMS

aneuploid	monosomy
concordance	nondisjunction
inborn errors of	polyploidy
metabolism	translocation
inversion	trisomy

3. Aneuploidy is caused by nondisjunction during meiosis, when either homologues during meiosis I or sister chromatids during meiosis II fail to separate, resulting in one daughter cell carrying an extra chromosome while the other is minus one (fig. 14.3).

4. Aneuploidy may also arise during mitosis early during embryonic development or if later, this results in a chromosomal mosaic.

H. Down syndrome—an extra autosome.
1. This is the most common of the known aneuploids.
2. It is usually caused by an extra chromosome 21.
3. It is characterized by slanted eyes, a flat face, short, straight hair, a tongue that protrudes through thick lips, poor muscle tone, slow development, and subnormal intelligence.
4. Fifty percent die before their first birthday due to heart and kidney defects, suppressed immune systems, and leukemia.
5. Many who reach adulthood develop amyloid (black tangles of protein) in their brains that resembles that of Alzheimer patients.
6. The association of the problem with an extra chromosome was first made in 1959, just 3 years after the normal number of chromosomes was identified.
7. The risk of having a Down syndrome child greatly increases with age (fig. 14.4).

I. Abnormal numbers of sex chromosomes.
1. Turner syndrome.
   a) This refers to a female who is lacking an X chromosome (XO).
   b) It is characterized by underdeveloped breasts and genitals; ovaries, fallopian tubes, and a uterus that are small and immature; and sometimes below-normal intelligence.
   c) It occurs in 1 in 3,000 females.
2. Triplo-X is an extra X chromosome in 1 in every 1,000 females that only causes menstrual irregularities.
3. Klinefelter syndrome.
   a) It refers to males with an extra X chromosome.
   b) Characterized by underdeveloped testes and prostate glands, no public and facial hair, very long arms and legs, large hands and feet, and mental impairment.
   c) It is found in approximately 1 in 500 males.
4. Males with an extra Y chromosome (XYY) (table 14.1).
   a) It is characterized by being very tall, prone to acne, aggressive and antisocial, and lower-than-average intelligence.
   b) Studies in Swedish mental hospitals and Scottish prisons found 1 in 7 compared with 1 in 1,000 in the normal population.
   c) It is presently believed that only the tall height should be attributed to the extra Y since tall boys are treated differently, perhaps leading to their aggressiveness.

J. Deletions.
1. This is when only a portion of a chromosome is missing.
2. Cri-du-chat, a syndrome in children who have pinched faces and a peculiar cry like a cat, have a deletion associated with chromosome number 5 (fig. 14.2*d*).

K. Translocations.
1. This may result in deletions or duplication when different chromosomes exchange parts.
2. The breaks may be caused by viruses, drugs, or radiation.
3. Nonreciprocal translocation (fig. 14.2*c*).
  a) This is when a piece of one chromosome breaks off and attaches to another chromosome.
  b) In 97% of the people with chronic myeloid leukemia, the tip of chromosome 22 is attached to chromosome 9.
4. Reciprocal translocation (fig. 14.5).
  a) When two different chromosomes exchange parts, the "carrier" has all the genetic material but it is arranged differently.
  b) Generally, the carrier does not have problems but when passing on genetic material, some gametes will have an imbalance of genetic material.

III. Abnormal Single Genes
A. Modes of inheritance.
1. Autosomal recessive.
  a) This occurs in most single-gene defects.
  b) Two normal people each pass a recessive trait to the child.
  c) Heterozygotes are referred to as carriers.

2. Sex-linked recessive.
  a) Since the gene is carried on the X chromosome, the trait is more likely to be expressed by the male, who only has one X.
  b) Sex-linked genes are passed from a carrier mother to son.
3. Autosomal dominant (fig. 14.6).
  a) They are passed directly to offspring since only one copy of the gene is needed to be expressed.
  b) Usually the symptoms are not expressed until adulthood.
4. Criteria of inheritance.
  a) The disorder appears in other family members in a pattern that can be explained by Mendel's laws.
  b) A chromosomal abnormality is evident.
  c) A biochemical defect is detected that is caused by a mutant allele.

B. Ethnic diseases.
1. A predominance of certain inherited diseases in particular populations is a result of the tendency to choose mates of the same ethnic background.
2. For example, among the Amish is a high incidence of six-fingered dwarfism and manic depression, since today's 12,500 people are mostly the descendents of the original 60 (fig. 14.7).

C. Inborn errors of metabolism (fig. 14.8).
1. In these disorders, a chemical reaction does not take place since the enzyme required to catalyze the reaction is missing.
2. Most are inherited as autosomal recessive diseases.
3. Heterozygotes, though normal, can sometimes be detected by measuring enzyme levels in affected tissues since they will have one-half the normal amount.
4. These errors were first discovered by Archibald Garrod in 1902, when he determined that individuals with alkaptonuria were missing the enzyme required to break down homogentisic acid and as a result had urine that turned black when exposed to the air.

D. Genetic defects of the blood.
1. There are 300 hemoglobin variants, many of which lead to types of anemia including the most common, which is sickle cell disease (fig. 14.10).
2. Hemoglobin is a molecule that transports oxygen in the blood and is composed of four chains wrapped around a heme group (fig. 14.9).
3. The variants are caused by various point mutations in one of the genes involved.
4. In sickle cell disease, a substitution of the amino acid lysine for valine caused the hemoglobin to crystallize when oxygen levels are low, resulting in blocked blood vessels, which causes muscle pain, pneumonia, or brain damage (fig. 14.11).

5. Beta-thalassemia.
 a) The hemoglobin in the fetus is different and produced by different genes than in the adult.
 b) Beta-thalassemia is caused when the adult form of hemoglobin is not produced because the genes are defective.
6. Some forms of anemia are caused by abnormal numbers of hemoglobin genes (fig. 14.12).

E. Orphan diseases.
1. This refers to rare genetic diseases that receive little financial or research support.
2. Tourette's syndrome results in involuntary tics of the face and limbs and progresses to uncontrollable grunting and barking, echoing of others' speech, and foul language.

IV. "It Runs in the Family"
A. Traits caused by more than a single gene.
1. This refers to genetic traits that appear in some relatives but not others and do not appear to follow Mendel's laws.
2. Breast cancer is higher among women who have mothers or sisters with breast cancer but not at the rate predicted by Mendelian genetics.
3. Neural tube defect.
 a) When the neural tube fails to close during development and the child is born with a portion of the spine exposed (spina bifida), it results in paralysis.
 b) Height and intelligence appear to be affected by both inheritance and environment (fig. 14.13).

B. Nature versus nurture—twin studies.
1. Identical twins (monozygotic, MZ) have identical genes, whereas fraternal twins (dizygotic, DZ) do not.
2. If a trait appears to occur more frequently in MZ than DZ twins it is assumed to be at least partially inherited.
3. Concordance is the number of pairs in which both twins express the trait divided by the number of pairs in which at least one twin expresses the trait.
4. A worldwide study of twins raised apart has been undertaken to reduce the variable of MZ twins being raised differently than DZ twins. (fig. 14.14).

LEARNING OBJECTIVES

After reading this chapter, the student should be able to answer these questions:

1. How do changes in chromosome number and chromosome structure occur?
2. Why do we not observe all possible chromosome abnormalities?
3. In what ways can chromosomal abnormalities affect health?
4. In what ways can single gene defects affect health?
5. Why are disorders resulting from abnormal autosomes more severe than those resulting from the sex chromosomes?
6. How can traits "run in families" yet not follow Mendel's laws of inheritance?
7. How can twins help us to understand the nature versus nurture question?

ANSWERS

"Questions"

1. a. Cri-du-chat syndrome is caused by a deletion of part of chromosome 5.
    b. Tay-Sachs disease is an autosomal recessive disease caused by a defective enzyme hexosaminidase.
    c. Sickle cell disease is an autosomal recessive disease caused by the substitution of glutamic acid by valine in one of the beta chains.
    d. Edward's syndrome is caused by trisomy 18.
    e. Huntington disease is an autosomal dominant trait.
    f. Down syndrome may be caused by trisomy 21 or translocation of chromosome 21 to another autosome.
    g. Alkaptonuria is an autosomal recessive disease caused by a missing enzyme.
    h. Tourette's syndrome is an autosomal dominant trait.
    i. Alpha thalassemia is a condition of anemia caused by the deletion of one or more genes that produce alpha globin chains.
    j. Klinefelter syndrome is found in males with an extra X chromosome.
    k. Hemophilia is usually a sex-linked condition where factor VIII clotting factor is absent, though there are autosomal gene disorders resulting in a lack of other clotting factors.
    l. Familial hypercholesterolemia is an autosomal dominant trait that results in the lack of a receptor for cholesterol on cells.

2. a. The student appears to have a reciprocal translocation between chromosomes 3 and 21.
   b. Though the gene has been rearranged, the student has a full complement of genes.
   c. Her children may receive a duplication or a deletion of chromosome 3 or 21 or normal chromosomes 3 and 21.
3. Sickle cell disease is an autosomal recessive disease. Individuals with the disease have two copies of the disease allele.
   Consequently 100% of the children from a marriage of two affected individuals will have the disease.
4. If the woman is not a carrier, then she can be sure that none of her children will have the disease regardless of her mate, since she will pass on a normal gene that could compensate for a defective gene.
5. Though intelligence has a genetic component, and an individual's ability to be successful is critically dependent upon loving and supportive parents, under these artificial conditions, it is questionable if the child will receive the same advantage as a child of lesser genetic stock for intelligence raised by both biological parents.

"To Think About"

1. In the case of multiple genes, extra gene products may or may not be produced. Consequently, normal protein function is still attainable. When a copy of a gene is missing, it may mean that not enough of a protein product is produced, leaving the cell unable to function.

2. Somatic mutations are errors that occur during mitosis. Depending on when in development the error occurs the resulting cell lines may differ in their genotypes.
3. Though there is no right answer, the student should consider how much evidence supports the claim that XYY results in high aggression. Even if the claim were true that behavior is affected by genes, does that justify misconduct?
4. Though there is no right answer, the student should consider an individual's right to meet his or her potential. The problem with control being exerted by someone other than the individual is that the expectations may be too great or too low.
5. Intelligence is believed to be a result of many genes and also greatly influenced by the environment. Consequently there is a lot that parents can do to enhance their child's intelligence.
6. Though height is inherited, it is affected by environmental factors. Consequently a person may never attain his or her possible genetic height because of poor nutrition at some point in development.

AUDIOVISUAL MATERIALS

Genetic Change, Time Life Film and Video, 1976.

Genetics: Human Heredity, Coronet Instructional Films, 1968.

15 Genetic Disease—Diagnosis and Treatment

KEY CONCEPTS

Most of the information regarding human heredity still comes from pedigrees, which are charts of family genetic characteristics.

Karyotypes are used to diagnose chromosomal disorders, exposure to environmental toxins, and evolutionary relationships between species.

Ultrasound is used in prenatal screening to detect physical deformities and to guide amniocentesis.

Amniocentesis is the removal of amniotic fluid for chromosome analysis and biochemical assays.

Chorionic villus sampling is the removal of sample cells surrounding the fetus for chromosome analysis.

Direct diagnosis of inherited disease is possible with DNA probes, which bind with known disease-causing alleles or a closely linked gene.

Treatment for genetic disease is directed toward treatment of the phenotype.

Nonheritable gene therapy corrects the genetic defect in the somatic tissue in which the defect is expressed.

In heritable gene therapy, defective genes in germ cells or fertilized ova are replaced with normal alleles.

In gene silencing, certain genes whose overproduction cause disease are blocked from being expressed.

KEY TERMS

amniocentesis	restriction fragment
genetic marker	length
karyotype	polymorphisms
pedigree	Southern blotting

CHAPTER OUTLINE

I. Studying Chromosomes and Genes
   A. Pedigrees (figs. 15.1 and 15.2).
      1. These are charts constructed on genetic traits of families.
      2. Interpretation may be difficult for various reasons.
         a) Not all information is given by family members.
         b) Adoptions and serial marriages result in blended families.
         c) The nature of the disorder is hard to trace.
   B. Chromosome charts.
      1. Karyotypes are charts of chromosomes arranged in pairs of decreasing size order.
      2. Production of karyotypes.
         a) White blood cells are extracted from a finger.
         b) Cells are treated with colchicine, a chemical that halts the movement of chromosomes in dividing cells.
         c) Cells are dropped on slides, causing the chromosomes to spread out.
         d) A photograph is taken and then the individual chromosomes are cut out and placed in order by size.
      3. The first karyotypes were constructed about 30 years ago.
      4. Differential staining results in banded chromosomes, each band consisting of about 20 genes (fig. 15.3).

5. Computerized karyotyping is rapid compared with the traditional method and also allows for a data base that can help identify genetic problems.
6. Karyotypes are used to diagnose aneuploidys, deletions, inversions, translocations, and fragile X syndrome (fig. 15.4).
7. They are also used to identify environmental toxins by looking for abnormalities such as broken chromosomes, which appear only in a group exposed to a particular contaminant such as the survivors of the atomic bomb in Japan.
8. Karyotypes are also used to determine evolutionary relationships, since the more recent the divergence of two species, the more alike are their karyotypes.

II. Prenatal Diagnosis
   A. This allows for prediction of genetic disease as well as understanding of the embryological process.
   B. Ultrasound (fig. 15.6).
      1. This is when sound waves are bounced off of the fetus and converted into an image on a screen.
      2. Body parts that are too large or too small for a particular point in gestation indicate a problem.
   C. Amniocentesis (fig. 15.7).
      1. In this procedure fetal cells and fluids are removed from the uterus with a needle guided by ultrasound.
      2. The cells are cultured for karyotypes and the fluid is examined for biochemicals that indicate metabolic disorders.
      3. Four hundred different chromosomal and biochemical problems can be detected.

   D. Chorionic villus sampling (fig. 15.8).
      1. A procedure that snips cells from the fingerlike projections that develop into the placenta.
      2. The chromosomes of the chorionic villus cells are derived from the fertilized egg, so they are the same as the fetus.
      3. Disadvantages.
         a) Sometimes a genetic mosaicism has occurred so genetic diagnosis may be in error.
         b) It does not allow for the sampling of biochemicals.
   E. Analyzing sampled cells.
      1. Since gender can be determined, the risk of a sex-linked disease can better be calculated.
      2. Fetuses that carry Tay-Sacs and spina bifida can be detected.

III. Molecular Approaches to Diagnosing Genetic Disease
   A. These allow for the examination of the DNA for a closer detection of disease-causing genes.
      1. DNA probes are pieces of DNA about 1,000 to 6,000 bases long that are complementary to part or all of a particular gene or any portion of a chromosome's DNA.
      2. They are grown in the presence of radioactive DNA.
      3. In a procedure called Southern blotting, probes pick out their complementary sequences in sample DNA and are visualized by their radioactivity (fig. 15.9).
   B. Direct genetic disease diagnosis—RFLPs within genes.
      1. A normal allele yields a different pattern of fragment sizes than the disease-causing allele, if each is cut with the same restriction enzymes.
      2. A childhood form of Gaucher's disease, which results in an enlarged liver and spleen, can be probed directly with a restriction enzyme.
   C. Genetic markers—RFLPs outside genes.

1. For disorders in which the gene has not been identified, restriction enzymes can be used for diagnosis if a piece of DNA located near the gene (genetic marker) has a characteristic restriction enzyme cutting site.
2. To be useful as a genetic marker, the linked DNA sequence must always be present in ill family members but never in healthy relatives.
3. This procedure is available for detection of Huntington disease where the G8 probe is about 96% accurate.
4. Presymptomatic tests for late-onset diseases like Huntington and Alzheimer's present ethical problems of knowing the future regarding health.
5. By using a process called chromosome jumping, researchers cut the chromosomal region containing the gene with several restriction enzymes and then sequence the resulting pieces in order to identify the actual gene.

IV. Genetic Screening
   A. Before Pregnancy: Couples who are carriers can determine the risk of producing an affected child.
   B. During Pregnancy: Genetic defects can be determined.
   C. Screening Children: For subtle diseases in infants such as sickle cell disease and phenylketonuria, this allows for changes in treatment that will extend the quality of life.
   D. Genetic screening in the workplace.
      1. Some chemical companies administer blood tests to identify workers who are genetically predisposed to becoming ill from exposure to certain chemicals.
      2. Preemployment tests compared with on-the-job tests can detect contact with potentially dangerous chemicals.

V. Genetic Counseling
   A. Genetic counselors are individuals who help families understand their problems and who evaluate pedigrees and test results to predict who in a family is likely to be affected by the condition.
   B. The ethical choice is up to the family.

VI. Medical Genetics, Ethics, and the Law
   A. There is concern regarding individuals who might have an abortion of a healthy fetus just because of its sex.
   B. The potential is there for companies to refuse to hire genetically susceptible people.
   C. Insurance companies could use genetic information to discriminate against certain groups of people.

VII. Gene Therapy
   A. Treating the phenotype—a short-term solution.
      1. Treating the symptoms does not alter the genetic material.
      2. These treatments may increase the frequency of disease-causing alleles in a population by allowing individuals to become parents.
   B. Nonheritable gene therapy.
      1. This is when only the affected somatic cells are treated.
      2. It will be implemented sooner than heritable gene therapy.
      3. Target diseases for this procedure are hereditary immunodeficiency disease and Lesch-Nyhan, where a missing enzyme causes mental retardation.
   C. Heritable gene therapy—a longer-term solution.
      1. A gene is altered or replaced at the germ cell or fertilized egg stage.
      2. Researchers have created a "supermouse" where a rat's growth hormone is implanted into the fertilized egg of a mouse (fig. 15.14).

3. Other researchers have "cured" beta-thalassemia in mice by this technique.
4. Gene silencing is another type of gene therapy where new genes indirectly may work to shut off oncogenes, which cause cancer.

LEARNING OBJECTIVES

After reading this chapter, the student should be able to answer these questions:

1. How are pedigrees, chromosomal charts, and biochemical tests used to detect or evaluate genetic disease?
2. What information is provided by prenatal diagnosis and genetic screening?
3. How are DNA probes and restriction enzymes used to diagnose genetic diseases directly?
4. How are genetic markers used to diagnose genetic disease before symptoms arise?
5. How can we alter genes and their expression?

ANSWERS

"Questions"

1. See figures 15.1 and 15.2.
   Since Huntington disease is a dominant trait, evidence of its presence in relatives would be known unless it arose through mutation.
2. The likely mode of inheritance of kinky white hair is X linked since it has only appeared in males and appears to have skipped generations. Through amniocentesis or chorionic villus sampling, the sex of the child can be determined. If it is a boy, the child would have a 50% chance of having the condition and if a girl a 50% chance of being a carrier.

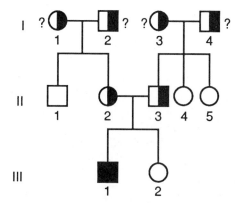

Figure 15.1 Cystic fibrosis

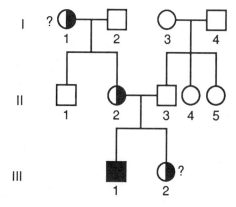

Figure 15.2 Hemophilia

3. A fetus with XX karyotype will be a girl unless a somatic mutation occurred early in development, resulting in a genetic inconsistency between the chorionic tissue and the fetus. The results of the alpha-fetoprotein test may have given a false negative if the amounts of the chemical were at undetectable levels due to the immature state of the embryo.

4. Though chorionic villus sampling allows for earlier genetic diagnosis, it has the drawback of a higher miscarriage rate and may be inaccurate due to somatic mutations that result in genetic inconsistencies between the chorionic tissue and the fetus. Amniocentesis does not often provide results until the 18th week of pregnancy, which increases the anxiety associated with deciding on an abortion for some individuals.

5. A karyotype of the Down syndrome child will provide clues as to the origin of the problem. If the condition is caused by trisomy 21, it is not likely that it is inheritable. If instead the X chromosome reveals the characteristic knob of fragile X syndrome, it is likely inheritable.

6. DNA probes are segments of DNA that bind with complementary strands of DNA of abnormal alleles or closely linked genes or with the DNA of an infectious agent.

7. Heritable gene therapy attempts to alter the genetic material in the gametes of an individual and does not affect the somatic cells of the person, which gave rise to the symptoms.

"To Think About"

1. Albinism is caused by having two copies of the recessive gene. In this case each parent was a carrier. They did not exhibit the trait because it was masked by the normal allele. Each pregnancy has a 25% chance of producing a child with the trait.

The counselor's job is to provide the best- and worst-case scenarios regarding children with Down syndrome for the prospective parents and to discuss the alternatives regarding abortion and childcare. The counselor can help the parents identify their feelings regarding the available choices but will not try to influence their decision.

The parents of the child with sickle cell disease should be told that the disease is caused by having two copies of the recessive gene. Each parent was a carrier of the gene in this case but did not exhibit the trait because they also had a normal functioning allele for hemoglobin. Each subsequent pregnancy would have a 25% chance of producing a child with the sickle cell disease trait.

2. To learn that you will get Alzheimer's disease in your 50s and have to look forward to the devastating symptoms might be difficult for an individual to handle. However, it would give you the knowledge whether or not you had the gene, which would pass to 50% of your children. On the other hand, you might discover that you did not have the gene and would be able to live without the fear of the unknown.

3. Though there is no correct answer, the student should consider the benefit to a couple of knowing about the defect early in pregnancy. It might allow for the choice of abortion or the psychological adjustment prior to birth. Is this benefit greater than the cost of the anxiety experienced by parents who must wait for the results of a second test to determine a false positive?

4. The information gained by sickle cell disease screening was widely misunderstood. It led to discrimination; tests were mandatory and the results were not kept confidential. Results were used by life insurance companies and employers to discriminate against heterozygotes who were unaffected by the gene. Its usefulness for parents is not definitive, since homozygotes show variable expression of the gene, from an early death to a long, useful life. Tay Sachs disease, which is caused by the homozygous condition, is lethal. Unlike sickle cell disease, Tay-Sachs disease was detectable by amniocentesis, so couples could conceive and elect for an abortion if present. The screening was voluntary and counseling was well provided. The test for cystic fibrosis should be voluntary, confidential, and accompanied by counseling.

5. Heritable gene therapy will be more difficult to accomplish because it requires alteration of the genetic material in the gametes. This is more risky since the technique requires insertion of genetic material by a virus into the normal DNA. Placement of the gene is not controllable by the experimenter and could result in the disruption of a normal gene or placement of the new gene next to a regulatory gene that alters its expression. This could result in the creation of even worse problems for the fetus. This has raised serious ethical questions. Nonheritable technology provides more control of the outcome because a few cells are removed from the individual and experimented on in the laboratory. The results can be better screened. Even the insertion of defective cells into the person would probably not lead to a worse condition. Also there is less of an ethical question because the affected individual can decide for himself or herself.

AUDIOVISUAL MATERIALS

Genetic Chance, Time Life Film and Video, 1976.

Genetics: Human Heredity, Coronet Instructional Films, 1968.

16 Neurons

KEY CONCEPTS

Neurons are cells with extensive processes that are specialized for communication.

Sensory neurons carry information to the central nervous system.

Motor neurons carry information from the central nervous system to muscles and glands for action.

Interneurons within the central nervous system process sensory information, initiate motor activities, and store information in the form of memory.

Neurons can communicate with other cells because of a charge across the cell membrane, termed the resting potential.

Signals are sent from neuron to neuron in the form of action potentials.

Synapses are connections between neurons where the action potential is converted to a chemical message for transmission to the next neuron.

Decision making is made at the synapse to either excite or inhibit the affected cell or neuron.

Alteration of the synaptic chemicals or neurotransmitters by drugs can alter behavior.

Some chemicals such as endorphins modify the activity of the nervous system and its response to pain.

Connections of neurons form networks for the production of complex behaviors.

KEY TERMS

acetylcholine	botulism
acetylcholinesterase	cell body
action potential	concentration gradient
(nerve impulse)	dendrites
acupuncture	depression
agonist	depolarization
antagonist	effector
agonist	electrical gradient
antagonist	endogenous
axon	endorphins

enkephalin	postsynaptic neuron
excitatory	presynaptic cell
gates	receptors
gray matter	repolarization
hyperpolarizes	resting potential
inhibitory	saltatory conduction
interneuron	Schwann cells
motor (efferent)	second messengers
neuron	selectively permeable
myelin sheath	sensory (afferent) neuron
nerve fibers	sodium-potassium pump
neural or synaptic	synapse
integration	synaptic cleft
neurotoxin	synaptic vesicles
neurotransmitter	synaptic knobs
node of Ranvier	white matter
polarized	

CHAPTER OUTLINE

I. The nervous system is composed of single cells termed neurons, or nerve cells, that make possible a variety of sensations, actions, emotions, and experiences.

II. The Anatomy of a Neuron
   A. A neuron has a rounded central portion from which many long, fine extensions emanate (fig. 16.1).
   B. The central portion of the neuron, called the cell body, contains the usual assortment of organelles.
   C. The extensions of a motor neuron.
      1. Dendrites.
         a) These are shorter, branched, and more numerous.
         b) They receive information from other neurons and transmit it toward the cell body.
      2. The second type of extension from the cell body is an axon.
         a) These conduct the message away from the cell body and transmit it to another cell.

b) They are sometimes called nerve fibers and are usually longer than a dendrite.
III. Types of Neurons (fig. 16.2)
  A. The shape of each type of neuron reflects its function.
  B. A sensory (or afferent) neuron has longer dendrites that carry its message from a body part, such as from the skin, toward the cell body, which is located just outside the spinal cord.
  C. A motor (or efferent) neuron conducts its message outward from the central nervous system toward a muscle or gland and has a long axon to reach the effector.
  D. A third type of neuron, an interneuron, connects one neuron to another to integrate information from many sources.
IV. A Neuron's Message
  A. A nerve impulse is actually an electrochemical change caused by ions moving across the cell membrane, and this change is called an action potential.
  B. The resting potential (fig. 16.3).
    1. Defined as when the neuron is resting and has an electrical potential that is slightly negative with respect to the outside.
    2. Cell membrane.
      a) A membrane in this condition is described as being polarized, and the differences across the membrane result from the unequal distribution of sodium ions (Na^+) and potassium ions (K^+).
      b) Selectively permeable cell membrane.
        (1) This is when the membrane admits some substances but not others.
        (2) Some channels in the membrane are always open, but others are opened or closed by the position of gates, which are proteins that change shape to block or clear the channel (fig. 16.4).
        (3) Some membrane channels are specific for Na^+ and others for K^+.
      c) The sodium-potassium pump.
        (1) This is a mechanism that uses cellular energy (ATP) to transport Na^+ out of the cell and K^+ into the cell.
        (2) The pump uses active transport, moving Na^+ and K^+ against their concentration gradients.
        (3) First, ions follow an electrical gradient: like charges tend to repel one another; unlike charges (negative and positive) attract one another.
        (4) Second, ion distribution is affected by a concentration gradient.
      d) Three basic mechanisms explain the resting potential.
        (1) First, the sodium-potassium pump, using ATP for energy, concentrates K^+ inside the cell and Na^+ outside: the pump ejects three Na^+ while pumping in two K^+.
        (2) Second, large, negatively charged proteins and other negative ions are trapped inside the cell because the cell membrane is not permeable to them.
        (3) Third, the membrane is 40 times more permeable to K^+ than to Na^+ in the resting state. As K^+ moves through the membrane to the outside of the cell, it carries a positive charge with it, leaving behind the

large, negatively charged molecules. A charge or potential is therefore established across the membrane; positive on the outside and negative on the inside.

C. The action potential.
   1. An action potential begins when a stimulus (a change is pH, a touch, or a signal from another neuron) changes the permeability of the membrane so that some Na^+ begins to leak into the cell (fig. 16.5).
   2. The membrane is said to become depolarized by the influx of Na^+.
   3. When enough Na^+ enters to depolarized the membrane to a certain point (the threshold), the sodium gates in that area of the membrane open, making the membrane even more permeable to Na^+.
   4. Na^+ floods the inside of the cell so that the interior becomes positively charged.
   5. A mass exodus of K^+ now begins, driven by both electrical and concentration gradients.
   6. The loss of positively charged K^+ restores the negative charge to the interior of the cell, and the outward flow of K^+ repolarizes the cell membrane.
   7. This slight increase in negative charge compared to the resting state is corrected by the sodium-potassium pump.
   8. While the Na^+ gates and then the K^+ gates are open, a second action potential cannot be initiated.
   9. The action potential spreads because some of the Na^+ rushing into the cell at a particular point moves to the neighboring part of the neuron.
   10. The action potential moves along the neuron just as surely as a line of properly spaced dominoes.
   11. An action potential is therefore an all-or-none phenomenon, since there are no degrees of action potentials.
   12. If all action potentials are of equal magnitude, how do we fell different degrees of stimulation?
      a) Sensitivity is transmitted by the frequency of action potentials.
      b) Our neurons also discern the type of stimulation.

D. The myelin sheath and saltatory conduction.
   1. The speed of conduction is determined by certain characteristics of the fiber; the greater the diameter, the faster it conducts an action potential.
   2. Myelin sheaths are formed by Schwann cells, which contain enormous amounts of lipid (fig. 16.6).
   3. Many Schwann cells form the sheath, each wrapping a small segment of the axon, and between each Schwann cell is a short region of exposed axon called node of Ranvier.
   4. Some neurons are wrapped in myelin produced by cells called oligodendrocytes.
   5. The action potential "jumps" from one node to the next, a type of transmission called saltatory conduction, which increases the speed of transmission 100 times faster than unmyelinated axons (fig. 16.7).
   6. Myelinated fibers are the white matter of the nervous system.
   7. The gray matter of the nervous system consists of cell bodies and interneurons that lack myelin and usually specialized in integration.

V. Synaptic Transmission
A. The action potential is converted into a chemical that travels from a "sending" cell to a "receiving" cell into a tiny space between the cells.
B. Once across this space, the neurotransmitter chemical alters the

permeability of the receiving cell's membrane either to provoke an action potential or prevent one.

C. The space between neurons is called a synapse.
    1. The end of an axon has tiny branches that form synaptic knobs.
    2. Within the synaptic knob are many small sacs, called synaptic vesicles, that contain neurotransmitter molecules.
    3. An action potential passes down the axon of the cell, sending the message, which is called the presynaptic cell.
    4. The calcium ions cause the vesicles containing neurotransmitter molecules to move toward the synaptic membrane, fuse with it, and dump their contents into the synaptic cleft by exocytosis (fig. 16.8).

D. Neurotransmitter molecules diffuse across the cleft and attach to protein receptors on the membrane of the receiving neuron (the postsynaptic neuron) (fig. 16.9).

E. When the neurotransmitter attaches to the receptor, the conformation (three-dimensional shape) of the receptor protein changes so that channels open in the postsynaptic membrane, allowing specific ions to flow through and changing the probability that an action potential will be generated.

F. In botulism, a bacterial toxin blocks vesicles in the presynaptic cell from releasing the neurotransmitter acetylcholine into the synapses to muscle cells, and if it lacks stimulation, the muscles cells cannot contract.

G. Disposal of neurotransmitters.
    1. Soon after a neurotransmitter is released, it is either destroyed by an enzyme or taken back into the axon that released it.
    2. Nerve gas and certain insecticides block the breakdown of acetylcholine by inhibiting acetylcholinesterase.

    3. This results in nerve endings stimulating skeletal muscle to contract continuously, and the victim convulses and dies.

H. Excitatory and inhibitory neurotransmitters.
    1. Excitatory synapses.
        a) They depolarize the postsynaptic membrane, and inhibitory synapses increase the polarization.
        b) They increase the probability that an action potential will be generated in the second neuron by slightly depolarizing it.
    2. Inhibitory synapses.
        a) It is more difficult for an impulse to be generated in the postsynaptic cell with an inhibitory neurotransmitter.
        b) The inhibitory neurotransmitter hyperpolarizes the postsynaptic membrane.

I. Synaptic integration—interprets its messages.
    1. Whether a neuron transmits an action potential depends on the sum of the excitatory and inhibitory impulses that it receives.
    2. Only when the postsynaptic membrane is depolarized to threshold level is an action potential generated.
    3. Whether an action potential is "fired" or not is termed neural or synaptic integration.
    4. We sleep when the activity of certain cells in a part of the brain called the reticular activation system is inhibited by the release of the neurotransmitter serotonin by other neurons.

VI. Neurotransmitters
A. Types of neurotransmitters (table 16.1).
    1. The peripheral nervous system (the part outside the brain and spinal cord) uses three

neurotransmitters: acetylcholine, noradrenaline, and adrenaline.

2. The central nervous system (the brain and spinal cord) has many additional transmitters such as dopamine, serotonin, the inhibitory transmitter, GABA (gamma amino butyric acid), and the fascinating internal opiates, endorphins.

3. Neurotransmitter levels are modulated by chemicals called second messengers such as cyclic adenosine monophosphate (cAMP).

B. Psychoactive drugs and neurotransmitters (table 16.1).
1. A drug that binds to a receptor, blocking a neurotransmitter from binding there, is called an antagonist.

2. A drug that activates the receptor, triggering an action potential, or helps a neurotransmitter to bind is called an agonist.

3. Amphetamine and cocaine enhance noradrenaline activity, thereby heightening alertness and mood.

4. Cocaine blocks the receptors on the membrane of the presynaptic cell that normally function as a reuptake gateway for noradrenaline resulting in the overabundance of noradrenaline in the synapses and producing the feelings of a cocaine high.

C. Disease and neurotransmitters (table 16.2).
1. Disturbances of neurotransmitter balance are thought to be behind a variety of medical problems, including epilepsy, insomnia, and sudden infant death syndrome.

2. It is often difficult to tell if the neurotransmitter abnormality is a result of a disorder or a cause of it.

3. Depression is a good example of a neurotransmitter imbalance.

D. The biochemistry of depression: Endogenous depression can often be traced to a biochemical abnormality in the brain such as deficiency of noradrenaline.

E. Opiates in the human body.
1. The effects of opiates are enjoyable, and animals can become physically addicted to these drugs.

2. In the 1970s the human body was found to produce its own opiates called endorphins.
   a) These influence mood and perception of pain.
   b) The first endorphin identified was a peptide consisting of five amino acids named enkephalin.
   c) A few months later, beta-endorphin, 30 amino acids long, was discovered.
   d) One particularly powerful endorphin is dynorphin, discovered in 1979 and it is 200 times as potent as morphine.
   e) Endorphins may also be responsible for the pain relief provided by the medical art of acupuncture, in which needles inserted at certain points in the body relieve pain in other areas.

LEARNING OBJECTIVES

After reading this chapter the student should be able to answer these questions:

1. What are the parts of a neuron?
2. What are the three types of neurons?
3. What happens during a nerve impulse?
4. How does a resting neuron differ from a neuron firing a nerve impulse?
5. How does the myelin sheath increase the speed of nerve transmission?
6. How do nerve cells communicate across a synapse?
7. How do nerve cells communicate with each other to integrate their function?

ANSWERS TO TEXT QUESTIONS

1. See figure 16.1a for a diagram of neuron.
2. Sensory neurons convey information from receptors to the central nervous system (CNS), motor neurons arise in the CNS and activate muscles and glands, interneurons lie within the CNS and integrate information from many sources and coordinate responses.
3. The inside of the neuron is slightly negative, approximately -65 millivolts, with respect to the outside, thus, the neuron is polarized in the resting state. The polarity is due to the unequal distribution of Na^+ and K^+ ions across the membrane. The membrane is relatively more permeable to K^+ than Na^+, so K^+ tends to follow its concentration gradient (established by the Na^+/K^+ pump) moving towards the outside of the cell. The large anions inside the cell contribute to the negative charge inside, and the positive charge outside is due to the small excess K^+. The movement of K^+ comes to equilibrium because the charge across the membrane is in opposition to the concentration gradient driving the movement of K^+.
4. The initial phase of the action potential, depolarization, is due to the inward movement of Na^+, which causes the cell to depolarize to zero potential, and then to become positive inside at which time the Na^+ gate closes. The repolarization is primarily due to the outward movement of K^+, which carries the positive charge from inside to the outside, returning the cell to its initial condition of negative inside and positive outside.
5. In myelinated neurons the rate of conduction is faster since the action potential "jumps" from node to node, rather than traveling as a wave along the length of the axon.
6. Action potentials are self-regenerating events ("all or none") and do not vary in magnitude. They are due to the movement of Na^+ and K^+ across the neural membrane. Synaptic transmission involves the release of a neurotransmitter from one neuron which alters the permeability to specific ions in a second neuron or

effector. This may or may not lead to an action potential, depending upon the type of neurotransmitter, the location of the synaptic connection, the nature of the receptor molecules, and other variables.
7. See figures 16.8 and 16.9 for a diagram of the synapse.
8. Parkinson's disease is caused by insufficient dopamine in the substantia nigra located in the base of the brain.

Schizophrenia is caused by excessive levels of dopamine in the brain and symptoms of schizophrenia may result from treating Parkinson's disease with L-dopa.

Epilepsy is caused by an imbalance in the central nervous system of neurotransmitters.

Botulism blocks the release of acetylcholine from motor neurons, activating muscle cells. The muscles cannot contract and breathing becomes paralyzed.

Heroin addiction alters the endorphin levels of the brain, reducing their synthesis. When heroin is removed the low level of endorphin is interpreted as pain.

Depression may be due to a low level of neurotransmitters such as noradrenaline in the brain.

Multiple sclerosis affects the myelination of neurons in the brain and spinal cord forming scar tissue around these neurons which are unable to transmit action potentials leading to muscle weakness and incoordination.

ANSWERS TO THE "TO THINK ABOUT" QUESTIONS

1. There are both positive and negative arguments for blanket testing of individuals for depression or other diseases. If the individual is found to have the chemical indicator for suicidal tendency, counseling and treatment for depression should be provided.
2. Beyond the epidemiological studies to link Parkinson's disease to herbicides, research could be conducted on animal models. Experimental animals could be

117

given certain classes of herbicides to see if they cause Parkinson's disease. You could also determine if the progress of the disease can be altered with L-dopa or other ways of increasing dopamine in the substantia nigra.

3. The advantage is that depression has an identifiable organic cause and can be treated with chemical therapy. Disadvantages may be that the form of depression is not treatable, or that drug therapy may include side effects that are intolerable to the individual.

4. It allows noradrenaline to accumulate in the brain by blocking the enzyme which degrades or inactivates this neurotransmitter.

AUDIOVISUAL MATERIALS

Nerve Cell, MacMillan Films, 17 min.
The Nerve Impulse, Encyclopaedia Britannica Educational Corp., 21 min., 1971.
Nerves at Work, Films for the Humanities, 26 min., 1985.
The Peripheral Nervous System, International Film Bureau, 19 min., 1977.

17 The Nervous System

KEY CONCEPTS

The brain and spinal cord form the central nervous system.

The spinal cord is capable of receiving sensory information, integrating it, and sending appropriate motor commands to muscles and glands which form the basis of spinal reflexes.

Sensory information is also transmitted to the brain for more complex levels of integration.

The brainstem controls many complex reflexes such as breathing, heartrate, blood pressure, body temperature, water balance, and certain drives and feelings.

The cerebrum is the portion of the brain associated with the higher functions of learning, memory, perception, and voluntary activity. The two halves of the cerebrum are specialized for different functions, but are connected by the corpus callosum which allows the two halves to work in harmony.

Short term memory is thought to be based on electrical circuits whereas long term memories are consolidated through specific synaptic connections.

The peripheral nervous system provides the linkage between sensory organs and receptors to the CNS through sensory neurons. It also provides the linkage of the CNS to muscles and glands through motor neurons.

The autonomic nervous system, with its two divisions, automatically controls the vegetative functions of the body.

KEY TERMS

amygdala
association areas
autonomic nervous
  system
basal ganglia

blood-brain barrier
brainstem
central nervous
  system (CNS)
cerebellum
cerebrospinal fluid

cerebrum
corpus callosum
cranial nerves
ganglia
hippocampus
hypothalamus
long term synaptic
  potentiation
medulla
memory
meninges
midbrain
motor areas
motor pathways
parasympathetic
  division
peripheral nervous
  system (PNS)

pons
reflex
reflex arc
reticular activating
  system
sensory areas
sensory receptor
sensory pathways
somatic system
spinal cord
spinal nerves
spinal reflex
sympathetic
  division
thalamus
tracts
vagus nerve
ventricles

CHAPTER OUTLINE

I. Organization of the Vertebrate Nervous System (fig. 17.1)
   A. The most general division of the vertebrate nervous system is the central nervous system (CNS), which consists of the brain and spinal cord.
   B. The peripheral nervous system (PNS) consists of nerves and ganglia (collections of cell bodies) that carry information to and from the CNS.

II. The Central Nervous System
   A. The spinal cord (fig. 17.2).
      1. The spinal cord runs about 17 inches from the base of the brain to about an inch below the last rib; it carries impulses to and from the brain and is a site of interaction between neurons involved in spinal reflex actions.
      2. A reflex is a rapid, involuntary response to a stimulus either from within the body or from the outside environment.

3. The spinal cord communicates with the rest of the body through spinal nerves.
   a) On each side of the cord is a pair of nerves; sensory information is delivered to the rear of the spinal cord, and instructions for activity of skeletal muscles pass outward from the front of the cord.
   b) The vertebral column of the lower back contains only spinal nerves and fluid.
   c) Spinal anesthetics ease childbirth pain (fig. 17.3).
4. The spinal cord conducts information to and from the brain via myelinated fibers, which form the white matter at the periphery of the cord.
   a) Ascending tracts carry sensory information to the brain, and descending tracts carry motor information from the brain to muscles and glands.
   b) The gray matter is motor neuron cell bodies, interneurons, and glial cells.
5. A spinal reflex usually involves several neurons but may require only two.
   a) A reflex arc is a neural pathway linking a sensory receptor and an effector such as a muscle (fig. 17.4).
   b) A dendrite of a sensory neuron in the skin is specialized as a sensory receptor.
   c) The sensory neuron's axon synapses with an interneuron, which synapses with a motor neuron.
   d) The motor neuron's axon exits the spinal cord on the ventral side, and its action potential stimulates a skeletal muscle cell to contract.
   e) The original sensory neuron synapses with other interneurons, some of which go to the brain.

B. The brain.
   1. Oversees the functioning of organs and also provides the qualities of "mind"—learning, reasoning, and memory.
   2. The activity of the brain accounts for 20% of the body's consumption of oxygen and 15% of its consumption of blood glucose.
   3. The human brain is built of three major regions and a few smaller structures (fig. 17.5).
      a) The brain stem.
         (1) The medulla.
            (a) This is the section of the brain closest to the spinal cord.
            (b) It regulates physiological processes that are essential to life, such as breathing and heartbeat.
            (c) It regulates blood pressure.
            (d) It contains reflex centers for vomiting, coughing, sneezing, urinating, defecating, swallowing, and hiccupping.
            (e) All messages entering or leaving the brain must pass through the medulla.
         (2) The pons.
            (a) The section above the medulla.

(b) White matter tracts in the pons connect the medulla and higher brain structures.

(c) Gray matter in the pons controls some aspects of respiration.

(d) A narrow region above the pons, called the midbrain, is also part of the brain stem.

(e) Gray matter in the brain stem contributes to seeing and hearing, and white matter connects the region to the cerebrum and other structures.

b) The cerebellum.
   (1) It is in back of the brain stem but connected to the cerebrum.
   (2) Its neurons refine motor messages, resulting in well-coordinated muscular movements.
   (3) It receives sensory input from the cerebral cortex (the outer portion) and the peripheral nervous system.

c) The thalamus.
   (1) A gray, tight package of nerve cell bodies and their associated glial cells located beneath the cerebrum.
   (2) A relay station for sensory input, processing incoming information and sending it to the appropriate part of the cerebrum.

d) The hypothalamus.
   (1) It helps to maintain homeostasis by regulating body temperature, heartbeat, water balance, and blood pressure.
   (2) It controls hunger, thirst, sexual arousal, and fear.
   (3) It also regulates the production of hormones in the pituitary gland at the base of the brain.

e) The reticular activating system.
   (1) Reticular means "little net," alluding to the RAS's role in filtering or screening sensory information so that only certain impulses reach the cerebrum.
   (2) If the RAS did not filter input from the environment, our senses would be overwhelmed with stimulation.
   (3) It is important in overall activation and arousal.

f) The cerebrum (fig. 17.6).
   (1) It controls the qualities of "mind," including intelligence, learning, perception, and emotion.
   (2) It consists of the gray matter that integrates or makes sense of incoming matter.

121

g) The cerebral cortex.
  (1) It is divided into sensory, motor, and association areas.
    (a) Sensory areas receive and interpret messages from sense organs about temperature, body movement, pain, touch, taste, smell, sight, and sound.
    (b) Motor areas send impulses to skeletal muscles.
    (c) Association areas, so named because they were once thought to connect the sensory and motor pathways, are the seats of learning and creative abilities.
  (2) A band of cerebral cortex extending from ear to ear across the top of the head controls voluntary muscles and is called the primary motor cortex; behind it is the primary sensory cortex, which receives input from the skin (fig. 17.7).
  (3) The surface area devoted to a particular body part is proportional to the degree of sensitivity and motor activity of the area (fig. 17.8).
  (4) Maps of the cerebral cortex that link certain areas to certain behavior have been constructed by intentionally damaging parts of the cortex in experimental animals or by studying loss of function in brain-damaged people (fig. 17.9).

h) Specializations of the cerebral hemispheres.
  (1) In most people, parts of the left hemisphere are associated with speech, linguistic skills, mathematical ability, and reasoning, while the right hemisphere specializes in spatial, intuitive, musical, and artistic abilities.
  (2) The corpus callosum is a thick band of hundreds of millions of axons, running between the cerebral hemispheres and enabling them to share information.
  (3) Epilepsy is an electrical disturbance in the brain resulting in seizures, loss of consciousness, and sensory disturbances that may be helped by drugs or cutting the corpus callosum.

C. Memory.
  1. Short-term memory.
    a) Thought to be electrical in nature.
    b) Neurons may be connected in a circuit so that the last in the series stimulates the first.
    c) Once an impulse begins, around and around it goes for seconds or hours.
    d) As long as the pattern of stimulation continues, you remember the thought.

2. Long-term memory.
    a) This involves some change in the structure or function of neurons that enhances synaptic transmission, perhaps by establishing certain patterns of synaptic connections.
    b) A certain pattern of synapses can remain unchanged for years.
    c) Structural changes such as an increase in the number of synapses and changes in the shape of neurons in rat brains have been noted following repeated electrical stimulation.
    d) Long-term synaptic potentiation.
        (1) It proposes that in an area of the cerebral cortex called the hippocampus, frequent and repeated stimulation of the same neurons strengthens their synaptic connections.
        (2) This strengthening could be in the form of a greater electrical change in the action potentials triggered in postsynaptic cells in response to the repeated stimuli.
    e) There seem to be two types of long-term memory.
        (1) Skill memories form in the cerebellum and in masses of nerve cell bodies within the cerebrum called the basal ganglia.
        (2) Factual memory is encoded in the hippocampus and in a nearby area called the amygdala.
            (a) One theory is that a rapid rate of action potentials opens calcium channels on postsynaptic membranes, and the influx of Ca^{2+} activates an enzyme that in turn alters proteins of the cytoskeleton.
            (b) The altered proteins change the shape of the neurons, enabling it to make new, and specific, synaptic connections.

D. Protection of the central nervous system.
    1. The bones of the skull and vertebral column shield the delicate nervous tissue from bumps and blows.
    2. Additional protection comes from trilayered membranes called meninges that jacket the central nervous system.
    3. A blood-brain barrier is created by specialized brain capillaries that fit so tightly against one another that only certain substances can cross into the cerebrospinal fluid that bathes and cushions the brain and spinal cord.

III. The Peripheral Nervous System (fig. 17.12).
    A. The PNS is subdivided into sensory pathways, which transmit impulses from a stimulus to the CNS, and motor pathways, which carry impulses from the CNS to effectors such as muscle or gland cells.
    B. The motor pathways are in turn subdivided into the somatic nervous system, which leads to skeletal muscles, and the autonomic nervous system, which goes to smooth muscle, cardiac muscle, and glands.

C. The somatic nervous system.
   1. The nerves send impulses to the muscles, the sense organs, and the sensory receptors in the skin, resulting in sensations such as light, sound, pain, body position, and contracting voluntary muscles.
   2. Twelve pairs of cranial nerves arise from the brain.
      a) Eleven of the cranial nerve pairs innervate portions of the head or neck.
      b) The vagus nerve leads to the internal organs.
   3. Thirty-one pairs of spinal nerves exit the spinal cord and emerge between the vertebrae.
D. The autonomic nervous system (fig. 17.13).
   1. The "automatic pilot" that keeps internal organs functioning properly without conscious awareness.
   2. The nerves transmit impulses to smooth muscle, cardiac muscle, and glands.
   3. It is subdivided into the sympathetic nervous system and parasympathetic nervous system.
      a) The sympathetic nervous system prepares the body to face emergencies—accelerating heart rate and breathing rate; shunting blood to the places that need it most, such as the heart, brain, and the skeletal muscle necessary for "fight or flight"; and dilating the airways so gas exchange can take place more easily.
      b) The parasympathetic nervous system is active while the body is at rest and returns heart rate, respiration, and digestion to normal levels after an emergency (fig. 17.14).

IV. When Nervous Tissue Is Damaged (table 17.1)
   A. Mature neurons usually cannot divide to repair damaged tissue.
   B. If a peripheral nerve is crushed, it can grow back, making the same synaptic connections as it originally did.
   C. In the mammalian central nervous system, heavily myelinated and damaged neurons cannot regenerate at all.
   D. Damaged unmyelinated neurons can sprout new axons, but these do not extend as far as the old ones.
   E. When a neuron dies, neighboring undamaged neurons extend new terminals to the cells that previously synapsed with the destroyed cell.
   F. In the aging brain, some synaptic connections deteriorate, some postsynaptic receptors become less sensitive to their particular neurotransmitters, and levels of some neurotransmitters fall.

LEARNING OBJECTIVES

After reading this chapter, the student should be able to answer these questions:

1. What are the major divisions of the human nervous system, and what is the rationale for this organizational framework?
2. What are the functions of the spinal cord?
3. What functions are provided by each of the major structures of the human brain?
4. What is thought to be the physical basis for memory?
5. How does the nervous system respond to a threat and then recover afterwards?

ANSWERS TO TEXT QUESTIONS

1. Reading—left cerebrum and amygdala; walking—cerebellum and basal ganglia; writing—left cerebrum; speaking—left cerebrum; driving—cerebrum and cerebellum; eating—hypothalamus and cerebrum; recognition of music—hippocampus and amygdala; taking tests—hippocampus and amygdala;

remembering a phone number long enough to dial the number—hippocampus and amygdala; riding a bicycle—basal ganglia and cerebrum.

2. Three methods of studying nervous system function are looking at the loss of function in brain-damaged individuals, intentionally damaging parts of the brain in experimental animals to determine loss of function, and observing behavioral responses to specific stimuli.

3. If the spinal cord is severed there would be loss of sensation and control of motor activities below the level of the lesion.

4. No, reflexes can protect the individual at the spinal cord level without the integrative properties of the brain.

5. The limited functions would be associated with the brain stem, heart beat, breathing, and perhaps other "vegetative" functions. The newborn would lack the higher functions associated with the cerebrum, consciousness, memory, sensory and motor control, etc. The organs would be normally developed, but with the lack of a functional brain, the individual could not survive.

6. The cerebellum is involved in trained patterns of movement and refinement of motor skills. It also integrates the position of the body in space and reacts to gravitational changes.

7. The hypothalamus is involved in temperature regulation, hunger, salt and water balance, emotions, and hormonal control through the pituitary gland. A tumor could affect some or all of these functions. For example, a tumor may cause an individual to grow abnormally large from the excessive production of growth hormone from the pituitary gland.

8. A person with narcolepsy probably has a defect in the reticular activating system.

9. The qualities of the "mind" are found in the cerebrum.

10. The different degrees of sensitivity depend upon the number of neuronal paths that project to a particular portion of the cerebral cortex and the mass of cells devoted to the processing of the information. For example, there is more sensory cortex devoted to interpretation of senses from the fingers than from the middle of the back.

11. Long-term memory is fixed in structural changes of the nervous system, by altered synaptic connections, and by neuronal facilitation. Short-term memory is the result of active neural circuits and may last only minutes or a few hours.

12. Visual recognition and association with the person's name may be the first in the process. Secondary associations of the personality traits, body shape, movements, voice, etc., with the visual image of the person could then occur. Loss of the amygdala and hippocampus would prevent the long-term memory of the characteristics and the ability to associate them with the visual image of the person.

13. The potentially dangerous situation will stimulate the sympathetic division of the autonomic nervous system through the hypothalamus to ready the body for the "fight or flight" response.

ANSWERS TO THE "TO THINK ABOUT" QUESTIONS

1. There are specific structures associated with tasks such as sensory interpretation, motor control of skeletal muscles, speech, etc. The processes of memory and learning may be coordinated in discrete structures, but probably they are more an interaction of many different areas and therefore appear to be a property of the whole brain. The subdivisions thus help in our understanding, but the interactions define the processes we term "the activity of the mind."

2. Following a period of sleep deprivation, the person will compensate by increasing both REM and non-REM sleep periods.

3. Each individual is unique due to the difference in patterns of stimulation and the resulting formation of unique connections between the billions of neurons in the central nervous system. Thus a given stimuli may cause a distinctly different response in different individuals.

125

4. The plasticity of the nervous system will mask the loss of some neurons. Adjacent cells in the different regions of the brain may take over the function of the lost cells.

5. One way would be to increase the variety of stimuli to the brain to enhance the formation of long-term memory. Saying an unfamiliar term out loud forms specific motor patterns of speech and provides auditory feedback since we hear the word spoken as well as see it written on the page. Writing new ideas or concepts down on paper as we hear them would also increase sensory input and may enhance memory.

6. Early detection of changes in brain chemistry may allow for psychological support to the individual and perhaps lead to treatment through chemical therapy. The disadvantage is that the disease may not be treatable, leading to depression and an inability to cope.

7. Design an intelligence test such as a maze capable of being used for a variety of different animal species. Test animals of varying brain size on their ability to master the maze.

AUDIOVISUAL MATERIALS

The Autonomic Nervous System, International Film Bureau, 17 min., 1973.

Biochemical Revolution: Moods of the Future, Washington University, 24 min., 1976.

Brain and Behavior, National Education Television Film Service, 30 min., 1962.

Drugs and the Nervous System, Churchill Films, 18 min., 1987.

The Hidden Universe: The Brain, CRM McGraw-Hill, 45 min., 1977.

Human Brain, Encyclopaedia Britannica Educational Corp., 16 min.

Miracle of the Mind, CRM McGraw-Hill, 26 min., 1967.

The Nervous System, Cornet Instructional Films, 22 min., 1980.

The Nervous System, Encyclopaedia Britannica Educational Corp., 18 min., 1980.

Nervous System of Animals, Association Films, 17 min., 1971.

Neurological: Cranial Nerves and Sensory Systems, J. B. Lippincott, 20 min., 1981.

18 The Senses

KEY CONCEPTS

Animals perceive the world in different ways because of differing sensory abilities.

Sensory receptors are specialized for different forms of energy.

Chemoreception depends upon receptor molecules that are found on the sensory cell membranes.

Photoreceptors contain pigments that absorb light resulting in action potentials of sensory neurons.

Basic organization of eye structure varies and can produce either mosaic or single images depending upon the nature of the lens.

The visual system is based on convergence of sensory neurons.

Color perception depends on different pigment molecules absorbing different wave lengths of light.

Mechanoreceptors respond to pressure, vibration, or touch with action potentials in sensory neurons.

The ear and vestibular apparatus convert sound and movement into mechanical stimulation of hair cells.

KEY TERMS

accommodation	hair cells
ampullae	hammer
anvil	iris
aqueous humor	lens
basilar membrane	middle ear
cerebral cortex	olfactory bulb
choroid coat	olfactory epithelium
ciliary body	olfactory receptor cells
ciliary muscle	ommatidia
cochlea	optic nerve
cochlear implant	otoliths
color blind	oval window
conductive deafness	Pacinian and Meissner's
cones	corpuscles
cornea	photoreceptors
free nerve endings	pupil
ganglion cells	receptor potential

retina	stirrup
rhodopsin	taste receptors
rods	tectorial membrane
saccule	thermoreceptors
sclera	tympanal organ
scotopsin	tympanic membrane
semicircular canals	utricle
sensory adaptation	vestibule
sensory or neural	vitreous humor
deafness	

CHAPTER OUTLINE

1. General Principles of Sensory Reception
   A. Sensory receptors detect certain environmental stimuli and pass the information on to sensory neurons, which in turn deliver nerve impulses to the brain.
   B. Sometimes sensory receptors are gathered into sense organs such as the eye or ear.
   C. Action potentials on sensory receptors are all the same, but they are selective in which types of energy they respond to, as in the eye only responds to light energy.
   D. Sensory receptors respond to environmental change with a receptor potential.
      1. Changes in membrane potential are caused by the redistribution of ions.
      2. The magnitude of receptor potential varies with the strength of the stimulus such that the louder the sound, the greater the depolarization.
      3. In some receptors, the strength and duration of the receptor potential determines the rate at which action potentials are generated and how long they last.

4. The variation in the number of sensory neurons carrying the message alters the message to the brain.
   E. Many sensory receptors detect changes in input and may ignore constant messages through sensory adaptation.
   F. Humans have multiple receptors for vision, hearing, taste, smell, and touch.

II. Chemoreception
   A. Smell.
      1. The detection of certain molecules is accomplished by specialized olfactory receptor cells in the nasal cavity (fig. 18.2).
      2. The olfactory epithelium.
         a) It contains cilia that have receptor sites for chemicals.
         b) It absorbs molecules with assistance of odorant-binding proteins that ferry the chemical to receptors.
      3. Each olfactory receptor cell synapses with neurons in the olfactory bulb of the brain.
      4. This message is relayed to the cerebral cortex, where the message is interpreted as smell.
      5. It has been proposed that distinction between odors may be due to certain combinations of receptors, such that garlic receptors are 1, 5, and 9, whereas banana receptors are 2, 4, and 7.
      6. Sensory information from olfactory receptors also stimulates the limbic system, where an emotional and memory response occurs.
   B. Taste.
      1. Chemicals are picked up by taste receptors.
      2. Some animals have them on tentacles, antennules, and even legs to detect food.
      3. Receptor cells live for about 1 week and then are replaced.
      4. Four primary taste sensations are picked up by sweet, sour, salty, and bitter receptors, which have different locations on the tongue.

III. Photoreception
   A. Photoreceptors, found in higher organisms, contain a pigment that alters its structure when it absorbs light.
   B. Invertebrate vision.
      1. Consists of the compound eye, which has many visual units called ommatidia (fig. 18.5).
      2. Each ommatidia contains a lens that transmits light to its own photoreceptor cell.
   C. The human visual system (table 18.1).
      1. The eye is constructed like a camera, with structures that focus the light like a lens in the front onto the retina, which acts like film.
      2. The retina consists of a sheet of photoreceptors.
      3. The sclera (fig. 18.6).
         a) The outer layer of the eyeball, which protects the inner structures of the eye.
         b) It becomes modified into the cornea, a transparent curved window that bends light rays to focus them on photoreceptors.
      4. The choroid coat.
         a) The middle layer of the eyeball is rich in blood vessels that nourish the eye.
         b) It contains a dark pigment that absorbs light and prevents it from reflecting off the retina.
         c) It thickens into the ciliary body, where the ciliary muscle alters the shape of the lens to adjust the focus of the image during near and far vision.
         d) The colored part of the eye, the iris regulates the amount of light by changes in the opening, called the pupil.

5. The retina (fig. 18.7).
    a) The first layer contains photoreceptors called rods and cones (fig. 18.8).
    b) Rods, which are concentrated around the edges of the retina, detect black-and-white vision in dim light.
    c) Cones detect color and are concentrated more centrally to the retina.
    d) The second layer is made up of bipolar neurons, which receive stimulation from the cones and rods.
    e) The third layer consists of ganglion cells, which receive input from the bipolar neurons and transmit them to the optic nerve, which leads to the visual cortex of the brain.
6. The aqueous humor.
    a) A jellylike substance that fills the cavity of the eyeball.
    b) It nourishes the eyeball, bends light rays, and helps maintain the shape of the eye.
    c) In the disease glaucoma, the aqueous humor builds up to levels that cause high pressure.
D. Focusing the light.
    1. Accommodation is the molding of the lens to suit the distance of the object being viewed.
    2. The ciliary muscles contract to focus on a very close object, and the lens flattens out to see an object far away.
    3. In farsightedness, the lens does not curve enough or is too short, and so the person has trouble seeing objects close at hand.
    4. In nearsightedness, the lens is too long so the person has trouble seeing objects far away.

E. Converting light energy to neural messages.
    1. Visual information is converted to receptor potentials by changes in pigment molecules within the rods and cones.
    2. Rods.
        a) In rods, the pigment rhodopsin is split by light into the chemicals scotopsin and retinal (fig. 18.10).
        b) The splitting of rhodopsin depolarizes the rod cell, which releases a neurotransmitter, passing the nerve impulse, via a bipolar neuron to the optic nerve.
        c) It takes a few minutes for rhodopsin to reform, so it takes a few minutes to make out shapes in a dark room.
    3. Cones.
        a) Cone cells contain different pigments, each of which absorbs light of different wavelengths.
        b) When light is absorbed by a cone, its pigment is chemically changed and a receptor potential results.
        c) Color blindness is when one or more of the cone types is missing.
    4. Depth perception is provided by the overlap of images from eyes that are close but separate.
IV. Mechanoreception
    A. Mechanoreceptors convert mechanical energy to action potentials.
    B. Hearing.
        1. Sound is created by pressure waves; the more cycles per second the higher the pitch of sound.
        2. Tympanal organ.
            a) A thin region of the outer body covering that vibrates

in response to sound and stimulates special receptor cells.
- b) The outer ear is the fleshy ear in humans, and it funnels the sound into the ear canal.
- c) The middle ear (fig. 18.12) contains the tympanic membrane, which vibrates with sound waves.
  - (1) Vibration of the tympanic membrane causes the three small bones, the hammer, anvil, and stirrup, to vibrate, resulting in amplification of the sound to 20 times.
  - (2) The vibrations hit the oval window, which is a membrane that opens into the inner ear.
- d) The inner ear.
  - (1) The first portion contains the semicircular canals and the vestibule, which are concerned with balance.
  - (2) The hindmost portion contains the cochlea, which consists of three fluid-contained canals called the vestibular, cochlear, and tympanic canals.
  - (3) The vibrations of the oval window are transferred to the fluid, which in turn initiates the change of mechanical energy to receptor potentials.
3. Transfer of mechanical energy into receptor potential.
- a) Within the cochlea, specialized hair cells lie between basilar membrane and the tectorial membrane above (fig. 18.14 and 18.15).

- b) Because of differences in width and flexibility of the basilar membrane, there are differences in the pitch.
- c) When a region of the basilar membrane vibrates, the hair cells are pushed against the tectorial membrane and action potentials are spiked in fibers of the auditory nerve, which sends signals to different regions of the brain (fig. 18.16).

C. Hearing loss.
1. Conductive deafness occurs when the transmission of sound through the middle ear is impaired; this is usually caused by infection of the middle ear or damage to the eardrum.
2. Sensory or neural deafness.
- a) This is caused by the inability to generate action potentials in the cochlea, a blocked communication network between the cochlea and the brain, or the brain's inability to make sense of the sensory message.
- b) It is commonly caused by aging or loud noises that eventually result in the permanent damage of hair cells (fig. 18.17).
3. Tinnitus, or ringing in the ears, is most often caused by loud noises.
4. Cochlear implants are devices that deliver an electronic stimulus directly to the auditory nerve, bypassing the function of the hair cells.

D. Balance.
1. Semicircular canals (fig. 18.19).
- a) They tell when the head is rotating.
- b) The enlarged base of the canal, called the ampullae, is lined with small, ciliated hair cells.

c) When the head moves, the fluid in the canal stimulates the hair cells, which cause receptor potentials in cranial nerves.

    2. Vestibule (fig. 18.13).

        a) It tells the position of the head with respect to gravity.

        b) It contains two pouches called the utricle and saccule, which are filled with jellylike fluid and ciliated hair cells.

        c) The fluid contains granules that move in response to vertical body movements and bend hair cells that trigger sensory impulses to the brain.

E. Touch (fig. 18.21).

    1. Pacinian corpuscles are stimulated by firm pressure like a bear hug.

    2. Meissner's corpuscles are stimulated by a light touch.

    3. Free nerve endings are sensitive to touch, pressure, and pain.

V. Thermoreception

A. Cold receptors are stimulated by temperatures between 50° F and 68° F.

B. Hot receptors are stimulated by temperatures between 77° F and 113° F.

VI. Detection of Magnetic Fields—A Sixth Sense?

A. A magnetic mineral called magnetite is found in many organisms and is believed to participate in migration.

B. Migrating birds and homing pigeons may also use the earth's magnetic field to determine which way is down.

LEARNING OBJECTIVES

After reading this chapter, the student should be able to answer these questions:

1. What is the difference between sensation and perception?

2. How do receptor cells in the different sense organs transmit information to the brain which enables us to sense and perceive the environment?

ANSWERS TO TEXT QUESTIONS

1. Approximately 75 to 80 percent of the flavor of food is determined by smell. Taste detects pleasant and unpleasant sensations while the food is being processed in the mouth. In addition the temperature, texture, and appearance may contribute to the flavor of food.

2. The receptor cells in the taste buds are replaced every 7 to 10 days.

3. An insects' view is based on a mosaic of images from multiple ommatidia, whereas the human eye forms a single image on the retina. The species of shrimp may be able to detect light from dark but the exact function is unknown at this time.

4. The ratio of activity of the three basic types of cones is interpreted by the brain as light of particular colors.

5. The major structures of the inner ear all have "hair" or ciliated cells in which the specific stimuli of sound or movement cause mechanical bending of the cilia, initiating receptor potentials activating the sensory neurons.

6. In the visual system there are many rods and cones that have direct connections to single ganglia cells in the retina. These cells process visual information before transmission of action potentials to the visual centers of the brain.

In the auditory system there are many hair cells on the basilar membrane that have direct input to specific auditory neurons which then project to the auditory centers of the brain.

On the tongue the taste buds have 60 to 100 receptors cells. Each of these receptor cells has sensory neurons projecting to the taste centers of the brain.

ANSWERS TO THE "TO THINK ABOUT" QUESTIONS

1. The loss of smell may be used to predict the possible onset of the disease and in the case of Parkinson's disease allow the beginning of early treatment with L-dopa.

2. People with an inability to smell certain substances probably lack a normal allele to produce a specific receptor protein found on the membranes of olfactory epithelial cells.

3. Vitamin A is the precursor for retinal, which when combined with scotopsin forms the visual pigment rhodopsin in the dark reaction of visual processing.

4. Deafness: the individual depends heavily on the visual system to communicate by lip reading, sign language, and written words. Blindness: the individual depends on sensory cues through hearing to communicate and also through touch to orient in the surrounding environment.

5. The damage may extend to the organs of balance and movement detection which are also part of the inner ear. The reduced information about movement reaching the brain reduces the probability of motion sickness.

6. Sensory perception is a blend of the five familiar senses by the brain into unique combinations. The smell of a familiar perfume may bring forth visual, auditory, and emotional memories associated with the odor.

AUDIOVISUAL MATERIALS

A Look at Sound, Time Life Multimedia, 30 min., 1971.

How The Eye Functions, Knowledge, 10 min., 1940.

Human Body Sense Organs, Coronet Instructional Films, 19 min., 1990.

The Human Ear, Association Films, 9 min.

The Human Eye, International Film Bureau, 14 min., 1978.

Nervous System, Coronet Instructional Films, 22 min., 1980.

The Nervous System, Encyclopaedia Britannica Educational Corp, 18 min., 1980.

Neurological: Cranial Nerves and Sensory System, J. B. Lippincott, 20 min., 1981.

The Sensory World, CRM McGraw-Hill, 33 min., 1971.

The Sensory World, Indiana University, 32 min., 1971.

The Skin as a Sense Organ, International Film Bureau, 12 min., 1974.

19 The Endocrine System

KEY CONCEPTS

Hormones control cellular reactions and are slower and longer-acting than neural systems.

Hormones bind to receptors on the target cells, which gives them their specificity.

Peptide hormones bind to the cell membrane and activate "second messengers" that alter biochemical reactions.

Steroids enter the target cell and activate genes for protein synthesis.

Hormone secretion is controlled by negative feedback loops.

The endocrine system is largely regulated by the nervous system through the hypothalamus and pituitary gland.

The anterior and posterior lobes of the pituitary gland have differing systems of control from the hypothalamus.

The gonadotropic hormones control sexual development and reproductive function through action on the gonads.

Oxytocin is involved in the birth process and milk production.

Growth hormone is essential for normal growth.

Antidiuretic hormone is essential for the conservation of water.

The adrenal cortex is essential in response to stress, control of glucose levels, and mineral metabolism.

The adrenal medulla plays a major role in the "fight or flight" response.

Thyroid hormones are necessary for normal growth and development and maintenance of normal metabolic function.

Calcitonin from the thyroid gland and parathyroid hormone control calcium levels of the body.

Pancreatic hormones control the level of blood glucose and nutrient absorption.

The gonads provide the sex hormones for secondary sex characteristics and the gametes.

Prostaglandins are fatty acids and play a major role in tissue repair as well as affecting many other systems.

Pheromones are substances that alter the physiology and behavior of other individuals of the same species.

KEY TERMS

acromegaly
adrenal cortex and medulla
adrenal glands
adrenocorticotropic hormone (ACTH)
aldosterone
antidiuretic hormone (ADH)
atrial natriuretic factor (ANF)
calcitonin
corpus luteum
cretinism
diabetes insipidus
diabetes mellitus
endocrine glands
epinephrine
estrogen
follicle
follicle stimulating hormone (FSH)
glucagon
glucocorticoids
goiter
gonadotrophic hormones
hormone
human chorionic gonadotropin (HCG)
hyperthyroidism
hypoglycemia
hypothalamus
hypothyroidism

insulin
insulin-dependent diabetes
interstitial cell stimulating hormone (ICSH)
islets of Langerhans
luteinizing hormone (LH)
luteinizing hormone releasing hormone (LHRH)
melanocyte stimulating hormone (MSH)
mineralocorticoids
myxedema
negative feedback loops
neurosecretory cells
norepinephrine
osmoreceptors
osteoporosis
ovaries
ovulation
oxytocin
parathyroid glands
parathyroid hormone
peptides
pheromone
pineal gland
pituitary
pituitary dwarfism
pituitary giant
positive feedback loops
progesterone
prolactin
releasing hormone
seasonal ovulators
second messenger

steroids
target cell
testes
thyroid stimulating
    hormone (TSH)

thyroxine
triiodothyronine
trophic hormone
vasopressin

CHAPTER OUTLINE

I. Hormones—Chemical Messengers and Regulators
   A. The endocrine system helps the nervous system coordinate the body's functions through the release of chemical messengers called hormones.
   B. Hormones are released by cells into the bloodstream which carries them to some distant part of the body where they alter cellular activity (fig. 19.1).
   C. Endocrine glands secrete hormones directly into the bloodstream, whereas exocrine glands such as salivary glands secrete their products into ducts (figs. 19.2 and 19.3).
   D. Only certain cells respond to a particular hormone and are known as target cells.
II. How Hormones Exert Their Effects (tables 19.1 and 19.2)
   A. Peptide Hormones.
      1. They are water soluble and so cannot pass into cells through the lipid bilayer.
      2. A peptide hormone binding to its receptor initiates the target cell's response, which might be the opening of specific ion channels in the membrane.
      3. The formation of a complex between a hormone and a membrane receptor activates another substance called the second messenger within the cell, which activates other enzymes in the cell.
      4. Cyclic adenosine monophosphate (cAMP) (fig. 19.4) serves as a second messenger.
         a) When the hormone binds to the receptor, the enzyme adenyl cyclase catalyzes the conversion of ATP into cAMP, which in turn activates certain enzymes within the cell.
         b) This occurs when parathyroid hormone secreted by parathyroid glands in the throat binds to specific kidney cells.
   B. Steroid hormones.
      1. Unlike peptide hormones, these are small and lipid soluble and so pass easily through the membranes of cells.
      2. Once inside, the steroid binds to a receptor in the cytoplasm and the complex enters the cell's nucleus, where a particular gene is activated.
III. Control of Hormone Levels
   A. Hormone levels are altered by changes in the level of ions or nutrients in the cellular environment, instructions from the nervous system, and directives from other hormones.
   B. Feedback loops.
      1. The level of a particular hormone in the blood can often maintain itself by means of a complex interaction between the hormone and its precursors through feedback loops.
      2. Negative feedback loops.
         a) These are the most common type found.
         b) When a certain biochemical accumulates to above-normal levels, its synthesis slows or temporarily halts.
         c) An example is the maintenance of blood-glucose levels.
            (1) Increased intake of carbohydrates increases blood-glucose levels.
            (2) Insulin levels increase, which causes increased uptake of glucose by the cells, thus removing it from the bloodstream.

(3) the drop in glucose levels causes a decline in insulin production (fig 19.6).

3. Positive feedback loops.
    a) When an accumulating biochemical increases its own production.
    b) An example is the onset of labor, when the uterus contracts and releases the hormone oxytocin and hormonelike substances called prostaglandins, which further intensify the uterine contractions.

C. Neuroendocrine control—the hypothalamus and the pituitary.
    1. Hypothalamus.
        a) A region of the brain that in addition to releasing its own hormones in response to nerve stimulation produces other hormones that control the output of several pituitary hormones.
        b) It releases hormones that are called tropic hormones since they cause the release of hormones by other glands.
        c) These tropic hormones are specifically known as releasing hormones.
        d) The hypothalamus is a link between the nervous and endocrine systems because it has both kinds of tissue.
        e) Neurosecretory cells in the hypothalamus function like neurons at one end, receiving neurotransmitters from other nerve cells and generating action potentials, but they act like endocrine cells at the other end, secreting a hormone.
    2. Pituitary gland (fig. 19.7).
        a) A small endocrine gland at the base of the brain.
        b) Under stimulation from the hypothalamus, it controls the endocrine function of other glands so that they work harmoniously.

IV. Pituitary Hormones
A. Anterior pituitary.
    1. The anterior portion is controlled by neurosecretory cells in the hypothalamus that secrete hormones that either stimulate or inhibit the production of anterior pituitary hormones (fig. 19.8).
    2. Growth hormone (GH).
        a) Promotes growth and development of all tissues by increasing rates of protein synthesis and cell division.
        b) It does this by increasing the uptake of amino acids, the mobilization of fat, and the release of glucose from the liver to supply energy.
        c) Pituitary dwarfism is caused by a deficiency of this hormone in childhood (fig. 19.9).
        d) Human growth hormone can now be manufactured using recombinant DNA technology to correct dwarfism.
        e) A pituitary giant results when a child has a gland that produces too much GH.
        f) Acromegaly, the thickening of bones noticed in the hands and feet, results when the adult has an overactive production of GH (fig. 19.10).

135

3. Prolactin.
   a) Stimulates milk production in a woman's breasts after she gives birth.
   b) Normal inhibition is prevented by sucking of the infant on the nipples.
4. Gonadotropic hormones (fig. 19.11).
   a) Follicle stimulating hormone (FSH) in females leads to the development of ovarian follicles, the maturation of oocytes, and the release of the hormone estrogen.
   b) FSH in males leads to the development of testes and the manufacture of sperm cells.
   c) Luteinizing hormone (LH) in females causes release of an oocyte from an ovary each month.
   d) LH in males is known as interstitial cell stimulating hormone (ICSH) and prompts the testes to produce the hormone testosterone.
5. Thyroid stimulating hormone (TSH) causes the thyroid gland in the throat to release its two hormones.
6. Adrenocorticotropic hormone (ACTH) prompts the release of the glucocorticoid hormones from the cortex of the adrenal glands.

B. Posterior pituitary.
1. Two hormones are produced in the hypothalamus but are stored and released by the pituitary gland.
2. Vasopressin, or antidiuretic hormone (ADH).
   a) It causes the smooth muscle cells lining the blood vessels to contract, thereby raising blood pressure.
   b) It causes the collecting ducts of the kidneys to be more permeable to water so more of it can be reabsorbed.
   c) Specialized cells in the brain called osmoreceptors sense the concentration of blood and release ADH if the blood is too concentrated, which results in the dilution of the blood.
   d) Diabetes insipidus results when there has been a disruption of the synthesis or release of ADH.
3. Oxytocin.
   a) This contracts the myoepithelial cells in the breasts, causing milk to be released when a baby nurses.
   b) It also contracts the uterine muscles, causing the force that delivers the newborn during labor.
   c) Intermediate region: releases melanocyte stimulating hormone (MSH), which causes pigment granules to disperse, darkening the skin (fig. 19.12).

V. The Thyroid Gland
A. A two-lobed gland in the front of the larynx and trachea in the throat (fig. 19.13).
B. The hormones thyroxine and triiodothyronine, released by the thyroid, increase the rate of cellular metabolism and require the use of iodine for production.
C. Calcitonin hormone decreases blood-calcium levels under certain conditions.
D. Under thyroid stimulation gas exchange in the lungs occurs faster, the small intestine absorbs nutrients more readily, and the fat levels in cells and in blood plasma are lowered.
E. Hypothyroidism refers to the effects of an underactive thyroid gland, where a person gains weight despite a poor appetite.

1. Cretinism is hypothyroidism before birth, resulting in a mentally retarded child.
2. Myxedema is hypothyroidism in the adult and is characterized by lethargy, a puffy face, and dry, sparse hair.
3. Goiters are caused by hypothyroidism due to lack of iodine in the diet (fig. 19.14).

F. Hyperthyroidism produces a swelling in the neck called a toxic goiter, and the person has a very short attention span and is irritable and hyperactive with elevated heart rate, blood pressure, and temperature.

VI. The Parathyroid Glands (fig. 19.13).
A. Four cell clusters embedded in the thyroid gland.
B. They control the level of calcium in the bones and blood.
C. Osteoporosis often occurs due to dropping estrogen levels in menopausal women, which results in excess parathyroid hormone.

VII. The Adrenal Glands
A. Paired glands that sit on top of the kidneys and have an outer region called the cortex and an inner portion called the medulla (fig. 19.15).
B. The adrenal medulla.
1. Secretes hormones called catecholamines, namely epinephrine (also known as adrenaline) and norepinephrine (also called noradrenaline).
2. In response to stress, the hormones are released in the "fight or flight" response, the heart and breathing rates increase, air passageways open, blood is shunted to the skeletal muscles and brain, and glucose is mobilized from the liver to provide energy.
C. The adrenal cortex.
1. Releases three types of steroid hormones.
2. Mineralocorticoids.
    a) They maintain blood volume and electrolyte balance by stimulating the kidney to return sodium ions and water to the blood but to excrete potassium ions.
    b) Specifically the major mineralocorticoid, aldosterone, maintains sodium ion levels in the blood and blood volume.
3. Glucocorticoids.
    a) Cortisol, which is the most important, is an essential component of the body's response to prolonged stress.
    b) They break down proteins into amino acids and then stimulate the liver to synthesize glucose from these freed amino acids.
    c) A deficiency of mineralocorticoids and glucocorticoids results in Addison's disease, with symptoms of weight loss, mental fatigue, weakness, and impaired resistance to stress.
    d) Excess cortical hormones causes Cushing's syndrome (fig. 19.16).
    e) Sex hormones— testosterone, estrogen, and progesterone—are secreted in small amounts.

VIII. The Pancreas
A. A large gland located between the spleen and the small intestine (fig. 19.17*a*).
B. Embedded in cells that produce digestive enzymes are cells called islets of Langerhans, which produce three hormones of different types (fig. 19.17*b*).
1. Insulin lowers the blood-sugar level by stimulating most of the body's cells to take up glucose from the blood and by stimulating reactions inside cells that metabolize glucose or store it as glycogen.
2. Glucagon, in contrast, breaks down glycogen into glucose,

thereby raising blood-glucose levels and providing energy between meals.

    3. Somatostatin controls the rate of nutrient absorption into the bloodstream.

C. Diabetes.
    1. Insulin-dependent diabetes.
        a) Also known as juvenile-onset diabetes because symptoms usually occur then.
        b) It is the most serious form caused by insufficient insulin production so a person must take daily injections of insulin.
        c) It is believed to be an autoimmune disease.
    2. Insulin-independent diabetes.
        a) Also known as adult-onset diabetes and results from the inability to use insulin.
        b) The source of this problem may be defective insulin receptors on cell surfaces starving the cells of glucose.

D. Hypoglycemia.
    a) A low level of glucose in the blood caused by excess insulin.
    b) The symptoms are also caused by many other conditions as well as stress.

IX. The Gonads
A. The ovary.
    1. Hormones.
        a) Estrogen increases cell division rate in the vagina, uterus, and breasts and triggers a buildup of fat beneath a female's skin.
        b) Progesterone controls secretion patterns associated with reproductive function.
    2. Menstrual cycle.
        a) First day of menstrual bleeding (fig. 19.19).
            (1) Low blood levels of estrogen and progesterone signal the hypothalamus to send luteinizing hormone releasing hormone (LHRH) to the anterior pituitary, stimulating it to release large amounts of follicle stimulating hormone (FSH) and small amounts of luteinizing hormone (LH).
            (2) The increase in FSH prompts the largest follicle in the ovary to produce estrogen, which stimulates growth of the lining of the uterus in preparation for pregnancy.
            (3) Rising estrogen stimulates the hypothalamus to produce more LHRH, which increases blood levels of LH.
        b) Day 14 of the cycle.
            (1) As LH levels peak the largest ovarian follicle releases an oocyte.
            (2) This is ovulation when a woman is most fertile.
        c) Day 14 to day 22.
            (1) The follicle cells that had surrounded the released oocyte send biochemical signals to lower blood-estrogen level.
            (2) Dropping estrogen results in lowering of LH levels.
            (3) The follicle now enlarges to form a gland, the corpus luteum, which produces estrogen and progesterone.
            (4) Increasing estrogen and progesterone

levels stimulate the buildup of the uterine lining.

    d) Day 23.

        (1) If a fertilized ovum has implanted in the uterine lining, a new hormone, HCG, is secreted by the fertilized ovum into the woman's blood, which prevents sloughing of the uterine lining.

        (2) If conception does not occur, the progesterone level in the blood rises, inhibiting LHRH production, estrogen and progesterone drop, and the uterine lining breaks down, beginning the cycle again.

B. The testes.

    1. Hormone levels are constant rather than fluctuating.

    2. LHRH stimulates the release of FSH and interstitial cell stimulating hormone (ICSH).

    3. FSH stimulates the early stages of sperm formation.

    4. ICSH completes sperm production and stimulates the testes to synthesize the male hormone testosterone.

    5. Testosterone promotes the development of male secondary sexual characteristics, including facial hair, deepening of the voice, and increased muscle growth.

X. Pineal Gland

A. A small oval structure located in the brain near the hypothalamus.

B. It produces melatonin hormone, which seems to help regulate reproduction in certain mammals by inhibiting the anterior pituitary hormones that regulate the activities of the gonads.

C. Light inhibits melatonin synthesis; darkness stimulates it.

D. Seasonal affective disorder (SAD) is a particular form of depression that is hypothesized as being caused by abnormal melatonin secretion patterns.

XI. Hormones Not Associated With Endocrine Glands

A. The mucus-rich lining of the stomach and intestines contains scattered hormone-secreting cells.

B. Cells in the atria secrete atrial natriuretic factor (ANF), which exert complex effects on several organs and endocrine glands.

C. Other organs, including the kidney and placenta, contain hormone-secreting cells.

XII. Hormonelike Molecules

A. Prostaglandins.

    1. Lipid molecules that affect various tissues and organs by altering hormone levels.

    2. When a cell membrane is disrupted by an injury, the binding of a hormone, or an attack by the immune system, certain fatty acids are released from the damaged membrane into the cytoplasm.

    3. They affect smooth muscle contraction, secretion, blood flow, reproduction, blood clotting, respiration, transmission of nerve impulses, fat metabolism, the immune response, and inflammation.

    4. Dysmenorrhea, a condition of painful menstrual cramps, results from elevated levels of two prostaglandins in the uterine fluid.

B. Pheromones.

    1. Substances secreted by an organism that stimulate a physiological or behavioral response in another individual of the same species.

    2. Pheromones are species-specific (fig. 19.21).

    3. The chemicals have been found in birds, fish, and mammals,

although a human pheromone has yet to be characterized.

LEARNING OBJECTIVES

After reading this chapter the student should be able to answer these questions:

1. What is a hormone?
2. What are the two major types of hormones, and how does each type exert its effect?
3. How does the hypothalamus control the activity of the pituitary gland?
4. What effects do the hormones secreted by the pituitary gland, the thyroid gland, the parathyroid glands, the adrenal glands, the pineal gland, the pancreas, and the gonads have on the human body?
5. What non-endocrine organs also secrete hormones?
6. What are prostaglandins and pheromones?

ANSWERS TO TEXT QUESTIONS

1. Because target cells have specific receptors that only bind to the appropriate hormone. This binding of the hormone to the receptor activates cellular functions.
2. Peptide hormones bind to receptors on the plasma membrane, activating a second messenger cAMP. Steroids are lipid soluble and pass through the membrane to the intracellular receptors which act on genes initiating protein synthesis.
3. Hormones are regulated through negative feedback loops. An example would be increased consumption of water. This acts on the osmoreceptors in the hypothalamus to inhibit the release of ADH. The low levels of ADH decrease the permeability of the collecting ducts of the kidney. Water is lost in the urine (diuresis) returning the body to normal concentration.
4. This results in an increase in growth and height before puberty and acromegalia after puberty if excessive growth hormone is released from the anterior lobe of the pituitary gland. A tumor in the posterior lobe may cause a decrease in the secretion of ADH leading to diabetes insipidus in which the individual drinks excessively and passes large quantities of water in the urine. Many different functions could be altered with a tumor in the hypothalamus since it directly controls the function of the pituitary gland. a) increase in growth (GH) b) water and electrolyte imbalance (ADH) c) blood pressure changes (ADH) d) metabolic changes (TSH and ACTH) e) reproductive disorders (gonadotropic hormones) f) blood glucose, protein metabolism, and fat mobilization (ACTH, glucocorticoids and mineralocorticoids).

5. Insulin causes blood glucose to fall, while glucagon causes blood glucose to rise; parathyroid hormone raises the level of blood calcium, while calcitonin lowers blood calcium.
6. A decrease in thyroid hormones would cause a decrease in cellular respiration depressing general body functions including lower metabolic rate, increase in weight, decrease in heart rate and blood pressure, and lower body temperature.
7. Insulin increases the uptake of glucose by body cells thus lowering blood glucose levels. Glucagon stimulates the breakdown of glycogen in the liver which releases glucose, raising blood sugar levels.
8. LHRH stimulates the anterior pituitary to release FSH which initiates follicle development in the ovary and the release of estrogen from the maturing follicle. Estrogen is the hormone that stimulates growth of the uterine lining and prepares the uterus for implantation of a fertilized ova. In the absence of sufficient estrogen the uterine lining will not grow so there is no subsequent breakdown and loss of tissue, which is menses.
9. The substance is probably a pheromone since it is passed between individuals of the same species. A hormone is a chemical released into the blood stream and a prostaglandin is a chemical released when cells are disturbed or injured. The prostaglandins then stimulate enzyme synthesis or other tissue responses.

ANSWERS TO THE "TO THINK ABOUT" QUESTIONS

1. One possible way would be to use extracts of the gland from cadavers and test the

140

effect on animal models. From these experiments you may be able to isolate the active hormone and its effects. The next step would be to look for the hormone in the human population by analysis of blood to see if it is a circulating hormone. You could also then do large population studies to see if individuals exhibit symptoms from either the lack of or too much of the hormone.

2. Either answer may be correct, depending upon value system severity of the deficiency, and family situation.

3. Yes, it may be stress induced or may mask more serious disorders of the pancreas or adrenal glands.

4. No, the negative effects to the individual are too great. Excess steroids may cause kidney damage, liver and heart damage, stunting of height, atherosclerosis, early baldness and possible infertility.

5. This seems to be the reaction to pheromones, possibly estrogen or progesterone.

6. This question might be argued either pro or con, but it seems that women should not be penalized for behavior that falls within the norm of human range. It may be considered a legal defense to a violent crime since the act is obviously outside the range of normal behavior and there is precedence for psychological and behavioral abnormalities in which the individual is not held responsible for the crime.

AUDIOVISUAL MATERIALS

Endocrine Glands and Metabolism, University of Texas at Dallas, 28 min., 1978.

The Endocrine System, Coronet Instructional Films, 15 min., 1980.

The Endocrine System, Encyclopaedia Britannica Educational Corp., 14 min., 1982.

Menstruation and Sexual Development, Washington University, 28 min., 1977.

Stress, Health and You, Time Life Multimedia, 18 min., 1978.

20 The Skeletal System

KEY CONCEPTS

Hydroskeletons provide shape and rigidity by fluid containment in flexible tissue.

Braced framework skeletons can be external, as seen in many invertebrate animals or internal endoskeletons, as seen in the vertebrates.

The human skeleton provides a lever system for movement, protection of internal structures, blood cell production, and mineral storage.

Cartilage forms the embryonic skeleton and forms smooth coatings on bone to reduce friction, it is both firm and flexible.

Bone is both strong and lightweight due to the presence of collagen and minerals.

Bone cells reside in lacunae and are supplied with nutrients and waste removal by capillaries that pass through canals in compact bone.

Ossification is the replacement of the cartilage model of the skeleton before birth with bone tissue.

During childhood, new bone forms at the epiphyseal plates of long bones, and continues through adolescence.

When a bone breaks, a blood clot forms and is replaced with dense connective tissue, then cartilage, then bone.

Bone is an active tissue constantly being broken down and built up in response to gravity, nutrition, and hormones.

The human skeleton is arranged into an axial skeleton that includes the skull, vertebral column, ribs and breast bone, and an appendicular skeleton which includes the limb bones and limb girdles.

Bones join together at movable and immovable joints.

Movable or synovial joints are capsules of fibrous connective tissue plus slippery cartilage on the bone ends that reduces friction on the moving surfaces.

KEY TERMS

appendicular skeleton	ligaments
arthritis	lordosis
axial skeleton	lumbar
bone	marrow cavity
braced framework	ossification
bursae	osteocytes
calcitonin	osteoporosis
canaliculi	parathyroid hormone
cartilage	
cervical	pectoral girdle
coccyx	pelvic girdle
endoskeletons	rib cage
epiphyseal plates	sacral vertebrae
estrogen	scoliosis
exoskeletons	spongy bone
Haversian canal	synovial joint
homeostasis	synovial membrane
hydrostatic skeletons	thoracic
	vertebral column

CHAPTER OUTLINE

I. Skeletal Diversity and Evolution
    A. A skeleton is a supporting structure or framework.
    B. Hydrostatic skeletons (fig. 20.2).
        1. The simplest of skeletons and consist of liquid surrounded by a layer of flexible tissue.
        2. Found in squid, sea anemones, slugs, and certain worms, including annelids such as earthworms.
        3. In earthworms, the circular muscles contract against the fluid in the body cavity; the worm lengthens, anchors itself to the ground with bristles, and then moves forward as muscles shorten.

C. Exoskeletons.
   1. A braced framework that protects an organism from the outside.
   2. They are found in many groups of invertebrates, including arthropods.
   3. Muscles attach to the inner surface of an exoskeleton and contract against it to bring about movement.
   4. The growing animal must shed or "molt" the outgrown exoskeleton and grow a new one that is slightly larger (fig. 20.3).
D. Endoskeletons.
   1. An internal scaffolding found in vertebrates.
   2. It does not restrain growth and consumes less of an organism's total body mass than an exoskeleton, so it increases mobility.
   3. Most vertebrates have skeletons composed primarily of bone.
II. The Human Skeletal System
   A. Skeletal functions.
      1. They support the body against gravity and give it its characteristic shape.
      2. Lever system for movement.
         a) Typically the two ends of a skeletal muscle are attached to different bones that are connected by a joint.
         b) When the muscle contracts, one bone is moved.
      3. They provide protection of internal structures such as the backbone, skull, and ribs.
      4. They produce red and white blood cells and platelets in the long bones of the arm and leg.
      5. They provide storage for the minerals calcium and phosphorus, which are used in important functions.
   B. Composition.
      1. Cartilage.
         a) During embryonic development, the skeleton first forms in cartilage, providing a mold for the bone that will later replace it.
         b) An elastic form of the tissue is found in the external ear and in the nose.
         c) It permits the expansion of the rib cage when the lungs inflate.
         d) Smooth cartilage coverings allow the adjoining bones of the limbs to slide past one another.
         e) Cartilage cells are distributed within the matrix in spaces called lacunae.
         f) The large nuclei and extensive endoplasmic reticula of chondrocytes reflect the cells' function of producing large amounts of protein.
            (1) Collagen forms strong networks of fiber that distribute weight, enabling the tissue to resist breakage and stretching.
            (2) Elastin provides flexibility.
            (3) Proteoglycans are long chains of disaccharides that interact with collagen and provide firmness and resilience (fig. 20.4).
      2. Bone.
         a) Its strength is due to both minerals and the protein collagen present in the matrix (fig. 20.5).
         b) The minerals calcium and phosphate precipitate out of body fluids, coating the collage fibers to provide strength.

143

c) Compact bone.
      (1) A matrix that is composed of spaces (lacunae) arranged in concentric rings called Haversian systems around a central portal containing the blood supply, the Haversian canal.
      (2) Osteocytes are located in spaces in Haversian systems and are responsible for growth and repair.
      (3) Narrow passageways called canaliculi connect the lacunae.
      (4) The entire structure is surrounded by a layer of connective tissue called periosteum.
   d) Spongy bone.
      (1) Found mostly in irregular bones and tips of long bones.
      (2) It has many large spaces between a web of boney struts that increase the bones' strength.
      (3) The spaces within spongy bone are filled with red marrow, where blood cells and platelets are produced.
      (4) The spongy bone is covered with compact bone.
      (5) Long bones also have yellow marrow, which contains fat cells and a few blood cells (fig. 20.6).
III. Bone Growth and Development
   A. Before birth.
      1. Most of the skeleton originates in the embryo as a cartilage model that is gradually replaced by bone tissue in a process called ossification (fig. 20.7).
      2. Each cartilage model has a shape that is similar to that of the bone it will become, and it is surrounded by a layer of connective tissue called the perichondrium.
      3. When the embryo is about 4 weeks old, cells just beneath the perichondrium secrete a "collar" of compact bone around the central shaft of the developing bone.
      4. The matrix becomes hardened with calcium salts, which block exchange with blood supply so cartilage cells die.
      5. Blood capillaries invade the pockets, and bone cells enter and secrete bone matrix to establish new bone.
   B. Bone elongation during childhood.
      1. Within a few months of birth, bone growth becomes centered near the ends of long bones in thin disks of cartilage called the epiphyseal plates.
      2. The cartilage cells in the plate divide, pushing daughter cells towards the shaft, where they calcify.
   C. Repair of fractures.
      1. Bleeding at the site of the fracture results in a blood clot, which is soon replaced by dense connective tissue fibers secreted by invading connective tissue cells (fig. 20.9).
      2. Cartilage cells enter the dense connective tissue to build a fibrous callus that fills the gap left by the injury.
IV. Bone as a Mineral Store
   A. Regulation of bone formation is involved with maintaining homeostasis for the organism.
   B. Calcium is constantly being shuttled between the bone and cells of the body via the blood, where it is used in activities of enzymes, muscle contraction, blood clotting, cell cohesion, and cell membrane permeability.

C. Parathyroid hormone raises the level of calcium in the blood by causing bone to release calcium into the blood, converting vitamin D into its active form, which increases absorption of dietary calcium from the intestine and decreases excretion of calcium from the kidneys.

D. Calcitonin lowers blood calcium by halting the release of calcium from bone.

E. Loss of estrogen in menopausal women can increase the loss of bone mass.

V. Skeletal Organization.

A. Axial skeleton.

1. Specialized for this role in shielding soft body parts.

2. The skull protects the brain and many of the sense organs (fig. 20.12).

   a) Sutures between the many bones of the skull help dissipate the shock of a blow.

   b) All the head bones are joined with immovable joints except those of the lower jaw and the middle ear, which are required to move for chewing, speech, and hearing.

   c) The regions between a baby's skull bones contain dense connective tissue so the skull will grow as the brain enlarges (fig. 20.13).

   d) The vertebral column protects the spine and is composed of 7 cervical vertebrae in the neck, 12 thoracic vertebrae in the upper back, 5 lumbar vertebrae in the small of the back, 5 fused pelvic vertebrae called the sacrum, and finally 4 vertebrae fused to form the coccyx (fig. 20.14).

   e) Scoliosis is a type of abnormal spinal curvature (fig. 20.15).

   f) Lordosis is curvature of the spine due to strengthening of back muscles only, as in runners.

   g) The rib cage protects the heart and lungs and is built with 12 pairs of ribs.

B. Appendicular skeleton.

1. It permits movement.

2. It consists of two limb girdles and their attached limbs.

3. The pectoral girdle forms the shoulders and consists of two clavicles and two scapulae.

4. The arms and hands.

   a) The humerus is the upper arm.

   b) The lower arm is make up of the ulna and radius.

   c) The wrist bones, called carpels, are joined to the metacarpals, which are the hand bones.

   d) The fingers are called phalanges.

5. The pelvic girdle.

   a) It consists of two hipbones, each of which is actually three separate bones.

   b) The hipbones join the sacrum in the rear and meet each other in front.

   c) It protects the lower digestive organs, the bladder, and some reproductive structures.

6. The legs and feet.

   a) The femur is the upper leg bone.

   b) The skin or tibia is the larger of the two bones of the lower leg and the fibula is the smaller.

   c) The kneecap is the patella.

   d) The ankle bones are called the tarsals.

   e) The foot is made up of bones called metatarsals, which are connected to toes called phalanges.

VI. Joints—Where Bone Meet Bone
   A. Bones of the skull are attached by a thin layer of fibrous connective tissue.
   B. Synovial joints (figs. 20.16 and 20.17).
      1. They are found between freely movable bones.
      2. They consist of a capsule of fibrous connective tissue called ligaments that is filled with synovial fluid.
   C. Arthritis is joint inflammation.
   D. Rheumatoid arthritis is an inflammation of the synovial membrane.

LEARNING OBJECTIVES

After reading this chapter the student should be able to answer these questions:

1. What are the functions of skeletal systems?
2. What are some of the forms of skeletal systems in different species?
3. What types of molecules and cells make up the human skeletal system, and how are they arranged in bone tissue?
4. How does the human skeleton develop and grow in the embryo, fetus, and child?
5. How does the human body repair bone fractures?
6. How is the mineral composition of bone regulated?
7. How are bones classified?
8. How are bones arranged in the human skeleton?
9. What is the structure of the synovial joint that connects some bones to each other?

ANSWERS TO THE TEXT QUESTIONS

1. A hydrostatic skeleton consists of a liquid surrounded by a layer of flexible tissue, found in sea anemones, slugs, and earthworms. Exoskeletons are braced frameworks on the outside of the animal. The structural components are strong enough to resist pressure without collapsing and are found in arthropods and mollusks. An endoskeleton is an internal scaffolding found in the vertebrate animals.
2. A jointed exoskeleton has the advantage over a shell in being movable and providing a lever system for muscle action. It allows many more options for movement and locomotion. The endoskeleton is lighter, stronger, and does not limit growth or need to be shed when growth occurs.
3. The functions of a human skeleton include: support, lever system for movement, protection of internal structures, production of blood cells, and mineral storage.
4. Cartilage is a precursor tissue in the formation of bone. It is flexible, with great strength and gives shape to soft tissue such as the ears and nose. It provides for shock absorption between bones at the joints, and allows for smooth gliding movements at the joints.
5. The major components of the bone matrix are protein, collagen, and minerals. The minerals give the property of hardness and the protein collagen gives the bone its great strength.
6. Bones cells are nourished by blood which reaches the cells through small canals that penetrate the matrix. The cells are in physical contact through small channels called canaliculi. Materials can be passed from the blood by diffusion and by cell to cell "bucket brigades" of up to 15 cells.
7. Spongy bone is found at the ends of long bones and also in flat bones. It is involved in blood cell production. Compact bone: found in the shafts of long bones and is involved in mineral storage and providing strength to the bone. Red marrow is found in spongy bone and plays a role in blood cell and platelet production. Yellow marrow is found in the shaft of long bones and is the site of fat storage.
8. The shaft consists of compact bone, which is very hard and strong due to the presence of protein collagen and minerals. The shaft is hollow which adds stiffness to the shaft.
9. In the embryo the skeleton originates as a cartilage model and is gradually replaced by bone tissue. Cells just beneath the outer layer of connective tissue termed the perichondrium secrete a collar of compact bone around the central shaft. The cartilage cells within the shaft enlarge squeezing the cartilage matrix into

146

thin spicules. The compressed matrix becomes hardened with deposits of calcium salts. The loss of fluid space and movement of water through the matrix kills the cartilage cells. The hardened cartilage matrix begins to degenerate, and blood capillaries invade the pockets left by the retreating cartilage. Bone cells then enter these spaces and secrete bone matrix to establish the internal region of new bone.

10. Cartilage disks at the ends of long bones called epiphyseal plates are the centers of growth. The cartilage cells in the plate divide, pushing daughter cells towards the shaft, where they calcify. The dividing cartilage cells keep the disk relatively constant in thickness, but the calcified daughter cells become part of the shaft elongating it. The growth at these plates continues until the late teens when they are replaced with bone tissue, and only a thin line remains marking the position of the cartilage disk in the adult.

11. The immediate reaction to a fracture is bleeding, followed by blood clot formation. The blood clot is replaced with dense connective tissue that is secreted by invading connective tissue cells. Cartilage cells enter the dense connective tissue to build a fibrous callus that fills the gap of the injured bone. Bone cells begin to manufacture new bone and the cartilage is replaced by spongy bone. The gap is closed. With exercise the bone cells are stimulated to secrete more collagen which compacts the newly formed bone.

12. Osteoporosis is the demineralization of bone. Pockets form with the loss of calcium salts and the mass of the bone decreases. It is more common in women than men because women have a lower bone mass than men so any demineralization with aging shows up in women before men. In addition women live longer than men so there is more time for the demineralization process to occur. In postmenopausal women the lower levels of estrogen cause an increase in sensitivity of bone cells to parathyroid hormone which raises the blood levels of calcium by releasing these salts causing demineralization.

ANSWERS TO THE "TO THINK ABOUT" QUESTIONS

1. The smaller skeletons are those of both female and male children. Those with the lower bone mass are the remains of females and the higher bone mass are males.

2. The cartilage chemical will prohibit the growth of capillaries to supply nutrients to the developing cancer tissue. Since the shark skeleton is entirely cartilage there would be more tissue to isolate the chemical than that of the rat.

3. Immobility prevents the movement of water through the cartilage matrix. The movement of water through the matrix supplies the nutrients and removes waste products of the cartilage cells which secrete the matrix.

4. In a hydroskeleton, the fluid in a confined space provides shape and rigidity to the organism. In cartilage the fluid moves through the matrix providing nutrients and removal of waste products of the cartilage cells. The fluid in synovial joints is very slippery, reducing friction at the joint. Very simple organisms have hydroskeletons. Exoskeletons provide sites for muscle attachment and increased mobility, but the mass of the tissue prevents the evolution of very large animals. Endoskeletons are seen in primitive vertebrate organisms such as the shark and the skeleton is entirely cartilage. In more advanced vertebrates the cartilage is replaced by bone tissue but cartilage remains in joints, and as supportive tissue seen in the ears and nose.

5. Spondylitis can reduce movement and the cushion of the cartilaginous pads between the vertebrae. The symptoms would be a lack of rotational movement and anterior-posterior curvature of the spine.

6. The hollow bones reduce the enormous mass of the bone as well as to make the bones stiffer and less likely to bend under the weight of the animal.

AUDIOVISUAL MATERIALS

Bones and Muscles, International Film Bureau,
        15 min., 1981.
Human Skeleton, United World Films, 11 min.,
        1951.
Skeletal System, Coronet Instructional Films,
        12 min., 1980.
The Skeleton, Encyclopaedia Britannica
        Educational Corp., 17 min., 1979.

21 The Muscular System

KEY CONCEPTS

Smooth, cardiac, and skeletal muscles differ in shape, location in the body, and neural control, but all utilize the sliding of thick and thin myofilaments to generate the force of contraction.

Muscle contraction requires four major proteins, calcium ions, and ATP as an energy source.

Muscle shortening results when thin actin filaments slide between the thick myosin filaments.

The myosin heads are able to swivel and attach at different points along the actin myofilament.

ATP is used as energy for the formation of cross bridges and the sliding of the myofilaments as well as the energy source to return calcium ions to the sarcoplasmic reticulum initiating relaxation.

ATP is supplied from creatine phosphate, aerobic metabolism, and anaerobic metabolism which is less efficient and leads to lactic acid accumulation.

Muscle fibers respond in an all or none manner.

The strength of muscle contraction depends upon how many muscle fibers a motor nerve cell stimulates.

Multiple stimulation of the muscle fiber leads to summation and tetanus.

The duration of muscle contraction depends upon the frequency of nerve stimulation to the muscle fibers.

The proportion of muscle fiber types (fast or slow twitch) within a muscle affects athletic performance.

Muscles and bones form lever systems, with bones the levers, joints as fulcrums, and force supplied by skeletal muscle.

Muscle spindles are receptors that control muscle tone in response to gravity.

Regular exercise leads to muscle hypertrophy, in which muscle cells increase in size, with enzymes and mitochondria being more numerous.

Tendons are frequent sites of injury, and muscles can tear in extreme conditions.

KEY TERMS

actin	oxygen debt
antagonistic	sarcolemma
atrophy	sarcoplasmic
cardiac muscle	reticulum
cells	skeletal muscle
creatine phosphate	cells
fast twitch-	sliding filament
fatigable	model
fibers	slow twitch-
fast twitch-	fatigue resistant
fatigue	fibers
resistant	smooth muscle
fibers	cells
insertion	summation
intercalated disks	tendon
lever systems	tetanus
motor unit	transverse (or T)
muscle tone	tubules
myofibrils	tropomyosin
myofilaments	troponin
myosin	twitch
origin	twitch types

CHAPTER OUTLINE

I. Muscle Function
   A. Locomotion.
   B. Sustained beating of the heart.
   C. Contractions move food along the digestive tract.
   D. Muscles shape the human form.
II. Muscle Cell Types
   A. Smooth muscle.
      1. They have long, tapered cells each with a single nucleus.
      2. They are under involuntary control.

3. They are found in the walls of certain blood vessels, where they regulate blood pressure and direct blood flow.
4. They are found in the walls of the intestines, where they push food through the digestive tract.
B. Cardiac muscle.
1. Their contraction creates the force that propels blood around the body.
2. They are multinucleated cells that appear striped or striated and are under involuntary control.
3. They are joined to one another by intercalated disks.
C. Skeletal muscle.
1. They are long, striated, multinucleated cells under voluntary control.
2. They are attached to bones and are responsible for moving them.
III. Skeletal Muscle Organization
A. Each muscle lies along a bone with opposite ends attached to different bones by tendons.
B. Each has a connective tissue sheath surrounding many bundles called muscle fasciculi.
C. Each fiber has its own connective tissue covering.
D. Blood vessels and nerves lie between bundles of fibers.
IV. Microscopic Structure and Function of Skeletal Muscle
A. Muscle fibers.
1. Each fiber contains hundred of thousands of units called myofibrils and regular cellular components.
2. The cell membrane of the muscle fibers is called sarcolemma, and the endoplasmic reticulum is called sarcoplasmic reticulum.
3. The parts that jut into the inner membrane are called the transverse, or T, tubules (fig. 21.3).
B. Myofibril composition.
1. Each myofibril contains two types of finer strings called myofilaments.

2. Thin myofilaments are composed primarily of the protein actin, which resembles a string of beads, and they also contain the proteins troponin and tropomyosin.
3. Thick myofilaments are composed of the molecules myosin, which is shaped like a golf club forming into bundles of several hundred molecules.
4. The myofilament assemblage is one thick myosin myofilament surrounded by six thin actin myofilaments, which allows cross-bridges to form, break, and reform between the myosin and actin.
5. Muscle appears striped because of the arrangement of thick and thin myofilaments.
C. How skeletal muscles contract.
1. The sliding filament model of muscle contraction describes how thin myofilaments slide between the thick ones.
2. The heads of the myosin molecules swivel back and forth, allowing actin to slide past myosin.
3. The four types of molecules along with ATP and Ca^+ are required for contraction.
4. Contraction is a multistep process.
   a) In a muscle at rest, troponin holds tropomyosin in such a way that cross-bridges do not form between actin and myosin.
   b) The directive to contract begins when a message from a motor nerve cell that extends from the brain or spinal cord to a muscle cell releases a chemical messenger, acetylcholine, at a special junction of nerve and muscle called a neuromuscular junction.

150

c) This messenger binds to the outer membrane of the muscle cell, inducing an electrical wave that is quickly sent to calcium storage sacs within the sarcoplasmic reticulum.

d) The sacs release calcium ions, which bind to the troponin molecules attached to the actin myofilaments.

e) This causes movement of troponin such that it can no longer block actin and myosin cross-bridging.

f) Meanwhile, ATP slips into its site on the myosin head and is catalyzed to ADP and inorganic phosphate.

g) This activates myosin so it attaches to actin, which releases the energy stored in myosin and causes the myosin head to swing back to its original position.

h) This causes the filaments to move past one another.

i) Soon afterwards, ADP and inorganic phosphate are released, myosin is free to bind to a new ATP, and the cross-bridges between actin and myosin is broken.

j) The head flips back to its original position and binds a new actin further along on the chain.

k) When no more calcium is available, tropomyosin goes back to blocking actin/myosin cross-bridging and the sarcomere relaxes.

5. Energy for muscle contraction.

a) As ATP is used for muscle activity, it is replaced by creatine phosphate, which is stored in muscle fibers.

b) When this supply is diminished, ATP is drawn from cellular respiration if oxygen is available.

c) If oxygen is unavailable, anaerobic respiration provides ATP through the breakdown of glycogen to lactic acid, which can build up and cause pain until oxygen debt is overcome.

V. Macroscopic Structure and Function of Skeletal Muscle

A. If nerve stimulation releases calcium within the fiber, then all of its sarcomeres contract and the entire fiber shortens, which is the all-or-none law.

B. If many of the muscle fibers encased in the same connective tissue sheath contract, then the sheath pulls on the attached tendon, which in turn pulls on the bone.

C. The strength of contraction depends upon the number and distribution of fibers in a muscle sheath.

D. A nerve cell and all of the muscle fibers it contacts are called a motor unit.

E. A twitch refers to a single stimulation, which causes the muscle to contract quickly and then relax.

F. Summation refers to two contractions that occur back to back without time for muscle relaxation between stimulations and results in an increased strength of contraction.

G. Tetanus refers to the smooth and continuous contraction that results from repeated strong stimulation without relaxation.

VI. Inborn Athletic Ability and Muscle Fiber Types

A. Slow twitch-fatigue resistant fibers.

1. These contract slowly because the myosin heads split ATP slowly.

2. They resist fatigue because they are well supplied with oxygen bound to myoglobin (a form of hemoglobin found in muscle cells).

3. They are called red or dark fibers.

151

B. Fast twitch-fatigue resistant fibers.
   1. They have abundant oxygen.
   2. Their myosin heads split ATP quickly, resulting in a twitch response that may be 10 times as fast as the slow twitch fiber.
C. Fast twitch-fatigable.
   1. It splits ATP quickly but does not have much of an oxygen supply due to lack of myoglobin.
   2. The muscle fibers are specialized for short bouts of rapid contraction.
D. The proportion of fast twitch to slow twitch fibers within particular muscles affects performance.
   1. People with a high proportion of slow twitch fibers tend to excel at endurance sports such as running and swimming.
   2. In some European nations, measurement of the ratio of fast twitch to slow twitch fibers is used as a predictor of athletic success in certain events.

VII. Muscles Working Together
A. Lever systems.
   1. A lever is a structure that can pivot around a point, called the fulcrum, when a force is applied to it.
   2. The bones function as levers, the joints as fulcrums, and the force that moves the levers is supplied by skeletal muscles.
   3. Muscle attachments to bones.
      a) Origin refers to the end of the muscle that is attached to the bone that does not move.
      b) Insertion refers to the end on the movable bone (fig. 21.13).
   4. Small differences in the placement of a muscle's insertion can make a big difference in strength.
B. Antagonistic pairs.
   1. Muscle groups that move a bone in opposite directions, for example, contraction of the bicep on the front of the upper arm bends the arm at the elbow, while the tricep straightens the arm.
   2. Generally one member of an antagonistic pair is stronger than the other.
   3. Overexercise of one antagonistic pair may result in pain.
C. Muscle tone.
   1. At any given time some fibers are contracted due to the effects of gravity.
   2. Sensory information from receptors called muscle spindles notifies the brain, which makes adjustments in muscle tone.

VIII. Effects of Exercise on Muscle
A. Regular exercise strengthens the muscular system and enables it to use available energy more efficiently.
   1. Hypertrophy, an increase in exercise-induced muscle mass, is due to an increase in the size of individual skeletal muscle cells rather than an increase in their number.
   2. In a trained runner, muscle fibers are both more active and more numerous and mitochondria are larger and more abundant, resulting in greater amounts of ATP.
   3. Athlete's muscles contain more blood and store more glycogen than do those of an untrained person.
B. Atrophy is the opposite condition of muscle degeneration resulting from the lack of use or immobilization in a cast.
C. The changes in muscles due to regular exercise disappear quickly if activity is discontinued.

IX. Injuries to the Muscular System
A. Injuries to tendons are fairly common since they are smaller than muscles and do not contract, resulting in inflammation as in Achilles tendinitis.
B. When a muscle is subjected to more stress that it can bear, muscle fibers tear, as is often seen in sprinters who tear hamstring muscles.

152

C. For suspected injuries of muscles, ligaments, joints, and bones, the RICE plan is suggested by the American College of Sports Medicine.
   1. **R**est to prevent further damage.
   2. **I**ce to contract blood vessels in the damaged area to stop bleeding and promote healing.
   3. **C**ompression to decrease swelling and promote healing.
   4. **E**levation of the legs above the level of the heart so gravity drains excess fluid.

LEARNING OBJECTIVES

After reading this chapter the student should be able to answer these questions:

1. How are the three types of contractile cells alike and how are they different?
2. What structures compose skeletal muscles, both macroscopic and microscopic?
3. How do thick and thin protein rods interact to cause the sliding action that underlies muscle contraction?
4. What factors contribute to the number, distribution, speed of contraction, and resistance to fatigue of muscle fibers within a skeletal muscle?
5. How do nerves affect a muscle's activity?
6. How do muscles, joints, and bones function together in lever systems?
7. How does exercise affect muscle tissue?

ANSWERS TO THE TEXT QUESTIONS

1. Smooth muscles are not under voluntary control and are single long and taped cells. They are found in internal organs, arterial walls, and in the respiratory system. Cardiac muscles are single striated cells that are under involuntary control. These cells are closely connected by intercalated disks and operate as a unit in the movement of blood. Skeletal muscles are striated and consist of long cells with many nuclei. They attach to the skeletal system and are involved in movement and heat production. They compose approximately 50% of the body mass and give form to the human body. All three types of muscles depend upon the sliding of actin and myosin filaments past each other to produce shortening or contraction.

2. a) The sarcomeres shorten. b) The muscle fibers shorten. c) The muscle fiber bundles shorten. d) The entire skeletal muscle shortens. e) The tendon pulls on the bone to which it is attached. f) The attached bone moves.

3. a) The nervous system controls the contraction of muscles by activating the muscle fibers. It controls the strength and duration of the contraction by altering the frequency of action potentials and the number of motor units activated. b) The skeletal system is moved by the muscular system producing locomotion. c) The circulatory system delivers nutrients and removes wastes from all tissues of the body including the muscular system. d) Connective tissue connects the muscular system to the bones and also holds together the cells of the muscular system.

4. Actin, slides past myosin in muscular contraction and is the site of the binding of the myosin head during contraction. Myosin, binds to actin and is activated by ATP. It generates the sliding action that causes shortening of the muscle tissue. Troponin, binds to calcium and moves tropomyosin, uncovering the sites of actin-myosin cross bridge formation. Tropomyosin, covers the site of cross bridge formation when the muscle is at rest. Dystrophin, relays the electrical message of the muscle membranes and T-tubule system to the sarcoplasmic reticulum which releases calcium thereby activating troponin.

5. The alignment of actin and myosin myofilaments in the sarcomeres. The alignment causes a light and dark band pattern to be seen in skeletal muscle fibers.

6. Tendons are smaller in diameter than muscles and cannot contract. The full force of the muscle is transmitted through the tendon attached to the bone.

7. ATP is supplied initially through transfer of high energy phosphate to ADP by creatine phosphate. ATP can be produced through aerobic metabolism in the

mitochondria or from anaerobic metabolism which is less efficient and leads to the accumulation of lactic acid.

8. Since muscles only generate force by shortening or contraction it takes two muscles to move a bone at its joint. For example, the biceps flex the forearm and the triceps extend it.

9. With regular exercise the leg muscles undergo hypertrophy in which the size of the individual muscle fibers increase due to more actin and myosin myofilaments. Myoglobin, glycogen, mitochondria, ATP, and the blood supply to the muscles all increase.

ANSWERS TO THE "TO THINK ABOUT" QUESTIONS

1. No, when muscles contract the diameter of the muscle increases and decreases when the length increases during relaxation.

2. This appears to contradict what occurs in humans in that only the size of the fibers increases following exercise, not the number of muscle cells.

3. Follow the "RICE" plan of rest, ice, compression, and elevation of the legs.

4. World-class athletes are routinely tested for percentage of fast and slow twitch muscle fibers, however, the apparent parental pressure for athletic superiority may have negative consequences.

5. Additional blood flow may help by delivering more oxygen to the muscle tissue for aerobic metabolism. The enhanced metabolic capacity would produce more ATP for sustained activity. However, the increase in numbers of red blood cells may increase the viscosity of the blood and put an additional workload on the heart.

6. Athletic stamina is not inborn but must be attained through consistent exercise. The entire body responds to the stress of exercise even though only specific muscles are utilized.

AUDIOVISUAL MATERIALS

Bones and Muscles, International Film Bureau, 15 min., 1981.
Muscle, CRM McGraw-Hill, 25 min., 1972.
Muscle: Chemistry of Contraction, Encyclopaedia Britannica Educational Corp., 15 min., 1969.
Muscle Contraction and Oxygen Debt, McGraw-Hill, 16 min., 1967.
Muscle: Dynamics of Contraction, Encyclopaedia Britannica Educational Corp., 21 min., 1969.
Muscles in Motion, Michigan Media Resources Center, 7 min., 1982.
Muscular System, Coronet Instructional Films, 11 min., 1980.
Musculoskeletal System, J. B. Lippincott, 16 min., 1981.

22 The Circulatory System

KEY CONCEPTS

Circulatory systems provide nutrients and remove wastes from cells.

In single cell organisms diffusion is effective in the exchange of materials with the environment.

In large animals the circulatory system links the cells to the environment through other organs for the exchange of gases, absorption of food, and removal of waste products.

Blood is the media for the transport of molecules within the circulatory system.

Blood helps to maintain homeostasis by regulating pH, water content, and protects from injury by clotting.

Blood is a complex tissue consisting of many different formed elements and cells suspended in a fluid matrix termed plasma.

Blood cells are specialized for transporting oxygen, protecting against injury, recognizing foreign agents, and destroying abnormal tissue cells, such as cancer.

There are many proteins in the plasma that are specialized to transport food and waste products, regulate osmotic pressure, serve as antigens against foreign agents, and react to form clots.

In closed circulatory systems there are specialized vessels (arteries) that carry blood from the heart to the tissues, thin vessels for the exchange of materials at the tissues (capillaries) and vessels to return blood to the heart (veins).

The heart contracts generating pressure to move the blood through the circulatory system.

The flow of blood is unidirectional because of the structure of the heart and its valves, and the one-way valves in the veins.

The path of blood flow in the human circulatory system is through two basic loops: the pulmonary loop to exchange gases in the lungs, and the systemic loop, which delivers blood to all the body tissues.

The heart is a pump which develops high pressure during contraction forcing blood into the large arteries, and during relaxation the chambers of the heart fill with blood returning from the body and lungs.

The heart has an internal system of regulating the sequence and timing of contraction and relaxation and is also controlled by the nervous system.

Exercise has positive advantages of developing the heart into a more efficient pump.

The lymphatic system returns excess fluid that leaks from the circulatory system into the tissue spaces back to the circulatory system.

The lymphatic system also plays a major role in protecting the body from infection, and in the absorption of fats from the digestive system.

KEY TERMS

aorta
aortic semilunar valve
arteries
arterioles
atria
atrioventricular (AV) node
atrioventricular valves
autonomic nervous system
bicuspid valve
blood
blood pressure
blood vessels
capillaries
cardiac cycle
cardioaccelerator area
cardioinhibition area

closed circulatory system
coagulation
coronary arteries
coronary circulation
deoxyhemoglobin
diastolic pressure
edema
erythropoietin
formed elements
heart
heart murmur
hepatic portal vein
hepatic portal system
inferior vena cava
intercalated disks
lymph
lymph capillaries
lymph nodes
lymph vessels
megakaryocyte

open circulatory system
oxyhemoglobin
pacemaker
parasympathetic nervous system
pericardium
plasma
platelets
pulmonary artery
pulmonary semilunar valve
pulmonary veins
Purkinje fibers
red blood cells
semilunar valves
sinoatrial node (SA node)

sphygmomanometer
spleen
stress test
superior vena cava
sympathetic nervous system
systolic pressure
thromboplastin
varicose veins
vasoconstriction
vasoconstriction area
vasodilation
vasodilation area
vasomotor center
veins
ventricles
venules
white blood cells

CHAPTER OUTLINE

I. Diversity Among Circulatory Systems
  A. Diffusion.
    1. In single-celled organisms and multicellular animals whose cells are close to the environment, diffusion across the body surface works to deliver raw materials and remove wastes.
    2. For protozoans, direct diffusion occurs, whereas in sponges a series of canals brings nutrients into inner cells.
    3. In flatworms, a central cavity branches so each cell can get nutrients.
  B. More complex animals.
    1. An increase in an animal's size decreases this surface-to-volume ratio, making diffusion inefficient.
    2. More complex animals require a circulatory system composed of a fluid (blood), a pump or heart to force the blood around the body, and blood vessels to serve as a network to deliver blood.
      a) Open circulatory systems.
        (1) Blood is not always contained in blood vessels, sometimes dumping into sinuses.
        (2) They are found in mollusks and arthropods.
      b) Closed circulatory systems.
        (1) Blood remains within blood vessels.
        (2) Large vessels conduct blood away from the heart and form smaller vessels that divide into even tinier vessels, where nutrients and wastes are exchanged.
        (3) Blood returns to the heart through gradually increasing vessels.

II. The Human Circulatory System
  A. Consists of a central pump, the heart, and an attached continuous network of tubes, the blood vessels.
  B. Also known as the cardiovascular system, cardio referring to heart and vascular to the vessels.

III. Functions of Blood
  A. Blood keeps cells and the abundant fluid around them "fresh" by bringing oxygen and nutrient molecules to cells and removing their wastes.
    1. Carbon dioxide is transported primarily in the form of bicarbonate ions (HCO_3^-).
    2. Urea, a product of protein metabolism, travels from the liver to the kidney.
  B. It carries hormones from their site of production to target cells.
  C. Blood maintains a constant pH of 7.4, with proteins that either release H^+ or absorb them.
  D. It regulates water balance with the aid of dissolved sodium ions and blood proteins, particularly albumin, which controls the movement of water from within the vessels to the tissues.

E. It protects from injury in two ways:
 1. Clotting involves cell fragments called platelets and inflammation, a swelling induced by white blood cells.
 2. White blood cells defend the body against foreign invaders.

IV. Blood Composition
 A. Blood is a matrix consisting of a liquid portion called plasma, in which molecules and blood cells are suspended.
 B. Plasma.
 1. It comprises approximately half of the blood's volume.
 2. It is 90% to 92% water.
 3. It contains various dissolved molecules, such as salts, nutrients, hormones, metabolic wastes, and gases.
 4. There are over 70 different blood proteins.
  a) Albumin regulates osmotic pressure.
  b) Globulins take part in immune response and transport lipids or metals.
  c) Fibrinogen is needed for blood clotting.
 C. Red blood cells.
 1. Also known as erythrocytes, they are packed with the protein hemoglobin, which transports oxygen.
  a) Oxyhemoglobin is bright red due to the oxygen it carries.
  b) Deoxyhemoglobin is the form after the oxygen is released and may pick up carbon dioxide to be carried back to the lungs.
 2. The mature cell is a biconcave disc lacking a nucleus.
 3. Carbon monoxide poisoning occurs because this molecule binds more strongly to heme groups than oxygen, so it cannot be picked up.
 4. When certain cells in the kidney do not receive enough oxygen they produce a hormone erythropoietin, which stimulates red blood cell production in the red bone marrow.
 5. Anemia is when there are too few red blood cells or too little hemoglobin.
 6. Individuals in high altitudes produce more red blood cells to compensate for low amounts of oxygen in the air.
 D. White blood cells.
 1. They are larger than red blood cells and retain their nucleus.
 2. They originate in the bone marrow and live about 1 year.
 3. There are five types of white blood cells or leukocytes.
  a) Neutrophils are the most abundant and surround and degrade bacteria at the site of inflammation.
  b) Eosinophils affect parasites that the neutrophils do not affect.
  c) Basophils release certain chemicals that are part of the inflammatory response.
  d) Monocytes attack bacteria that are resistant to neutrophils.
  e) Lymphocytes produce antibodies that protect against infection.
 4. Certain diseases are associated with white blood cell imbalances.
  a) In tuberculosis and whooping cough, the proportion of lymphocytes is elevated.
  b) High levels of eosinophils are seen in hookworm and tapeworm infections.
  c) Excess monocytes are found in typhoid and malaria sufferers as well as in individuals with mononucleosis.
  d) In leukemia, one type of cell greatly predominates because it divides more often.

    e) In AIDS (acquired immune deficiency syndrome) a subcategory of cells called T-lymphocytes are killed by the virus.

E. Platelets.
1. Small, colorless cell fragments originating from huge bone marrow cells called megakaryocytes.
2. They initiate clotting by being broken and collecting near a wound.
3. They release chemicals that trigger a chemical cascade resulting in clot formation.
    a) The injured blood vessel constricts.
    b) The blood protein thromboplastin converts prothrombin into thrombin, which converts fibrinogen into fibrin.
    c) Fibrin forms a meshwork of fibrils that entrap red blood cells and additional platelets.
4. Poor clotting can lead to hemorrhage.
5. Abnormally rapid clotting can lead to a thrombus or an embolus.

V. Blood Vessels
A. Large elastic blood vessels called arteries leave the heart and branch into smaller arteries, which in turn branch into even smaller arterioles, finally forming the tiniest vessels of all, the capillaries.
B. On the blood's return voyage, capillaries flow into small venules, then small veins, and finally large veins that return the blood to the heart.
C. Arteries.
1. They have thick walls that are able to expand to take up the surge in pressure with each wave that comes from the pumping action of the heart.
2. The outermost layer consists of connective tissue, the middle

layer is built of elastic tissue and smooth muscle, and the inner layer is a smooth one-cell-thick lining called endothelium.
D. Capillaries.
1. The site of nutrient and waste exchange.
2. They are built entirely of endothelium.
3. They are arranged in networks called capillary beds.
E. Veins.
1. They carry blood back to the heart from the capillaries.
2. They are thinner and less elastic than arteries because they conduct blood that is low in pressure.
3. Veins must move blood back against gravity when bringing it back from the legs and need assistance.
    a) Flaps called venous valves allow blood to move only in the direction of the heart.
    b) Contraction of skeletal muscles also helps move blood in veins.
F. Systemic circulation.
1. The aorta is the first vessel to leave the heart and branches to feed the heart itself, kidneys, muscles, abdominal organs, liver, brain, and other parts of the body.
2. Hepatic portal system.
    a) A special division of the circulatory system where nutrients are metabolized for quick energy and toxins and microbes are removed.
    b) Capillaries from the stomach, small intestine, pancreas, and spleen converge into the hepatic portal vein, which leads to the liver.
    c) Blood leaves the liver through the hepatic vein

and enters the heart through the inferior vena cava.
G. Blood pressure.
1. Blood pressure is a result of the heart's pumping action.
2. Systolic pressure is the pressure at its maximum when the heart chambers called the ventricles contract.
3. Diastolic pressure is blood pressure at a minimum when the ventricles relax.
4. A pressure gauge called a sphygmomanometer gives a reading of the systolic pressure over the diastolic pressure as in the average reading of 120/80.
5. Pressure drops the further away from the heart due to the increasing cross-sectional of blood vessels.
6. Dropped pressure ensures adequate time for pickup and delivery of wastes and nutrients.
7. Pressure can be increased by the narrowing of blood vessels, called vasoconstriction, or lowered when they are opened by vasodilation.
8. Hypotension refers to lower than normal blood pressure, which may cause fainting, and hypertension refers to higher than normal pressure, which may cause death.

VI. The Heart
A. The structure of the pump.
1. The heart is composed of cardiac muscle, which contracts to propel blood.
2. It is enclosed in a tough connective tissue called the pericardium.
3. The heart consists of four chambers.
   a) The atria receive blood returning from the heart and pump it into ventricles.
   b) The ventricles are much larger and have thick walls to generate enough force to push blood to the entire body.

B. Heart valves.
1. The atrioventricular valves (AV valves) prevent backflow of blood from the ventricles into the atria.
2. The semilunar valves prevent blood from flowing back into the ventricles from the arteries.
3. Heart sounds.
   a) The "lub-dub" sound is caused by the closing of the AV valves (("lub") and the semilunar valves ("dub").
   b) A heart murmur is caused by an abnormal function of the valves.
C. The journey of blood.
1. The right side of the heart.
   a) Dark red, oxygen-poor blood is delivered into the right atrium from the body through the superior vena cava and the inferior vena cava and from the heart itself through the coronary sinus.
   b) It passes into the right ventricle through the AV valve and then into the pulmonary semilunar valve to the lungs.
2. The left side of the heart.
   a) Bright red, oxygen-rich blood returns from the lungs to the left atrium of the heart.
   b) It passes through the AV valve to the left ventricle and then through the aortic semilunar valve to the aorta, which takes the blood to the body.
D. Cardiac cycle.
1. Consists of a sequence of contractions (systole) and relaxation (diastole).
2. Muscle cells.
   a) Each muscle cell contracts on its own.
   b) When two or more are in contract they synchronize

the contractions due to tight junctions between the cells called intercalated disks.

3. Coordination of heartbeats.

    a) The Purkinje system is a network of muscle fibers permeating the heart that trigger contraction of the ventricles.

    b) Heartbeat begins with specialized cells in the sinoatrial node (SA node) located in the wall of the right atrium and are known as the pacemaker.

    c) They trigger specialized cells in the atrioventricular node, which then branches into cardiac muscle fibers called Purkinje fibers.

    d) During quiet times the parasympathetic nervous system inhibits the SA node and slows the heart, whereas when life is threatened, the sympathetic nervous system is dominant.

E. Coronary circulation.

    1. Two branches from the aorta are the left and right coronary arteries, which branch to surround and infuse the heart with blood vessels that nourish the active muscle cells.

    2. Vasomotor center in the brain.

        a) The vasoconstriction area and cardioaccelerator areas stimulate circulation by constricting blood flow and speeding the heart, respectively.

        b) The vasodilation area and cardioinhibition area dilate vessels and slow the heart.

VII. Exercise and the Circulatory System

A. It lowers heart rate so there is less workload on the heart.

B. It increases the number of red blood cells and the quantity of hemoglobin in each cell.

C. It lowers blood pressure.

D. It elevates the level of high density lipoproteins (HDLs), which may reduce cholesterol in the blood.

E. Development of extra blood vessels occurs.

VIII. The Lymphatic System

A. Another transport system that carries fluid that has leaked from the blood vessels back to the heart.

B. It contains lymph, which is plasma minus the large proteins and which cannot leave the blood vessels.

C. Lymph is propelled only by contraction of skeletal muscles.

D. Edema is a breakdown in lymphatic function resulting in excess fluid in the tissues.

E. Lymph nodes are fibrous enlargements along the lymph system and contain white blood cells that protect against infection.

F. Lymphocytes are manufactured in the spleen, the thymus, and possibly the tonsils.

LEARNING OBJECTIVES

After reading this chapter the student should be able to answer these questions:

1. How do circulatory systems in diverse organisms work?

2. How do the structures of the human circulatory system provide oxygen and nutrients to the body's cells, and remove wastes from them?

3. What are the functions of each of the blood's components?

4. What is the structure of the blood vessels and heart?

5. What is the pathway of the blood to and from the lungs, and to and from the other body tissues?

6. At what level of the circulatory system does exchange of oxygen and nutrients and wastes occur?

7. What factors determine and control heartbeat?

8. What effects does exercise have on the circulatory system?

9. What are the structures and functions of the lymphatic system?

ANSWERS TO TEXT QUESTIONS

1. Circulatory systems are needed in large animals because the cells and tissues buried deep within the body must be connected with the environment to survive. Nutrients must be brought in and waste products removed, while at the same time the internal environment must remain nearly constant in order to sustain life.

2. In an open circulatory system the blood is not always contained in blood vessels. There is usually a heart and blood vessels that lead to spaces called sinuses, in which blood directly bathes cells before returning to the heart. In closed systems the blood remains within blood vessels. Materials move between cells and blood across the walls of the capillaries.

3. Too few red blood cells would reduce oxygen carrying capacity and too many will increase the viscosity of the blood adding an additional workload to the heart. With too few white cells the individual may not be able to combat disease organisms or viral infections. Too many white cells will crowd the red blood cells and reduce the oxygen carrying capacity of the blood. Too few platelets may inhibit coagulation or blood clotting. With too many platelets, clots may form abnormally leading to the possibility of coronary thrombosis, strokes, or pulmonary embolisms.

4. The red blood cell is biconcave giving it a large surface area for gas exchange. It is easily bent and can squeeze through the narrow passageways of the capillaries. It has no nucleus, mitochondria, or ribosomes but is literally packed with hemoglobin, the oxygen carrying molecule.

5. It is difficult to find appropriate chemicals that mimic the action of hemoglobin in the blood. Many of the chemicals investigated provoke an immune attack on the "foreign" tissue and cause severe reactions.

6. The lack of vitamin K impairs the ability of blood to form clots and prevents the excessive loss of blood from the cardiovascular system.

7. The young man's blood lacks the normal ability to clot preventing blood loss.

The acquired immune deficiency syndrome is caused by a virus that kills lymphocytes called T cells that are essential in fighting viral infections.

8. Arteries are thick walled vessels with an outer connective tissue layer, a layer of smooth muscle and elastic tissue, and an inner smooth one-cell layer of endothelium. Veins are thinner and less elastic than arteries. The middle layer is reduced or even absent. In addition they may have one-way valves to prevent the back flow of blood.

9. Smooth muscle in an arteriole can contract and reduce the blood flow to the capillary bed it feeds, or relax and open the flood gates to that area. The pattern of dilation and constriction of the arterioles provides adaptation to exercise, temperature changes, and local metabolic needs of the tissues.

10. Blood pressure is the force generated by the contraction of the ventricles of the heart and large arteries, which drives blood through the circulatory system.

11. The heartbeat begins within specialized cells in the sinoatrial node located in the wall of the right atrium. The SA node sets the tempo of beating and is called the pacemaker. The electrical impulses travel over the atrial cells to the atrioventricular node where the signal is delayed before branching out through the ventricles via the Purkinje fibers. The heart delivers blood through two different circulatory systems. The right side of the heart pumps blood through the pulmonary circulation for the exchange of gases in the lungs. Oxygen rich blood returns to the left side of the heart which then pumps this blood through the systemic circulation to all parts of the body.

12. The individual has high blood pressure or hypertension, and anemia indicated by the low red blood cell count. The angina of exertion would be due to lack of adequate oxygen delivery to the heart muscle with exertion, indicating a sedentary person. He does not have atherosclerosis or diseases of the white blood cells such as leukemia. Moderate exercise may increase the red blood cell count and reduce the

hypertension relieving the angina during exertion.

13. The flow of blood from the lungs to the big toe and its return to the heart follows: pulmonary veins, left atria, left ventricle, aorta, descending aorta, iliac artery, capillary of the toe, iliac vein, inferior vena cava, right atria, right ventricle, pulmonary artery, lungs.

14. Lymph is made up of blood plasma minus certain large proteins that are too large to diffuse out of the blood capillaries. Lymph is propelled by contractions of the surrounding skeletal muscles, and is assisted by valves in the lymph vessels.

ANSWERS TO THE "TO THINK ABOUT" QUESTIONS

1. Platelets will adhere to the damaged wall and initiate a clot reaction. The clot may remain in place blocking blood flow through the vessel which could cause a coronary thrombosis if it occurs in the heart or a stroke if it occurs in the brain. The clot may be released and pass to the lung causing a pulmonary embolism.

2. Machines that give inaccurate low reading may mask the presence of hypertension which strains the heart, raising the risk of heart attack or failure. Too high a reading may create a false sense of "being at risk," causing unnecessary worry and leading to true hypertension.

3. The individual could obtain a more accurate reading by taking his or her blood pressure at home or in a more comfortable setting by an individual dressed more casually than in the nurse or doctor "uniform."

4. When the coronary arteries are narrowed by disease such as atherosclerosis, the individual has symptoms of pressure, fullness, squeezing sensation, sweating, shortness of breath, dizziness, or feeling nauseated. In general coronary heart disease may result in angina pectoris, arrhythmia, and congestive heart failure.

5. The doctors recognized that in trained athletes the heart rate is slowed to well below normal levels and arrhythmias are common in low heart rates.

6. The cardiovascular system responds to many different factors which are part of the complex seen in athletes. Excessive weight, cigarette smoking, and stress, all increase the risk of high blood pressure affecting the performance of the cardiovascular system. It is difficult to look at just the effect of exercise on the cardiovascular system in athletes since they tend not to smoke cigarettes, have low blood pressure, and use effective methods to reduce stress. In the general population these factors may be present and the single effect of exercise could be observed more easily.

7. A plan of moderate exercise, no smoking, reduction of excess salt in the diet, reduced cholesterol in the diet, maintenance of adequate diet in terms of minerals, vitamins, proteins, carbohydrates, and fats, reduced weight, and avoidance of stress or development of effective means of reducing stress, will maintain a healthy cardiovascular system.

AUDIOVISUAL MATERIALS

Anemia, Squibb, 25 min.

The Blood, Encyclopaedia Britannica Educational Corp., 16 min., 1961.

Body Defenses Against Disease, Encyclopaedia Britannica Educational Corp., 11 min., 1937.

Circulation, Boston Broadcasters, 22 min., 1980.

Incredible Voyage, McGraw-Hill, 26 min., 1968.

The Lymphatic System, International Film Bureau, 15 min., 1979.

Secret of the White Cell, Indiana University, 30 min., 1966.

Work of the Blood, Encyclopaedia Britannica Educational Corp., 13 min., 1957.

23 The Respiratory System

KEY CONCEPTS

Oxygen is needed by cells to extract maximal energy from nutrient molecules, and carbon dioxide is produced as a waste product.

A moist membrane is required for the exchange of gases between cells and the environment.

The respiratory surface must be extensive enough to meet the animal's metabolic needs and have a way to transfer the oxygen to the cells.

The process of diffusion is sufficient for single-cell organisms and simple multicellular organisms with a flat body form.

More complex organisms use a circulatory system to transport oxygen.

In vertebrate animals the nose filters the air, warms it, and adds moisture before entering the lungs.

The respiratory system is shaped like a tree with many branches that end in small sacs, the alveoli, where the exchange of gases occurs.

Surfactant prevents the collapse of lungs by lowering the surface tension of water.

Air moves into and out of the respiratory system because of pressure differences between the atmosphere and the thoracic cavity.

The pressure differences created by changes in thoracic volume are brought about by muscles acting on the ribs and the action of the diaphragm.

Oxygen binds to hemoglobin in red blood cells when the concentration of oxygen is high in the lungs and is released when the oxygen concentration is low at the tissues.

Carbon dioxide is transported bound to the hemoglobin molecule, dissolved in the plasma, and as the bicarbonate ion.

Breathing is controlled by the rhythmicity center in the medulla; neurons that initiate inspiration are active during quiet and heavy breathing.

Neurons that initiate expiration are only active during heavy breathing.

Detection of blood carbon dioxide level by chemoreceptors is the most important factor in regulating breathing.

Environmental pollutants may cause lung disease.

The respiratory, circulatory, nervous, skeletal, endocrine, and muscular systems all function together to enable you to breathe.

KEY TERMS

alveolar ducts	heme
alveoli	human lung
alveolus	surfactant
analgesics	hyperventilation
antihistamines	inspiration
apneustic center	larynx
asthma	lungs
bronchi	Mycobacterium
bronchioles	tuberculosis
bronchitis	nasal conchae
carbonic anhydrase	operculum
cerebral cortex	oxyhemoglobin
countercurrent flow	pharynx
dead space	residual air
decongestants	rhinoviruses
diaphragm	tidal volume
emphysema	trachea
epiglottis	tracheae
expiration	tubercle
gills	vital capacity
glottis	vocal cords
Heimlich maneuver	

CHAPTER OUTLINE

I.   Why We Need Oxygen
  A.   Oxygen is used by cells to extract the maximal amount of energy from nutrient molecules and to store that energy in the bonds of ATP.
  B.   The energy in ATP can then be tapped for a variety of biological processes—maintaining the potential

difference across a nerve cell membrane, muscle contraction, active transport across membranes, and hormone synthesis.

C. When cells lack oxygen, they can produce only 2 to 3 ATP molecules from a single glucose molecule.

D. In contrast, the total metabolism of the same molecule of glucose in the presence of oxygen can yield at least 36 ATP molecules.

E. Oxygen is the final acceptor of the low-energy electrons at the end of the chain of acceptors; if oxygen is not present, the electrons cannot pass along this series of acceptors, so the breakdown of organic molecules stops.

II. Diversity Among Respiratory Surfaces

A. Respiration is the exchange of gases between the cells and the environment across a moist membrane.

B. The larger and more active animals need a greater respiratory surface and also require a circulatory system.

C. The amount of water that an aquatic organism must move over its respiratory surface is greater than the volume of air that a land dweller must move across its respiratory surface.

D. Aquatic organisms require a larger respiratory surface than a terrestrial animal of similar size and energy requirements, and land dwellers have to keep the respiratory membrane from drying out.

E. Body surface.
   1. A single-celled organism, for instance, can use this method because its cell membrane provides enough surface area to transport sufficient oxygen into the cell.
   2. Multicellular organisms have a flat body form, such as Hydra.
   3. Flatworms are thin enough for both gas exchange and the distribution of gases to occur by diffusion.
   4. Earthworms and amphibians still use the body surface for gas exchange but employ a

circulatory system to transport the gases between the cells and the environment.

F. Tracheal systems.
   1. Some organisms have tough, waterproof exoskeletons that are indented to form an extensively branching system of tubules, called tracheae, that bring the outside environment in close enough contact with every cell for gases to be exchanged by diffusion.
   2. Examples include arthropods such as insects, centipedes, millipedes, and some spiders.

G. Gills.
   1. In aquatic organisms, the respiratory surface usually takes the form of gills, which are extensions of the body wall that are so highly folded and branched that they may superficially resemble feathers.
   2. Gases are exchanged across a very thin respiratory membrane and the single layer of cells of capillary and transported by the circulatory system.
   3. They are protected by a flap called the operculum and require a continuous supply of fresh water from the mouth over the gills.
   4. Countercurrent flow.
      a) This is the flow of blood through the capillaries in the opposite direction of the flow of water over the gill membrane.
      b) It maximizes the amount of oxygen that can be extracted.
      c) The blood leaving the gills has a relatively high oxygen concentration and encounters water entering the gills, which has the highest available oxygen concentration.
      d) Thus oxygen diffuses into the blood and the oxygen

164

load of the blood is topped off as it leaves the gills.

H. Lungs.
1. A respiratory surface within the body that is usually in close association with the circulatory system and can be kept moist.
2. The more complex the folding in a lung, the greater its surface area for gas exchange.

III. The Human Respiratory System
A. The nose.
1. Cells lining the nose purify and warm incoming air.
2. Nasal cavity.
   a) Each nasal cavity is partitioned into channels by three shelflike bones, the nasal conchae.
   b) The entire surface is covered with a mucous membrane containing many blood vessels (fig. 23.5).
3. The air is cleaned by hairs, mucus, and cilia.
B. The pharynx and the larynx.
1. The pharynx, or throat, conducts food and air.
2. The larynx, or voice, is where inhaled air passes though the opening called the glottis.
3. A piece of cartilage called the epiglottis flips down and covers the glottis during swallowing.
4. Stretched over the glottis are two elastic bands of tissue, the vocal cords, that vibrate when air rushes past them, producing sound that can be molded into speech.
C. The trachea, bronchi, and bronchioles.
1. Atop the larynx sits the trachea, also known as the windpipe.
2. Cartilage hold the trachea open in spite of the negative pressure created by exhalation (fig. 23.7).
3. The trachea branches into two bronchi.
4. The bronchi branch repeatedly, each branch decreasing in diameter until it consists of only a thin layer of smooth muscle and elastic tissue, and at this microscopic level, the tubes are called bronchioles (little bronchi).
5. The respiratory passageways within the lung are called the "bronchial tree" (fig. 23.8).
   a) The bronchioles have no cartilage, but their walls do contain smooth muscle.
   b) During a stressful or emergency situation, the sympathetic nervous system causes the bronchioles to constrict, reducing their diameter and then returning them to normal.
   c) Asthma occurs when the bronchial muscles go into a spasm, making air flow exceedingly difficult (fig. 23.9).
      (1) Most asthma attacks are triggered by an allergy to pollen, dog or cat dander, or tiny mites in house dust.
      (2) Some aerosol inhalers spray epinephrine onto the bronchial walls.
D. The alveoli.
1. Each bronchiole narrows into several alveolar ducts, each of which opens into a grapelike cluster of alveoli.
2. Surrounding each cluster of alveoli is a vast network of capillaries.
3. Gas exchange occurs through the thin walls of the alveoli and those of the neighboring capillaries by diffusion.
4. The alveoli must be inflated for their extensive surface area to be used efficiently, which is accomplished by a mixture of phospholipid molecules called human lung surfactant.

165

5. The lack of surfactant in the newborn is called infantile respiratory distress syndrome.
E. The lungs.
   1. The bronchial tree and the alveoli are housed within the paired lungs.
   2. The left lung is divided into two regions, called "lobes," and the right is divided into three.

IV. Mechanism of Breathing
A. Air moves between the atmosphere and the lungs in response to pressure gradients.
   1. The thoracic cavity is separated from the abdominal cavity by a broad sheet of muscle, the diaphragm.
   2. Changing the size of the thoracic cavity creates pressure changes that move air in and out of the lungs.
   3. Air is drawn into the lungs when the size of the thoracic cavity increases due to the contraction of the muscles of the rib cage and the diaphragm a process called inspiration, or inhalation.
      a) Rib muscles contract and pull the rib cage upward and forward.
      b) Meanwhile, the contraction of the diaphragm elongates the thoracic cavity.
      c) The increase in the size of the thoracic cavity causes a drop in the pressure between the lungs and the outer wall of the cavity (this area is called the plural cavity), and air is drawn in (fig. 23.12).
   4. When the muscles of the rib cage and the diaphragm relax, the thoracic cavity is decreased in size, causing expiration, or exhalation; when pressure within the lungs exceeds atmospheric pressure, air is pushed out.
B. Not all of the respiratory tract actively participates in gas exchange.

   1. The part that is not used in gas exchange is called dead space.
   2. The air in the bottom third of the lungs is not exchanged with each breath; because most of this air remains in the lungs it is called residual air.
   3. The amount of air inhaled or exhaled during a normal breath is called the tidal volume.
   4. Vital capacity is the maximal amount of air that can be moved in and out of the lungs during forceful breathing.

V. Transport of Gases
A. Oxygen.
   1. Oxygen is carried by the blood throughout the body and is bound to hemoglobin in the red blood cells.
   2. Each hemoglobin molecule has four subunits, which include a protein chain, the globin, and a heme group (fig. 14.9).
   3. The heme group includes an iron atom that bonds to the oxygen; all four of the protein chains have a heme group and, therefore, each hemoglobin molecule can carry four molecules of oxygen.
   4. The compound formed when hemoglobin binds with oxygen is called oxyhemoglobin.
   5. As cells use oxygen, they lower the partial pressure of oxygen in the blood there and oxygen is released from the hemoglobin.
   6. Exercise increases the acidity in the blood, and the hemoglobin releases more of its oxygen load to the cells.
   7. Active tissues also produce heat as a by-product of energy metabolism, and blood flowing through warmer parts of the body releases more of its oxygen.

B. Carbon dioxide.
   1. A small amount of carbon dioxide, about 7%, is transported dissolved in the plasma.
   2. Slightly less than a quarter (23%) of the transported carbon dioxide is carried by hemoglobin molecules, where it is bonded to the protein portion.
   3. About 70% of the carbon dioxide is transported as a bicarbonate ion dissolved in the plasma.
      a) The bicarbonate ion forms when carbon dioxide produced by the cells diffuses into the blood and into the red blood cells.
      b) The carbon dioxide forms carbonic acid, which dissociates to form hydrogen ions(H^+) and bicarbonate ions (HCO_3^-).
      c) The H^+ combines with hemoglobin, keeping the pH of the blood fairly stable.
C. Carbon monoxide poisoning.
   1. This is a by-product of combustion.
   2. Carbon monoxide is poisonous because it binds to hemoglobin more readily than does oxygen.
   3. It is less likely to leave the hemoglobin molecule so the unsuspecting person begins to experience symptoms of oxygen deprivation.

VI. Control of Respiration
A. Twelve times a minute is considered steady breathing and is controlled by a rhythmicity center in the brain's medulla (fig. 23.14).
   1. The impulses sent by these neurons contract muscles used in inhalation, increasing the size of the chest cavity and drawing air into the lungs.
   2. The neurons in the inspiratory center rest and you passively exhale.
   3. During strenuous activity, the respiratory cycle is changed so that exhalation is no longer a passive event.
   4. The inspiratory center also sends impulses to a part of the medulla's rhythmicity center called the expiratory center.
B. The apneustic center allows the forceful inspiration of a deep breath.
C. In the cerebral cortex you can control breathing for speech and voluntarily pant like a dog, sigh, and hold your breath.
D. The most important factor in regulating breathing rate is the blood CO_2 level, or more precisely, the number of H^+ formed when the CO_2 goes into solution and forms carbonic acid.
E. Chemoreceptors in the medulla, and to some extent those in the aortic bodies and carotid bodies, monitor the blood CO_2 levels.
F. When the blood CO_2 level increases, thereby raising the H^+ concentration, chemosensitive areas are stimulated and send messages to the inspiratory center in the medulla and the increased breathing rate, called hyperventilation, decreases the level of CO_2.

VII. The Unhealthy Respiratory System
A. The common cold.
   1. The most common respiratory problem is the viral infection known as the cold and can be due to any one of a family of viruses, the rhinoviruses.
   2. Many of these viruses can alter their arrangement of surface molecules, which hinders development of a vaccine that relies on the immune system's recognition of viral surfaces.
   3. Cold remedies.
      a) Decongestants shrink nasal membranes and ease breathing.
      b) Antihistamines decrease mucus secretion and combat watery eyes and sneezing.
      c) Analgesics relieve pain and discomfort.

B. Influenza.
   1. Although the flu may start out with a stuffy nose and therefore feel like a cold, the onset of fever, joint aches, fatigue, and weakness indicates the flu.
   2. Flu is viral in origin, and the causative viruses change their surface characteristics so frequently that vaccines cannot be developed fast enough to keep up with them.
C. Bronchitis.
   1. Bronchitis is an inflammation of the mucous membrane of the bronchi and can be caused by viruses or bacteria.
   2. Bronchitis can be acute or chronic.
      a) Acute bronchitis often follows a cold and may lower the body's resistance to bacteria.
      b) Chronic bronchitis is a more serious long-term or recurring condition with symptoms similar to acute bronchitis.
      c) The poor drainage predisposes the patient to lung infections such as pneumonia, which can be fatal, and to additional degenerative changes in the lungs, such as emphysema.
D. Emphysema.
   1. The alveoli become overinflated and burst and create fewer and larger alveoli (fig. 23.17).
   2. The surface area available for gas exchange decreases.
   3. The volume of residual air in the lungs increases, and more forceful inspiration is necessary to ventilate the lungs adequately.
   4. The main symptom of emphysema is shortness of breath.
E. Cystic fibrosis.
   1. The protein for cystic fibrosis lacks only one amino acid.
   2. In the lungs mucus is not sufficiently mixed with water, causing lung congestion that encourages infection, and in the pancreas, thick mucus blocks release of digestive enzymes and the person becomes very thin.
F. Tuberculosis.
   1. TB is caused by Mycobacterium tuberculosis bacteria inhaled into the lungs.
   2. This is highly contagious and as the bacteria multiply, they cause a small area of inflammation.
   3. The body resists the bacteria by killing them or sealing them in a fibrous connective tissue capsule called a tubercle.
   4. The formation of tubercles slows the spread of the disease, but it does not actually kill the bacteria.

LEARNING OBJECTIVES

After reading this chapter, the student should be able to answer these questions:

1. Why do we require oxygen?
2. What kinds of structures have different species evolved to respire?
3. How do we breathe?
4. How are gases exchanged between the respiratory system, the circulatory system, and the body's tissues?
5. How is breathing rate controlled?
6. What are some disorders of the respiratory system?

ANSWERS TO TEXT QUESTIONS

1. The color would be pink because of the lower level of airborne pollutants such as smoke.
2. In small animals or those with a flat body form, diffusion can effectively deliver oxygen to the cells. In larger animals diffusion is ineffective, and a circulatory system is necessary to transport oxygen from the respiratory membrane to the body cells. In other organisms the respiratory system consists of branching tracheal tubes that deliver air directly to the cellular level. In these organisms the circulatory

system functions for the transport of nutrients and waste products and is not involved in gas exchange. the respiratory membrane in large animals is many times larger than the body surface. This is accomplished in gills by branching and folding of the body wall and in lungs by extensive branching of the respiratory tree, ending in small sacs for gas exchange. More active animals require a very large respiratory surface and effective circulatory systems to deliver oxygen to the tissues. In aquatic environments the respiratory surface can be external because it is supported by the density of water. In terrestrial animals the respiratory surface is internal for support and to prevent drying out.

3. The pathway air follows is through the nose, pharynx, larynx, trachea, bronchi, bronchioles, alveolar ducts, alveoli, and respiratory membrane to the blood.

4. In the nose, hair acts as a filter for large particles suspended in air. Smaller particles are trapped in the mucus of the nose or in lower portions of the respiratory tract, where they are swept out by the action of cilia.

5. The muscles of the rib cage raise the ribs and sternum, and contraction of the domed diaphragm lowers it towards the abdomen, increasing the volume of the thorax. This lowers the pressure in the thorax compared to the environment, so air rushes into the lungs due to this pressure difference.

6. Oxygen binds to hemoglobin in the red blood cells forming oxyhemoglobin in the lungs. The blood returns to the heart through the pulmonary veins and is then pumped to the cells of the body through systemic circulation.

7. Normally hemoglobin releases 27% of the oxygen bound to it, but when the oxygen level in the tissues decreases because of metabolic activity, approximately three times more oxygen can be released. Metabolically produced acids increase the release of oxygen from hemoglobin (Bohr effect). The higher temperatures of active tissues also causes more oxygen to be released from hemoglobin.

8. Seventy percent of the generated carbon dioxide is transported as bicarbonate ion dissolved in the plasma. The enzyme carbonic anhydrase found in red blood cells catalyzes the reaction of carbon dioxide and water to form the bicarbonate ion. Carbon dioxide is also transported on the hemoglobin molecule bound to the protein component and simply dissolved in the plasma.

9. When at rest the inspiratory center in the medulla sends a burst of activity (2 seconds in duration) to the inspiratory muscles causing inspiration. The inspiratory center is quiet for the following 3 seconds, which allows for passive expiration. When activity increases, the inspiratory center causes inspiration and the expiratory center triggers forced expiration by initiating contraction in the muscles, which pull down the rib cage, actively forcing air from the lungs.

10. The most important factor is the concentration of carbon dioxide, or more precisely, the level of $H+$ ions in the blood. Chemoreceptors in the medulla, carotid, and aortic bodies respond to increasing $H+$, which stimulates the rate of respiration. Under unusual conditions, for example, high altitude, low oxygen levels in the blood can trigger an increase respiratory rate.

ANSWERS TO THE "TO THINK ABOUT" QUESTIONS

1. The lungs do not freeze at low air temperatures because the inspired air is warmed and moistened as it enters the nose, pharynx, and upper portion of the respiratory tree.

2. During mouth breathing, moisture is not added to the air from the nasal mucosa, so water vapor comes from the throat and upper respiratory tract, drying out these structures.

3. The epiglottis protects the respiratory tract from the entrance of food or liquids during swallowing. These materials pass through the pharynx on their way to the stomach. Swelling of this tissue could close off the glottis or opening to the larynx. This would decrease oxygen levels and cause difficulty in breathing and loss of

voice. The tracheostomy would have saved him by allowing air to enter the respiratory system through the trachea below the larynx.

4. Breath holding decreases the level of oxygen and increases the level of carbon dioxide. With low oxygen levels the brain is unable to function and the individual loses consciousness. The high carbon dioxide level stimulates the respiratory centers in the medulla to resume normal respiration.

5. Hyperventilation causes excess carbon dioxide to be given off, altering the respiratory pattern. Breathing in a paper bag restores the normal levels of carbon dioxide and a normal breathing pattern.

6. Not all individuals exposed to asbestos would know of its presence, so it seems reasonable to remove it from the workplace altogether rather than have people wear masks.

7. The endurance times were slow due to the low level of oxygen in the air at the relatively high altitude of Mexico City.

8. Mucus would accumulate in the lungs since the cilia are no longer active. The accumulation of mucus could block the movement of air in the respiratory tract or lead to lung infections such as pneumonia.

AUDIOVISUAL MATERIALS

Mechanisms of Breathing, Encyclopaedia Britannica Educational Corp., 11 min. 1954.

Respiration, Boston Broadcasters, MTI, 22 min., 1980.

Respiration, United World Films, 12 min., 1953.

Respiration in Man, Encyclopaedia Britannica Educational Corp., 26 min., 1968.

Respiratory System, Coronet Instructional Films, 15 min., 1980.

Smoking/Emphysema: A Fight for Breath, CRM McGraw-Hill, 12 min., 1975.

24 The Digestive System

KEY CONCEPTS

Macromolecules are too large to be absorbed by the circulatory system so they are hydrolyzed to micronutrients in the process of digestion.

Carbohydrates are digested to glucose, fats to fatty acids, and glycerol and proteins to amino acids.

Animals eat for a supply of energy and building blocks for growth, repair, and maintenance of tissues.

The human gastrointestinal tract is designed to secrete mucus, water, and enzymes for digestion.

Muscles pulverize food mechanically and keep it moving down the tract.

Nervous and hormonal responses to the presence of food regulate digestion.

The circulatory system maintains the vitality of the organs while picking up the products of digestion.

Mechanical digestion increases the surface area so chemicals can attack bonds in the macromolecules.

Starch digestion begins in the mouth and is completed in the small intestine.

Digestion in the stomach involves mainly the breakdown of protein by pepsin under pH conditions provided by hydrochloric acid.

The gastrointestinal tract protects itself from digestion by secreting mucus, replacing its cell lining frequently, and secreting enzymes usually only when food is present.

Digestion is completed in the small intestine, and most of the nutrients are absorbed into the circulatory system.

Water, minerals, and salts are absorbed by the large intestine; normal bacteria that reside there extract needed vitamins from undigested food.

The pancreas produces digestive enzymes and sodium bicarbonate, which are secreted into the small intestine.

The liver supplies bile, which emulsifies fats.

Bile is stored in the gallbladder.

KEY TERMS

anal canal	hydrochloric acid
anus	ileum
appendix	intracellular
bile	digestion
body	jaundice
bolus	jejunum
carbohydrase	lactase
cardia	lacteal
cecum	lactose intolerance
cementum	lipases
cholecystokinin	liver
(CCK)	microvilli
chyme	pancreas
chymotrypsin	pepsin
colon	pepsinogen
colorectal cancer	peptidase
colostomy	periodontal membrane
constipation	peristalsis
crown	pharynx
dentine	pulp
diarrhea	pyloric stenosis
diverticulosis	pylorus
duodenum	rectum
enamel	root
esophagus	rugae
extracellular	saliva
digestion	salivary amylase
fundus	salivary glands
gallbladder	secretin
gastric juice	segmentation
gastric lipase	sphincter
gastrin	trypsin
gastrointestinal	ulcer
tract	ulcerative colitis
heartburn	villi
hemocault test	
hepatitis	

CHAPTER OUTLINE

I.   Eating and Digesting
      A.   The body dismantles nutrient molecules into smaller molecules—proteins to amino acids,

complex carbohydrates (starch) to simple carbohydrates (sugar), and fats to fatty acids and glycerol.
- B. Only small molecules can enter cells lining the digestive tract, from which they enter the circulation.
- C. Once inside cells, nutrient molecules are further broken down to release the energy in their chemical bonds, or they are built up into human versions of macromolecules.
- D. Digestion is chemically carried out by hydrolytic (water-splitting) enzymes (fig. 24.1).

II. Types of Digestive Systems
- A. Digestion occurs within a specialized compartment.
- B. Intracellular digestion.
  1. Lower invertebrates, such as rotifers, brachiopods, and simple sea dwellers, separate digestion from other activities by sequestering food in food vacuoles.
  2. The lysosomes in the cells of higher organisms carry out intracellular digestion to break down fats and carbohydrates of worn-out cell parts.
- C. Extracellular digestion.
  1. Food particles are dismantled by hydrolytic enzymes in a cavity.
  2. When nutrient molecules are small enough, they enter cells lining the cavity.
  3. Hydra and the flatworm planaria have digestive systems with a single opening.
  4. A major advantage of a two-opening digestive system is that regions of the tube can become specialized for different functions: breaking food into smaller particles, storage, chemical digestion, and absorption.

III. An Overview of the Human Digestive System
- A. The gastrointestinal (stomach-intestine) tract includes the mouth, esophagus, stomach, small intestine, large intestine, and accessory structures.
- B. These structures aid the breakdown of food either mechanically (the teeth and tongue) or chemically by secreting digestive enzymes (fig. 24.3).
- C. The digestive organs and glands are supported by the mesentery, which is an epithelial sheet reinforced by connective tissue.
- D. The mucus lining protects underlying cells from rough materials in food and from digestive enzymes.
- E. Waves of contraction called peristalsis propel the food through the digestive system and mix it with enzymes (fig. 24.4).

IV. Structures of the Human Digestive System
- A. The mouth and the esophagus.
  1. Chemical digestion begins with secretion of saliva from three pairs of salivary glands near the mouth, which produce the enzyme salivary amylase, which breaks down starch.
  2. Mechanical digestion (fig. 24.6).
     - a) Teeth tear food into small pieces, increasing the surface area upon which chemical digestion can begin.
     - b) Tooth structure.
       (1) A thick enamel covers teeth.
       (2) Beneath the enamel is the bonelike dentine, and then the soft inner pulp, which contains connective tissue, blood vessels, and nerves.
       (3) two layers on the outside of the tooth, the periodontal membrane and the cementum, anchor the tooth to the gum and jawbone.

(4) the visible part of a tooth is the crown, and the part below the surface is the root.

3. Swallowing.
   a) the tongue rolls chewed food into a lump called a bolus.
   b) as the bolus of food is swallowed, it passes first through the pharynx (the throat) and then to the esophagus, a muscular tube leading to the stomach.

B. The stomach.
   1. Its functions are storage, some digestion, and regulation of the flow of food into the small intestine (fig. 24.7).
   2. The stomach can expand due to folds in the stomach's mucosa, called rugae.
   3. Entry to and exit from the stomach are controlled by muscular rings called sphincters.
   4. the stomach has four regions: the entry of the stomach is the cardia, the domelike top is the fundus, the midsection is the body, and the bottom is the pylorus.
   5. Churning action mixes the food with gastric juice secreted by stomach cells to produce a semifluid mass called chyme.
   6. Gastric juice.
      a) The hydrochloric acid creates a highly acidic environment, which activates an enzyme called pepsin from its precursor, pepsinogen (fig. 24.9).
      b) Another stomach enzyme is gastric lipase, which splits butterfat molecules found in milk.
      c) Secretion of gastric juice is regulated by nerves and hormones.
      d) Gastric cells release a hormone, gastrin, that

causes gastric juice to be secreted (fig. 24.10).
      e) Protection of the stomach lining.
         (1) Most gastric juice is not secreted until food is presented for it to work on.
         (2) Some of the stomach cells secrete mucus, which coats and protects the stomach lining from the corrosive action of gastric juice.
         (3) Pepsin is produced in an inactive form, pepsinogen, and cannot digest protein until hydrochloric acid is present.

   7. Some water and salts, a few drugs (such as aspirin), and alcohol can be absorbed.
   8. When the upper part of the small intestine is full of chyme, receptor cells on the outside of the intestine are stimulated, which in turn temporarily shuts the pyloric sphincter.
   9. When the sphincter at the upper end of the stomach is unable to hold back the stomach contents, acidic chyme will squeeze out of the esophagus, a process known as heartburn.
   10. Pyloric stenosis is when children are born with an enlarged circular muscle in the pyloric sphincter, which as a result cannot open properly.

C. The small intestine.
   1. The duodenum is the first 10 inches, the next two-thirds is the jejunum, and the remainder is the ileum.
   2. Segmentation is localized muscle contractions (fig. 24.12).
   3. Chemical digestion.
      a) Protein digestion is accomplished by peptidase enzymes manufactured by

173

intestinal cells and by the enzymes trypsin and chymotrypsin, which are made in the pancreas.

b) Fat digestion.
   (1) Lipases are the enzymes that chemically digest fats.
   (2) Bile, produced by the liver, is needed to emulsify fat, breaking it into droplets small enough to remain in suspension, and works as a detergent (fig. 24.13).
   (3) Pancreatic lipase can chemically digest fats.
   (4) Most fats in our diet are triglycerides, which are digested into fatty acids and glycerol.

c) Carbohydrate digestion.
   (1) The small intestine produces carbohydrase enzymes, which chemically break down certain disaccharides into specific monosaccharides.
   (2) Lactose intolerance results from the absence of the enzyme lactase in the small intestine, causing abdominal pain, gas, diarrhea, bloating, and cramps.

4. Regulation.
   a) The hormone secretin triggers the release of bicarbonate from the pancreas and stimulates the liver to secrete bile.
   b) Cholecystokinin (CCK) signals the release of substances needed for fat digestion.
      (1) It stimulates bile secretion and contraction of the gallbladder.

      (2) It releases pancreatic enzymes, including lipase, into the small intestine.
   c) Other glands in the small intestine secrete mucus, which protects the intestinal wall from digestive juices and assists in neutralizing stomach acid.

5. Absorption of nutrients.
   a) there is a maximization of surface area in the small intestine (fig. 24.15).
      (1) Due to additional folds, there are about 6 million tiny projections called villi (fig. 24.16).
      (2) Villi bristle with projections of their own called microvilli, which increase surface area.
   b) The capillary network surrounds a lymph vessel, called a lacteal, that absorbs fat molecules.
   c) Monosaccharides are absorbed by facilitated diffusion, and active transport and amino acids are taken up by active transport.
   d) The products of fat digestion move into cells by passive diffusion, which is possible because they are soluble in the lipids of the intestinal cell membranes.

D. The large intestine.
   1. A pouch at the start of the large intestine is the cecum and dangling from it is the appendix (fig. 24.18).
   2. The large intestine absorbs most of the remaining water, plus salts and some minerals, from chyme, producing a solid or semisolid mass known as feces.

3. The "intestinal flora" produce vitamins B_1, B_2, B_6, B_{12}, K, folic acid, and biotin and breaks down bile and foreign chemical such as those in drugs.
E. Associated glands and organs.
   1. Pancreas (fig. 24.20).
      a) This gland contains fluid with trypsin and chymotrypsin to digest polypeptides, pancreatic amylase to digest carbohydrates, pancreatic lipase to further break down emulsified fats, and nucleases to degrade DNA and RNA.
      b) Sodium bicarbonate neutralizes the acidity of chyme.
   2. Liver.
      a) Its functions include detoxification of harmful substances in the blood, storage of glycogen and fat-soluble vitamins, synthesis of blood proteins, and monitoring the blood-sugar level.
      b) Bile is stored in the gallbladder and consists of pigments derived from the breakdown products of hemoglobin, salts, and cholesterol.

LEARNING OBJECTIVES

After reading this chapter, the student should be able to answer these questions:

1. Why do animals need to eat?
2. What are some different types of digestive systems?
3. How do the structures and biochemicals of the human digestive system mechanically and chemically degrade food into nutrient molecules small enough to be absorbed and enter the bloodstream?
4. How are the activities of the human digestive system regulated?

5. What are the building blocks into which proteins, fats, and carbohydrates are broken down?

ANSWERS TO TEXT QUESTIONS

1. The circulatory system picks up nutrients from the digestive tract as well as servicing the needs of the digestive organs. Muscle action results in mechanical digestion and also keeps the food moving through the digestive tract. Nerves control muscle action and the release of digestive hormones. Hormones in turn cause the release of gastric and intestinal juices.
2. Starches break down into the disaccharides, fructose, sucrose, and maltose, and the monosaccharide glucose.
3. Proteins are broken down into amino acids, while fats are digested into glycerol and fatty acids.
   a. Heartburn is caused by the improper function of the pyloric valve, resulting in the backup of chyme into the esophagus.
   b. Diarrhea may be caused by the buildup of harmful bacteria in the colon.
   c. Hepatitis is a viral infection of the liver that causes overwhelming fatigue, jaundice, dark urine, and pale-colored feces.
   d. Ulcers may be caused by excessive hydrochloric acid or insufficient mucus.
4. Mechanical digestion involves the physical breakdown of food into smaller particles by chewing and muscle contractions, whereas chemical digestion involves the breakdown of food by the secretion of digestive enzymes from the salivary glands and pancreas. Mechanical digestion increases the surface area so enzymes and water can attack chemical bonds.
5. The gastrointestinal tract protects itself from digestion by secreting mucus. It also has a high turnover of epithelial cells and secretes enzymes only in the presence of food.
6. The rugae of the stomach and the villi of the small intestine greatly increase the surface of these two organs. This allows

for maximum chemical and mechanical digestion and for absorption.

7. The normal flora of bacteria in the large intestine assists the individual by extracting certain vitamins from undigested nutrients. If the bacteria are destroyed by antibodies, harmful bacteria may start to grow, which can cause diarrhea.

ANSWERS TO THE "TO THINK ABOUT" QUESTIONS

1. Since all foods contain at least some of each macronutrient, diets can widely vary.
2. Without amylase for the breakdown of starch, the pasta was not digested. Since pasta is mostly starch, the bulk of it would pass into the colon undigested.
3. Since the meat would still be in the form of protein, it would be too large to leave the circulatory system.
4. It is likely that the children were exposed to hepatitis A through food preparation by a day-care worker.
5. The most sophisticated organ or gland of the digestive system is the small intestine. Most digestion and absorption of nutrients into the circulatory system takes place here.

AUDIOVISUAL MATERIALS

Digestion, Body Work Series, Boston Broadcasters, 22 min., 1979.

Digestion, Open University Educational System Media Guild, 25 min., 1980.

Digestive System, Coronet Instructional Films, 15 min., 1980.

The Digestive System, Encyclopaedia Britannica Educational Corp., 16 min., 1980.

The Digestive Tract, University of Texas at Dallas, 28 min., 1978.

25 Nutrition

KEY CONCEPTS

Animals require nutrients to promote growth and to maintain and repair tissues of the body.

Nutrition is the study of nutrients and what happens to them in the body.

The macronutrients carbohydrates, fats, and proteins provide energy for the body.

Micronutrients include vitamins and minerals.

Recommended Dietary Allowances (RDA) indicate the mean number of kilocalories consumed daily and the amounts of vitamins and minerals that are sufficient for most adults.

Nutrient deficiencies may be caused by diet or metabolic disorders.

Starvation occurs under a variety of conditions and is the result of inefficient calories to maintain the normal function of the body.

Obesity is defined as being 20% over ideal body weight, and it increases the risk of disease.

KEY TERMS

anorexia nervosa
bomb calorimeter
bulimia
carbohydrate loading
energy nutrients
energy RDA
essential nutrients
exchange system
food group plan
kilocalories
kwashiorkor
lean tissue
macronutrients

marasmus
micronutrients
nonessential nutrients
nutrient
nutrient dense
obese
primary nutrient
    deficiencies
Recommended Dietary
    Allowances (RDA)
secondary nutrient
    deficiencies
subclinical

CHAPTER OUTLINE

I. Human Nutrition—From a Prehistoric Meal to Fast Food
    A. A substance obtained from food and used in the body to promote growth, maintenance, and repair is a nutrient.
    B. Nutrition is the study of nutrients and their fate in the body, including their ingestion, digestion, absorption, transport, metabolism, interaction, storage, and excretion.

II. The Nutrients
    A. Macronutrients include carbohydrates, proteins, and fats.
        1. Complex carbohydrates and disaccharides are broken down to glucose during digestion, and the glucose is metabolized to yield energy in the form of ATP.
        2. Energy can also be derived from proteins and fats under certain conditions, but these nutrients have other major functions. (tables 25.2 through 25.4).
    B. Micronutrients are vitamins and minerals needed in very small amounts.
    C. Water is also a vital part of the diet.
    D. Essential nutrients are those that must be obtained from the diet because the body cannot synthesize them.
    E. Nonessential nutrients, such as 11 of the 20 amino acids, are found in foods.
    F. The amount of energy that a nutrient provides is measured in units called kilocalories, which are determined by burning food in a bomb calorimeter. (fig. 25.2).
    G. Good nutrition is a matter of balance; taking in more kilocalories than are expended in basal metabolism can lead to overweight, and not consuming sufficient kilocalories to support the activities of life can lead to underweight.
    H. The RDAs (United States Recommended Daily Allowances) are directed at the "generalized human adult," which does not include children, the elderly, and women who are pregnant or breast-feeding.

III. Planning a Balanced Diet
  A. The food group plan.
    1. To make nutrient watching easier, the United States government in the 1940s introduced the food group plan.
    2. The groups include meat and meat substitutes, milk and milk products, fruits and vegetables, and grains.
  B. Dietary guidelines.
    1. Eat a variety of foods.
    2. Maintain a desirable weight.
    3. Cut down on fats and cholesterol and eat more fish and beans, trim the fat from meat, broil or bake foods instead of frying them, and eat less oil and butter.
    4. Increase intake of fiber and starch by eating more fruits and vegetables, grains, beans, peas, and nuts.
    5. Reduce sugar intake, because sugar contributes to tooth decay.
    6. Lower intake of sodium (found in many processed foods), which can contribute to the development of hypertension.
  C. The exchange system.
    1. A more specific way than the food group plan to select nutritious foods is the exchange system.
    2. Foods are grouped according to their proportions of carbohydrate, protein, and fat.
    3. The food exchange lists include only foods that are nutrient dense—that is, foods that offer a maximum amount of nutrients with minimum number of kilocalories.
IV. Nutrient Deficiencies
  A. Primary nutrient efficiencies are those caused by the diet.
  B. Secondary nutrient deficiencies.
    1. These result from an inborn metabolic condition in which a particular nutrient is not absorbed sufficiently in the small or large intestine, is excreted too readily, or is destroyed.
    2. Menkes disease is where copper is not adequately absorbed from food in the small intestine.
    3. Micronutrient deficiencies develop slowly as in the case of anemia, where not enough iron in the diet results in a low number of red blood cells.
    4. A compulsive disorder that may result from mineral deficiency is pica, in which people consume huge amounts of substances such as ice chips, dirt, sand, laundry starch, and clay and plaster.
V. Starvation
  A. A healthy human can stay alive for 50 to 70 days without food.
    1. After only 1 day, the body's reserves of sugar and starch are gone.
    2. The body extracts energy from fat and then muscle protein.
    3. Gradually, metabolism slows to conserve energy, blood pressure drops, and the pulse slows.
  B. Marasmus and Kwashiorkor.
    1. Profound nutrient deficiency is called marasmus (fig. 25.6a).
    2. Protein starvation is called Kwashiorkor and typically appears in a child who has recently been weaned from the breast (fig. 25.6b).
  C. Anorexia nervosa (figs. 25.7 and 25.8).
    1. When starvation is self-imposed because the person is terrified of any weight gain.
    2. The individual develops low blood pressure, a slowed or irregular heartbeat, constipation, and constant chilliness, and she stops menstruating as her body fat level plunges.
    3. The anorexic's eating behavior is ritualized.
    4. Anorexia nervosa has no known physical cause.
  D. Bulimia.
    1. A person suffering from bulimia is often of normal weight.

2. They rid their bodies of the thousands of extra kilocalories by vomiting, taking laxatives, or exercising frantically.
VI. Overweight
    A. Obese is defined as 20% above "ideal" weight based on population statistics considering age, sex, and build; an obese person is at higher risk for diabetes, digestive disorders, heart disease, kidney failure, hypertension, stroke, and cancers of the female reproductive organs and the gallbladder.
    B. The additional body weight is primarily muscle, bone, connective tissue, and water, known as lean tissue.
    C. Obesity is caused by both heredity and the environment; we inherit genes that control metabolism, but studies on twins show that 70% of weight is influenced by genes and 30% by environment.
    D. A safe goal of weight loss is 1 pound of fat per week; this can be lost by eating 500 kilocalories less per day or exercising off 500 kilocalories each day.
    E. A rule of thumb is to leave the proportion of protein kilocalories about the same or slightly increased, cut fat kilocalories in half, and cut carbohydrates by a third.
VII. Gaining Weight
    A. One weight-gain diet for a convalescent patient offers 3,000 to 3,500 kilocalories per day, with approximately 14% of kilocalories as protein, 24% as fat, and 62% as carbohydrate.
    B. An infant also needs to gain weight rapidly.

LEARNING OBJECTIVES

After reading this chapter, the student should be able to answer the following questions:

1. What are the nutrients for humans?
2. How is the energy content of foods measured?
3. What are some ways to plan a diet?

4. What forms of starvation are seen in developing nations?
5. What are eating disorders?
6. Why is being overweight dangerous?

ANSWERS TO TEXT QUESTIONS

1. A nutrient is a substance obtained from food and used in the body to promote growth, maintenance, and repair.
2. A kilocalorie is equal to the amount of energy required to raise 1 gram of water 1 degree Celsius.
3. Micronutrients are required in such small quantities that given ample calories of food, one is likely to meet the RDA for minerals and vitamins.
4. Margarine could be substituted for butter, skim milk for whole milk, lean meats instead of fatty meats, and reduce the number of eggs per week.
5.   a. Determine your recommended energy intake per day from table 25.9 For example a twenty year old make requires on the average 2900 calories. to lose one half pound of fat per week he should reduce his daily caloric intake by 250 calories and exercise.
    b. 30% = fat; 12% = protein; 48% = complex carbohydrates; 10% = sugar
    c. See page 37.
6. Human milk is rich in the lipids needed for rapid brain growth and contains the protein lactoferrin, which binds with microorganisms. It also contains a biochemical called bifidus factor, which encourages the growth of beneficial bacterial in the digestive tract, which keeps out harmful bacteria.
7. The teenager suffers from a primary nutrient deficiency and the other person from secondary nutrient deficiency.
8. The teenager suffers from a primary nutrient deficiency and the other person from secondary nutrient deficiency.

ANSWERS TO THE "TO THINK ABOUT" QUESTIONS

1. a. The bikini diet of less than 1,200 calories per day is a starvation diet and is not recommended.
   b. The Cambridge diet is below 1,200 calories per day and also provides little variety in the kind of food taken in.
   c. The Beverly Hills diet is not balanced in the intake of carbohydrates, fats, and protein per day. Also it is restrictive in what can be eaten at certain times, which makes it more difficult to follow.
   d. The Weight Loss Clinic diet is a starvation diet and is high in protein and low in carbohydrates and fats.
   e. The macrobiotic diet is high in carbohydrates and low in protein and fat. The food choices are not very palatable.
   f. The No Aging diet is low in fats and carbohydrates and high in protein. Also it is very restrictive in what can be eaten.
2. RDAs are based on the nutritional needs of healthy adults.
3. Vitamins do not provide calories and so cannot supply energy.
4. The players are taking in high amounts of protein and fat when they should be taking in high amounts of carbohydrates to provide a ready supply of energy to the muscles.
5. Anorexia nervosa is a psychological disorder found most often in young females with low self-esteem. Since there is no physical cause, psychotherapy is in order.
6. Low caloric intake robs the body of building blocks needed by the immune system, so mounting an immune response against infection is inhibited.

AUDIOVISUAL MATERIALS

Dieting, Washington University, 20 min.
Dieting: The Danger Point, CRM McGraw-Hill, 20 min., 1979.
Food Revolution, CRM McGraw-Hill, 26 min., 1968.
For Tomorrow We Shall Diet, Churchill Films, 24 min., 1976.
Look Before You Eat, Washington University, 22 min., 1979.
Nitrite Saga, Washington University, 28 min.
Nutrition: What's In It for Me?, Document Associates, 24 min.
The Sugar Film, Pyramid FIlms, 27 min., 1980.
To Catch a Meal, Association Films, 12 min.
Vitamins: What Do They Do, Indiana University, 21 min., 1979.

26 Homeostasis

KEY CONCEPTS

Homeostasis is the ability to maintain constant conditions in the body, such as temperature, fluid balance, and chemistry.

Temperature regulation is important because biochemicals, including proteins, work optimally at certain temperatures.

Ectotherms regulate body temperature primarily by their behavior, whereas endotherms do so by utilizing metabolic energy to generate heat.

Many organisms use both ectothermic and endothermic strategies.

Some species maintain a constant body temperature at all times, some alter their body temperature at certain times, and some can alter the temperature only in certain body parts.

Humans retain heat by shivering, vasoconstriction, and increased metabolism, and they cool through the evaporation of sweat.

Body temperature is controlled by the hypothalamus, which elicits appropriate physiological responses.

Nitrogenous wastes form from the metabolic breakdown of proteins and nucleic acids.

The amino groups of amino acids gain a hydrogen ion to form ammonia, which is excreted in dilute form by some aquatic animals, but it is converted to urea by amphibians and mammals and to uric acid by birds, reptiles, and insects.

Water gain must be balanced against water loss to keep tissues sufficiently hydrated.

The human excretory system consists of the kidneys, ureters, urinary bladder, and urethra.

The kidney is built of microscopic nephrons, and the urine formed along the nephrons drains into the innermost kidney region, the pelvis.

The kidney extracts wastes and toxins from the blood and recycles valuable substances by filtering, reabsorbing, and then secreting.

Urine formation begins as fluid, and small molecules cross from the glomerulus into the Bowman's capsule by filtration.

Nutrient substances are reabsorbed from the renal tubule back into the peritubular capillaries.

Waste products that are not filtered are secreted into the renal tubules from the peritubular capillaries.

The loop of Henle creates an osmotic gradient from the retention of water and concentration of urine.

The blood's solute concentration and volume are regulated by osmoreceptor cells in the hypothalamus, which control secretion of ADH, which acts on the collecting duct.

Aldosterone controls the sodium ion concentration and indirectly the blood volume and pressure.

Renin, a hormone released by special cells in the afferent arterioles, controls aldosterone secretion.

afferent arteriole
aldosterone
ammonia
antidiuretic
    hormone (ADH)
ascending limb
Bowman's capsule
cortex
countercurrent
    multiplier
    system
descending limb
distal convoluted
    tubule
ectotherms
efferent arteriole
endotherms
extracorporeal
    shock wave
    lithotripsy
glomerular filtrate
glomerulus
gout

homeostasis
kidney failure
kidneys
kidney stones
loop of Henle
medulla
nephrons
pelvis
peritubular
    capillaries
piloerection
proximal
    convoluted
    tubule
renal tubule
sphincters
urea
ureter
urethra
uric acid
urinary bladder
urinary tract
    infection

CHAPTER OUTLINE

I.   The ability to maintain constancy of body temperature, fluid balance, and chemistry is termed homeostasis.

II. Temperature Regulation
     A. If cellular temperatures vary, enzymes function less efficiently and vital biochemical reactions occur more slowly.
     B. Extreme temperatures alter biological molecules.
     C. Ectothermy and endothermy.
         1. Ectotherms.
             a) They lose or gain heat to their surroundings by moving into areas where the temperature is suitable.
             b) When an ectotherm cannot warm itself, activity slows and the animal seeks shelter, offering some safety from predation.

             c) An ectotherm thus regulates its body temperature by its behavior.
         2. Endotherm.
             a) Their source of heat is internal, that is, the source is its own metabolic heat.
             b) The metabolic rate is generally five times that of an ectotherm of similar size and body temperature.
             c) Layers of insulation, feathers, or fur help retain body heat.
         3. Ectothermic and endothermic ways of life each have advantages and disadvantages.
             a) The ectotherm is dependent upon a continuous ability to escape or take advantage of the environment.
             b) An endotherm is able to maintain body heat even in the middle of the night.
             c) The endotherm must eat large amounts of food compared to the ectotherm, and 80% of the energy from that food is used to maintain body temperature (fig. 26.1).
                 (1) In some species, body temperature fluctuates with the temperature of their surroundings, peaking when the animal is active on a hot sunny day but dropping quite low on a cold night.
                 (2) Other animals seem to compromise, maintaining a constant body temperature most of the time but sometimes varying it or doing so only in a part of the body.
                 (3) Hummingbirds at night, when food is

no longer being gathered, enter a sleeplike state called torpor in which the body temperature falls toward that of the surroundings.

    (4) Some large fish such as the mako shark and tuna conserve energy by heating only parts of the body.

D. Temperature regulation in humans.
1. We consciously control our temperature by dressing appropriately for weather conditions, seeking comfortable locations, or altering the environment by heating or cooling the air.
2. We also have physiological responses that adjust body temperature.
    a) Human response to cold is shivering.
    b) Other musclelike cells in the skin also contract in the cold, forming goose bumps and making our hair "stand on end."
    c) Finally, when body temperature begins to drop, the hypothalamus in the brain triggers release of hormones that increase metabolic rate, thereby increasing heat production (fig. 26.3).
    d) A very dramatic response to sudden, extreme, and prolonged exposure to cold (hypothermia) is seen in young drowning victims when blood flow is temporarily rerouted so that vital functions are maintained for a short time.
    e) The human body cools itself by perspiring; this can upset the body's balance of water and salts (fig. 26.4).

III. Regulation of Body Fluids
A. Two important ways in which the composition of body fluids is regulated are by removal of nitrogenous wastes and osmoregulation (water balance).
B. Nitrogenous waste removal.
1. Three types of nitrogenous wastes are generated by protein destruction: ammonia, urea, and uric acid.
2. Ammonia is very toxic and must be excreted in a very dilute solution (fig. 26.5).
3. Birds, amphibians, and mammals, including humans, convert ammonia to urea as their primary nitrogenous waste.
4. Birds, reptiles, and insects convert much of their nitrogenous wastes to uric acid, which can be excreted in an almost solid form (fig. 26.6).
5. In humans, our uric acid comes from the breakdown of the purine bases of DNA and RNA.
6. Gout is when too much uric acid is produced and it accumulates in the joints, causing pain.
C. Osmoregulation (fig. 26.7).
1. The control of water and salt balance in the body.
2. If water loss reaches 10%, the person becomes deaf, delirious, and can no longer feel pain.
3. Water loss of 12% or more of the body weight is usually fatal because muscles, including those that control swallowing, cease to function; a person can die of thirst in only 3 days without water from food and drink.
4. The most important process in regulating water loss is excretion.

IV. The Human Excretory System—An Overview
A. The paired kidneys are the major organs responsible for both excretion and osmoregulation (fig. 26.8).

B. Urine is formed within each kidney and drains into a muscular tube called the ureter.

C. Waves of muscle contraction squeeze the urine along the ureters and squirt it into a saclike muscular urinary bladder.

D. Urine drains from the bladder and exits the body through the urethra.

E. The exit from the bladder is guarded by two rings of muscle, called sphincters.

F. Stretch receptors in the bladder send impulses to the spinal cord, which stimulate nerves that contract the bladder muscles, and these generate strong urges to urinate.

V. The Kidney

A. Structure.

1. Each kidney is packed with 1.3 million microscopic tubules called nephrons.

2. An individual nephron is built of a continuous renal tubule plus a blood supply, the peritubular capillaries, that entwines around it (fig. 26.9).

3. The outermost part of the kidney is called the cortex, and it is grainy in appearance.

4. The middle section, called the medulla, looks like it is composed of many aligned strings, and this region corresponds to the long sections of the renal tubules.

5. The innermost portion of the kidney, the pelvis, is where the urine produced from each nephron is collected before it leaves the kidneys to enter the ureter.

B. The entire blood supply courses through the kidney's blood vessels every 5 minutes, and most of the material that enters the renal tubules from the blood is reabsorbed back into the blood rather than excreted.

C. Nephrons extract wastes and recycle valuable nutrients and salts to the bloodstream.

VI. Activities Along the Nephron

A. Bowman's capsule.

1. Blood approaches a nephron in one of many branches of the renal artery known as afferent arterioles.

2. Each arteriole narrows into a ball of capillaries called the glomerulus.

3. These capillaries than come together to form an efferent arteriole; the glomerulus is surrounded by the cup-shaped end of the renal tubule, which is known as Bowman's capsule (figs. 26.10 and 26.11).

4. Anything that fits through the pores in the glomerulus and is not repelled by the charge on the membrane passes into Bowman's capsule.

5. Large molecules (such as plasma proteins) and formed elements of the blood (cells and platelets) remain behind in the bloodstream and exit the nephron in the efferent arteriole, which ultimately leads into the capillary network surrounding the remaining tubules of the nephron.

6. These capillaries then empty into a venule, which runs into the renal vein, which finally joins the inferior vena cava.

7. The material passing into the Bowman's capsule is called the glomerular filtrate.

B. The proximal convoluted tubule.

1. The glomerular filtrate passes from the Bowman's capsule into the proximal convoluted tubule.

2. Here, specialized cells actively reabsorb all the glucose and vitamins and about 75% of the amino acids and important ions.

C. The loop of Henle.

1. After the proximal convoluted tubule, the renal tubule of a mammal forms the loop of Henle, where water is conserved and the urine concentrated.

2. This consists of a descending limb and an ascending limb, and it dips into the medulla region of the kidney.

3. Differing permeabilities of the descending and ascending limbs create a situation in which the bottom of the loop sits in a salty brew, which forces water to diffuse out of the renal tubule and into the surrounding capillaries (fig. 26.13).

4. The movement of Na^+ and water between the limbs of the loop of Henle and the medullary space is called a countercurrent multiplier system.

   a) The "countercurrent" refers to the fact that fluid movement direction is opposite in the two limbs (down and then up).

   b) The effect is multiple because of both the active extrusion of $Na+$ from the ascending limb and the passage of $Na+$ into the descending limb.

D. The distal convoluted tubule.

1. The next region of the renal tubule, located in the kidney's cortex.

   a) Sodium is reabsorbed into the peritubular capillaries.

   b) The accumulation of $Na+$ outside the distal convoluted tubule stimulates water to diffuse passively out of the tubule into the capillaries.

2. The distal convoluted tubule also helps to maintain the pH.

   a) If the pH of the blood falls too low, the distal convoluted tubule can raise it by secreting $H+$ into the urine.

   b) A blood level that is too high can be lowered by inhibiting the secretion of hydrogen ions by the tubules.

E. The collecting duct: After adjustment by reabsorption and secretion, the filtrate is urine.

VII. Control of Kidney Function

A. The somatic pressure of the blood, that is , how concentrated the plasma solutes are, affects many cellular activities, particularly the exchange of materials between the cells and the blood.

B. The volume of blood is a factor determining blood pressure and, therefore, cardiovascular health.

C. Antidiuretic hormone (ADH).

1. When the blood plasma becomes too concentrated, increasing osmotic pressure, the osmoreceptor cells send impulses to the posterior pituitary gland in the brain, which then secretes (ADH).

2. This increases the water permeability of both the distal convoluted tubule and the collecting duct.

3. It also increases the reabsorption of water into the blood.

4. The distal convoluted tubule and collecting ducts become less permeable to water as more water is retained.

5. Ethyl alcohol stimulates urine production by decreasing production of ADH.

6. Aldosterone synthesized in the adrenal glands enhances the reabsorption of $Na+$ in the distal convoluted tubules, as well as in the salivary glands, sweat glands, and large intestine.

7. Renin is a response to lowered blood volume and pressure and initiates a series of chemical reactions that eventually boost aldosterone levels.

VIII. The Unhealthy Excretory System

A. Urinary tract infections.

1. These most often affect women.

2. The placement of the urethra near the vaginal and anal

openings provides an open pathway for bacteria to enter.

B. Kidney stones.
   1. These form when salts, usually calcium salts or uric acid, precipitate out of the newly formed urine and accumulate in the kidney tubules and pelvis (fig. 26.14).
   2. They can be caused by excessive milk consumption, infection, dehydration, hormonal problems, or the inherited metabolic disorder gout.
   3. They cause fever, chills, frequent urination, and kidney pain.

C. Kidney failure.
   1. This is the result of damaged renal tubules.
   2. One theory attributes kidney failure to elevated blood pressure in the glomerulus, which in itself could have a number of causes.
   3. High blood pressure strains the filtering capabilities of the glomerular capillaries.
   4. Malfunctioning kidneys lead to a rapid buildup of toxins in the blood, altered ion concentrations in the tissues, and water retention that causes painful swelling of tissues.
   5. A dialysis machine is used to pass a patient's blood through a tube that is separated from a balanced salt solution by an artificial semipermeable membrane.
   6. Newer forms of dialysis use the peritoneal membrane lining the body cavity.
      a) The balanced salt solution is placed into the patient's abdominal cavity, where waste products and excess water diffuse from the person's blood into the fluid.
      b) The fluid is suctioned out after about a half an hour.
   7. A harder-to-find alternative to dialysis is a kidney transplant, but because the body usually rejects tissues that it does not recognize as "self," this procedure is most successful when the donor is a blood relative.

LEARNING OBJECTIVES

After reading this chapter, the student should be able to answer these questions:

1. How are viruses different from cells?
2. How are the two basic types of cells (prokaryotic and eukaryotic) alike and how are they different?
3. How did compartmentalization and division of labor within the cell become ways to maintain functions as cells grew larger?
4. Which organelles carry out the processes of secretion, waste removal, inheritance, and obtaining energy? How are secretion and waste removal carried out in eukaryotic cells?
5. What are the four tissue types in humans? What are their functions?
6. How might eukaryotic cells have evolved from prokaryotic cells?

ANSWERS TO TEXT QUESTIONS

1. Homeostasis is the ability to maintain constancy of body temperature, fluid balance, and body chemistry. In other words, maintenance of a stable internal environment.
2. Iguana lizards are ectotherms and gain heat from outside sources. They depend upon their behaviors to maintain optimal body temperatures for life. Humans are endotherms with internal sources of heat production. We have five times the metabolic rate of equivalent ectotherms. Up to 80% of our metabolism is used to maintain internal body temperature at a constant level. Endotherms also use changes in behavior to maintain body temperature, such as seeking shade when too warm or basking in the sun when cold.
3. Water loss can have drastic effects on human health. A loss of 1% body weight causes dry mouth and thirst. If the loss reaches 10%, the person becomes deaf,

delirious, and no longer can feel pain. A water loss in excess of 12% body weight causes death because muscles can no longer function.

4. The nitrogenous wastes are ammonia from amino acids, urea from the conversion of ammonia, and uric acid from the breakdown of purine bases of nucleic acids.

5. The nephron is diagrammed in figure 26.12. Filtration occurs in Bowman's capsule, secretion in the distal convoluted tubule, and reabsorption in the proximal convoluted tubule.

6. The loop of Henle functions to conserve water and concentrate urine. The ascending and descending limbs of the loop have different permeabilities to salts and water, which create an osmotic gradient for water to move from the renal tubule into the peritubular capillaries. The ascending limb is impermeable to water and actively transports sodium ions into the medullary space. Water in the descending limb diffuses outward because of the high concentration of sodium ions in the medulla and then into the peritubular capillaries and back into circulation. Sodium also diffuses into the descending limb and travels down the loop back to the ascending limb, setting up a cycle of passive and active transport. This movement of sodium ions and water is the "countercurrent multiplier" system, which creates a very concentrated salt solution at the tip of the loop of Henle. The collecting ducts pass through this region and depending upon the presence or absence of ADH, additional water can be removed from the renal tubule, further concentrating the urine.

7. Food and drink affect the osmotic concentration of blood, which is filtered into the renal tubular system. Dilute solutions will pass through and be lost as dilute urine. Food or drink high in salt concentration will result in water conservation to maintain the normal osmotic concentration of the blood. In this case the urine will be concentrated to remove excess salts and conserve water.

8. Many diseases can be detected by examining the urine, including diabetes insipidus (large volume of urine), urinary tract infections (cloudy and foul-smelling urine), kidney stones, and gout (uric acid crystals in the urine).

9. Antidiuretic hormone (ADH) is released from the posterior pituitary gland when the osmotic concentration of the blood increases. This results in conservation of water, accompanied by a decrease in urinary volume. The collecting ducts become more permeable to water in the presence of ADH, which allows water to diffuse out of the tubule and back into the peritubular capillaries. Release of aldosterone from the adrenal glands responds to decreasing sodium concentrations in the blood or a decrease in blood pressure or volume. The hormone stimulates the kidney to reabsorb sodium ions, thereby retaining water osmotically, which increases the cardiovascular volume and raises blood pressure.

10. With loss of kidney function there is a rapid buildup of toxins in the blood, such as ammonia, urea, and uric acid. In addition, the ion concentration of the blood is altered and water retention causes painful swelling of the tissues. Death results in a short period of time unless the composition of the body fluids is regulated within narrow limits.

ANSWERS TO THE "TO THINK ABOUT" QUESTIONS

1. Drinking seawater would result in increased urinary volume to rid the body of the excess salts. This would lead to further dehydration, creating greater thirst.

2. In infants excessive fluid loss through vomiting and diarrhea can lead to water loss in excess of 10% body weight, which may lead to convulsions and death.

3. Urinary tract infections usually accompany sexually transmitted diseases because of the close proximity of the urethral opening to the vagina and anus, allowing bacteria to enter.

4. Excess secretion of renin acts to produce the release of aldosterone from the adrenal glands. This hormone promotes the reabsorption of sodium ions from the renal tubules. The increase in blood-sodium levels leads to water retention. The water retention increases the blood volume and raises blood pressure, the symptom of hypertension.

5. Large molecules such as proteins are not normally filtered in the glomerulus into Bowman's capsule. Therefore, if proteins appear in the urine there is probably inflammation or another disease process that allowed them to leak out of the blood into the tubular system.

6. Coffee is a diuretic and increases the urinary output by altering filtration at the glomerulus. The symptoms of a "hangover" are the results of dehydration caused by the consumption of alcohol. Coffee compounds the problem of dehydration; it does not rehydrate the body.

7. Individuals with high-protein diets are advised to drink large quantities of water because protein metabolism leads to excess nitrogenous waste compounds such as ammonia and urea. It takes additional water to flush these waste products out of the body through the urine and prevent dehydration.

8. Low blood pressure would reduce the filtration rate across the glomerulus so waste products would accumulate in the blood. The accumulation of waste products would have serious effects on all cells of the body and interfere with normal function.

AUDIOVISUAL MATERIALS

The Basic Principles of Dialysis, Schering, 20 min.

The Dynamic Kidney: I. Structure and Function, Lilly, 29 min.

Excretion, McGraw-Hill, 28 min., 1960.

Excretory System in Animals, Association Films, 16 min., 1971.

Kidney Function in Health, Lilly, 38 min.

Regulating Body Temperature, Encyclopaedia Britannica Educational Corp., 22 min., 1972.

Temperature Regulation, International Film Bureau, 12 min., 1979.

Work of the Kidneys, Encyclopaedia Britannica Educational Corp., 10 min. 1940.

27 The Immune System

KEY CONCEPTS

Viruses and bacteria entering the human body encounter a variety of nonspecific defenses.

Inflammatory response dilutes toxins and sends in white blood cells to stem infection.

Interferons inhibit viral replication.

Fever fights viral infections by helping interferon, and it fights bacterial infections by reducing blood-iron levels.

The immune system distinguishes between self and nonself antigens with specialized white blood cells and the biochemicals they produce.

The immune response is fast, highly specific, and very diverse and has memory.

The immune system is a grouping of components of the lymphatic system and the blood circulatory system.

Lymphocytes produce a variety of biochemicals called lymphokines that are essential to the immune response.

Lymphocytes are made in the bone marrow and then sent to the lymph nodes, spleen, tonsils, and thymus gland.

Lymphocytes circulate in the blood and tissue fluid.

The skin is a vital outpost for lymphocytes, for this is where many infectious agents enter the body.

Producing antibodies constitutes the humoral immune response.

Antibody binding to antigen inactivates or detoxifies invaders, makes them more visible to macrophages, or stimulates the complement system.

The cellular immune response is carried out by T cells, which are made in the bone marrow and then pass through the thymus.

The fetal immune system learns to distinguish self from nonself while not rejecting maternal cells and biochemicals.

Infants are protected by antibodies transmitted first through the placenta and then through breast milk.

The immune system begins to decline early in life.

AIDS is an assault on helper T cells and other cells by a retrovirus, HIV.

In autoimmune disorders, autoantibodies attack a person's own tissues.

In allergies, cells release allergy mediators in response to nonthreatening substances.

Allergies may have been an adaptation to life-threatening substances.

A vaccine is a portion of or an entire inactivated virus or bacterium that evokes a secondary immune response when the disease-causing agent is encountered.

The key to a successful transplant is to suppress immunity against foreign tissue but to retain immunity against infection and cancer.

Monoclonal antibodies can be used to diagnose a wide range of infections, pinpoint hormonal events, and detect and possibly treat cancer.

KEY TERMS

active immunity
AIDS-related
     complex (ARC)
allergens
allergy
allergy mediators
anaphylactic
     shock
antigen binding
     sites
antigens
autoantibodies
B cells
complement system
constant regions
cyclosporin
desensitization
endogenous pyrogen
heavy chains
histamine

humoral immune
     response
hybridoma
idiotypes
inflammation
interferons
light chains
lymphokines
macrophages
major histo-
     compatibility
     complex (MHC)
mast cells
memory cells
monoclonal
     antibodies
natural killer
     cells
neutrophils
passive immunity

CHAPTER OUTLINE

I.   Non-specific Defenses
   A.   Barriers to infection.
      1.   Unpunctured skin prevents invasion.
      2.   Mucus in the nose traps inhaled dust particles.
      3.   Tears wash chemical irritants from the eyes and contain lysozyme, a biochemical that kills bacteria.
      4.   Wax traps dust particles in the ears.
      5.   Acidic stomach secretions destroy microorganisms in food.
      6.   Airways of the respiratory system sweep out dust by use of cilia.
   B.   Phagocytosis.
      1.   Intruders are met by phagocytes that engulf and digest them (fig. 27.1).
      2.   The phagocytic neutrophils and microphages roam about.
   C.   Inflammation.
      1.   Plasma accumulates at the wound site, which dilutes toxins secreted by bacteria.
      2.   Increased blood flow warms the environment.
   D.   Antimicrobial substances.
      1.   Interferons.
         a)   The interferons are polypeptides produced by a cell infected with a virus.
         b)   These stimulate cells to produce biochemicals that block viral replication.
      2.   Complement system.
         a)   A group of proteins that assists, or complements, several of the body's other defense mechanisms.
         b)   Some complement proteins trigger a chain reaction that punctures the cell membranes of microbes, bursting them.
         c)   Mast cells release a biochemical called histamine.
         d)   Histamine widens blood vessels, easing entry of white blood cells.
   E.   Fever.
      1.   First, white blood cells are stimulated to proliferate.
      2.   Cells then secrete a protein called endogenous pyrogen, which resets the thermoregulatory center in the brain's hypothalamus to maintain a higher body temperature.
      3.   A fever-reducing medication may actually counteract an effective biological defense.
      4.   Microbial infections are also stemmed by fever because of the reduction of the level of iron in the blood.
II. Specific Defenses
   A.   The human response includes protection against specific "foreign" structures or entities and distinguishes "self" (the body's cells) from "nonself" surfaces (e.g., a disease-causing microorganism).
      1.   It consists of an army of lymphocytes plus specialized biochemicals that some of these cells synthesize.
      2.   These biochemicals are a response to any antigen that is a foreign substance.
   B.   The immune system is fighting such agents as a virus, microbes, a fungal infection, and food poisoning, and the components of the immune system carry out different battle strategies.
III. Cells and Chemicals of the Immune System
   A.   Lymphocytes.
      1.   These produce a variety of biochemicals called lymphokines.
      2.   They are made in bone marrow and then sent to the lymph nodes, spleen tonsils, and thymus gland.

B. Macrophages.
   1. These engulf foreign matter and cellular debris and activate lymphocytes.
   2. The antigen is held to the surface of the macrophage by a protein badge called the major histocompatibility complex (MHC) and travels to the nearest lymph node.
   3. The macrophage secretes a lymphokine called interleukin-1, which stimulates the T cell to begin replicating and causes a fever, which may slow microbial activity.
C. The humoral immune response—B cells produce antibodies.
   1. Lymphocytes called B cells destroy antigens by secreting proteins called antibodies into the bloodstream, and this release is called the humoral immune response.
   2. Several cells interact to manufacture antibodies and the alerted macrophage then activates a helper T cell, which stimulates a B cell to produce antibodies specific for that antigen.
   3. The activated B cell divides, producing a clone, and the stimulated B cells develop into plasma cells and memory cells.
   4. Plasma B cells are antibody factories, each specific for the detected antigen.
   5. The immune system's reaction to its first meeting is a primary immune response and it takes a few days, during which the B cells divide and mature into plasma cells and produce the appropriate antibodies.
   6. Subsequent encounters are called a secondary immune response and the defense is quicker.
D. Antibody structure.
   1. Antibodies are large, complex molecules built of four polypeptides.
      a) The two larger polypeptide chains are called heavy chains.
      b) The other two are light chains.
   2. Constant regions are the lower portions of each antibody polypeptide chain, consisting of a sequence of amino acids that tends to be the same or very similar in all antibody molecules.
   3. The variable regions vary between individual antibody molecules.
   4. The specialized ends of the antibody molecule are called antigen binding sites, and the particular parts that actually bind the antigen are called idiotypes (fig. 27.7).
   5. The binding of an antigen inactivates the microbe or neutralizes the toxins it produces and can activate the complement system.
   6. Antibody molecules can also aggregate in groups of two or five to form antibody complexes (fig. 27.8).
   7. Five classes of antibodies are distinguished according to their locations in the body and their functions.
   8. During the early development of B cells, sections of their antibody genes are randomly moved to other locations among the chromosomes, creating new gene sequences and, consequently, instructions for producing different antibodies.
E. The cellular immune response—T cells.
   1. This is carried out by lymphocytes called T cells.
   2. T cells travel to where they are needed and originate in the bone marrow and then pass through the thymus.
   3. T cells are said to "mature" or be "educated" in the thymus, where they acquire the ability

191

to recognize particular nonself cell surfaces and molecules.

    4.   Kind of T cells.

        a)   Helper T cells stimulate B cells to produce antibodies and secrete lymphokines and activate another type of T cell, called a killer T cell.

        b)   Killer T cells attack nonself cells by attaching to them and releasing chemicals.

        c)   Suppressor T cells inhibit the response of all lymphocytes to foreign antigens, shutting off the immune response when an infection is controlled.

        d)   Natural killer cells cause cells to burst and differ from killer t cells in that they kill without interaction with other lymphocytes or antigen and are a first line of defense against viruses and perhaps other microbes.

IV.   Development of the Immune System

    A.   Active immunity is to mount your own immune defense.

    B.   Passive immunity.

        1.   When the newborn receives some antibodies from the mother during fetal existence, which continues to provide temporary immunity.

        2.   Mammary glands secrete a yellow substance called colostrum, which is rich in antibodies that protect the newborn against certain digestive and respiratory infections.

        3.   The immune system begins to lose effectiveness early in life, and the thymus gland reaches its maximal size in early adolescence and then slowly degenerates.

V.   When Immunity Breaks Down

    A.   AIDS.

        1.   Individuals die from opportunistic infections, meaning infections that take advantage of a weakened immune system, such as the appearance of Kaposi's sarcoma.

        2.   AIDS-related complex (ARC) is characterized by weakness, swollen glands in the neck, and frequent fever.

        3.   Modes of transmission.

            a)   It can be transferred through anal or vaginal intercourse.

            b)   Intravenous drug users spread AIDS when they share needles.

            c)   It was also passed in blood transfusions prior to 1985.

            d)   Another source is during birth from infected mothers.

        4.   Mode of action.

            a)   AIDS is caused by a virus, HIV, that attacks the immune system (fig. 4.2).

            b)   HIV is found in blood, sperm, and to a lesser extent in milk, tears, and saliva.

            c)   It is a retrovirus, in which its genetic material is ribonucleic acid (RNA).

            d)   It attaches to a receptor called CD4 on helper T cells and sends in its RNA, including instructions to manufacture the enzyme reverse transcriptase.

            e)   Transcriptase builds a DNA strand complementary to the viral RNA and then replicates to form a DNA double helix representing the vial genetic material (fig. 27.13).

            f)   The viral DNA then enters the T cell's nucleus.

            g)   When some other virus activates a helper T cell, it also activates HIV.

            h)   The helper T cell can no longer release lymphokines or stimulate B cells, and as it dies it unleashes HIV

particles, which suppress immunity.
   5. Treatment so far has taken three approaches: inhibiting the ability of the virus to infect and replicate; boosting the functions of the immune system; and treating the various opportunistic infections.
B. Severe combined immune deficiency.
   1. This is when neither T nor B cells function.
   2. "David" was one such youngster who had no thymus gland and spent 12 years of his life in a vinyl bubble.
C. Autoimmunity.
   1. The immune system manufactures antibodies that attack the body's own cells and are called autoantibodies.
   2. One theory of its source is that a virus, while replicating within a human cell, "borrows" protein from the host cell's surface and incorporates them onto its own surface; when the immune system "learns" the surface of the virus to destroy it, it also learns to attack the human cells that normally bear the particular protein.
   3. Types of autoimmune disease.
      a) Myasthenia gravis is when neuromuscular junctions are destroyed, resulting in muscular weakness in the arms and legs.
      b) Multiple sclerosis is when the myelin coat around neurons is attacked.
      c) In rheumatoid arthritis, the synovial membranes of the joints become inflamed.
D. Allergies.
   1. An allergy is an inappropriate response, such as attacking substances that are not a threat to health.
   2. The offending substances, called allergens, activate antibodies of type IgE.

   3. These antibodies in turn prompt mast cells to release substances called allergy mediators, which include histamine and heparin (fig. 27.15).
   4. The most common allergens are foods, dust mites, pollen, and fur.
   5. Many people are allergic to more than one substance, and allergies run in families (fig. 27.16).
   6. Desensitization is when small amounts of allergen are periodically injected under the skin, and although the doses are not enough to stimulate production of IgE, they do stimulate the production of IgG.
   7. Anaphylactic shock.
      a) A reaction to certain stimuli that is frightening and potentially life-threatening.
      b) Mast cells release mediators throughout the body, and the individual will lose consciousness and die within 5 minutes if he or she does not receive an injection of epinephrine or a tracheotomy.
VI. Altering Immune Function
   A. Vaccines—augmenting immunity.
      1. A vaccine is a killed or weakened form of a virus or bacterium or merely the part of the infectious agent that starts the immune response.
      2. They can be alive or killed by heat or a toxic chemical.
      3. A safer vaccine uses only the precise part of the virus's or bacterium's surface.
      4. A "super vaccine" can be constructed by inserting genes for surface proteins from various disease-causing viruses into the well-studied vaccinia virus such as herpes simplex type I, hepatitis B, influenza, and smallpox.

B. Organ transplants—suppressing immunity.
  1. Because blood types are determined by surface characteristics of blood cells, the more blood types two people share, the more likely a transplant is to "take" between them.
  2. Cyclosporin suppresses the T cells that attack transplanted tissue but not those that attack cancer cells or stimulate B cells to produce antibodies.
  3. Continuing challenges to successful transplants are the lack of donor organs and rejection of transplants by the recipient's immune system.
  4. A form of autologous (from oneself) transplantation is the practice of storing one's own blood for use in a future transfusion.
C. Monoclonal antibodies—targeting immunity.
  1. Monoclonal antibodies are antibodies that descend from a single B cell.
  2. Cesar Milstein and Georges Köhler received the Nobel Prize in medicine in 1984 for their contribution of monoclonal antibody (MAb) technology.
  3. They isolated a single B cell from a mouse's spleen and fused it with a cancerous white blood cell called a myeloma to produce a resulting fused cell called hybridoma.
  4. Combinatorial libraries engineer bacteria to harbor antibody genes from a higher organism, such as a mouse.
  5. Uses of MAbs.
    a) Cell biologists can use pure antibodies to localize and isolate each of the thousands of proteins in a cell.
    b) They can be used to diagnose viral or bacteria diseases and diagnose cancer.
    c) They can be used to treat cancer.
    d) MAbs bind to cancer cells, macrophages are attached, and they destroy the MAb cancer cell complex.
    e) MAbs can ferry conventional cancer treatments to where they are needed.
    f) There are magnetic beads, each coated with MAbs specific to cancer cells, and when the marrow is passed through a magnet, the cancer cells are pulled out.
D. Biotherapy—using immunity to treat cancer.
  1. Biotherapy is the use of substances naturally made in the body, such as immune system cells and biochemicals.
  2. Monoclonal antibodies could be used to deliver the biochemicals selectively, and recombinant DNA technology could provide vast amounts of pure proteins.
  3. Interferon was the first substance proven effective.
  4. Another form of biotherapy.
    a) A patient is given interleukin-2, and then killer T cells are removed from samples of tumors and then incubated in the laboratory in interleukin-2 as the patient receives even more of it.
    b) The activated killer T cells, along with more interleukin-2, are injected into the patient and this is called "lymphokine activated killer cells."

194

LEARNING OBJECTIVES

After reading this chapter, the student should be able to answer these questions:

1. What is the overall function of the immune system?
2. What are some ways that the body protects itself nonspecifically?
3. How is the immune response fast, specific, and diverse?
4. How does the immune system remember?
5. How do the cells and biochemicals of the immune system interact with each other to protect against specific invaders?
6. How does immune function change over a lifetime?
7. How is immune function altered by AIDS, autoimmune disorders, and allergies?
8. How can the immune system be altered to prevent, diagnose, and treat disease?

ANSWERS TO THE TEXT QUESTIONS

1. a. The HIV retrovirus attaches to helper T cells. When a second virus activates the helper T cell it also activates HIV replication, which then infects additional helper T cells.
    b. Myasthenia gravis is an autoimmune disease that attacks the neuromuscular junctions of skeletal muscles.
    c. The hay fever allergen activates B cells to produce IgE antibodies, which promote mast cells to release allergy mediators.
    d. In anaphylactic shock, mast cells release mediators throughout the body causing hives, itching, and rapid drop in blood pressure.

2. a. Mast cells are activated by IgE antibodies and release allergy mediators.
    b. Macrophages engulf foreign material and transport the antigen on protein "badges" on their cell membranes to helper T cells. They also secrete interleukin-1, which stimulates T cell replication.
    c. Killer T cells attack nonself cells and cancer cells by releasing cytolysin,

which destroys the cell membrane of the attacked cell.
    d. Helper T cells stimulate B cells to produce antibodies.
    e. B cells destroy antigens by producing antibodies. They become committed to a specific antigen termed plasma cells and memory cells.

3. A killed vaccine can cause illness when a virus or bacterium somehow survives a killing treatment and mutates so that it becomes pathogenic.

4. The young man suffers from a severe allergic reaction termed "anaphylactic shock," in which mast cells release allergic mediators throughout the body causing itching, hives, diarrhea, and difficulty in breathing.

5. The HIV attaches to helper T cells and replicates through reverse transcriptase and inserts the viral DNA in the helper T cell nucleus. Other viruses activate the helper T cell, which then turns on the replication of the HIV. The helper cells burst and infect other T cells. The helper T cells can no longer release lymphokines when infected and can no longer activate B cells to manufacture antibodies.

ANSWERS TO THE "TO THINK ABOUT" QUESTIONS

1. AIDS vaccines could be supplied to volunteers infected with the virus.
2. Exposure to ultraviolet light may help by depressing the activation of helper T cells, which harbor the HIV. When the helper T cells are activated by other viruses the HIV is replicated and released to infect additional helper cells.
3. The intentional conception of a child to serve as an organ or tissue donor is unethical. The life of the newborn is given secondary consideration to the role the individual would serve as a donor.
4. By injecting allergens before the T cell recognizes self from nonself the allergen would be considered a normal part of the body and would not produce allergic responses. Foreign tissue would also be recognized as self. Later this foreign tissue could be transplanted into the

individual as an organ and not be rejected as nonself.

5. Answers will vary but it may be in society's interest to notify the individual so that they may take precautions to prevent the spread of the HIV.

6. The individual risk, although low, cannot be discounted. However, the risk to society is greater in that many more cases of whooping cough will develop in the absence of widespread vaccinations in the population.

7. Monoclonal antibodies could be used to detect the protein myosin in the blood. The quantification of the myosin-monoclonal antibody complex in the blood would indicate the extent of damage to the cardiac muscle.

AUDIOVISUAL MATERIALS

Blueprints in the Bloodstream ("NOVA" Series), Time Life Video, 57 min., 1978.

The Genetics of Transplantation, Milner-Fenwick Incorporated, 19 min., 1975.

Infectious Diseases and Natural Body Defense, Coronet Instructional Films, 11 min., 1961.

The Immune Response, Indiana University, 30 min., 1962.

Immunocompetence in the First Year of Life, Schering, 28 min.

Secret of the White Cell, Indiana University, 30 min., 1966.

Work of the Blood, Encyclopaedia Britannica Educational Corp., 13 min., 1957.

28 Plants through History

KEY TERMS

legumes, so eating these two foods together provides a good protein balance.

   b) Spices can also preserve foods.

C. Cereals—staples of the human diet.

   1. Cereals.

      a) Members of the grass family that have seeds that can be stored for long periods of time.

      b) The seed is the kernel or grain, and it consists of the embryo, a large starch supply called the endosperm, and outer protective layers called the aleurone and the pericarp, which are rich in protein, lipid, and vitamins (fig. 28.4).

   2. Wheat.

      a) The grain of the "hard" bread wheats contains 11% to 15% protein, mostly of a type called gluten, which when mixed with water forms an elastic dough.

      b) Ancient wheat called Einkorn was crossbred with another type of grass.

      c) The hybrid grass then underwent a genetic accident (nondisjunction) that prevented the separation of chromosomes in some of the developing germ cells (fig. 28.5).

      d) The result was a new type of plant that had twice the number of chromosomes and larger grains; it was called Emmer wheat and was a far better source of food than the parent wheat.

      e) Emmer wheat crossed with another weed, goat grass, and after another fortuitous "accident" of chromosome doubling led to bread wheat, which has 42 chromosomes.

   3. Corn.

      a) The tasty, sweet, or starchy kernels of the corn plant *Zea mays* have sustained the Incas, Aztecs, and Mayan Indians of South America and are used to manufacture food products, drugs, and industrial chemicals (fig. 28.7).

      b) Modern corn comes from a grass called teosinte (*Zea mexicana*) (fig. 28.8).

      c) About 7,500 years ago, a population of wild teosinte faced an environmental stress that was overcome only by individual plants whose tassels included some female structures.

      d) Part of the teosinte tassel might have enlarged to evolve into the corn tassel.

      e) Modern corn, with female ears and male tassels, was artificially selected by farmers seeking plants with plump and tasty kernels.

      f) Charles Darwin found that inbred plants were more prone to disease and produced ears with fewer rows of kernels.

      g) Gregor Mendel demonstrated the genetic phenomenon of hybrid vigor; plants with unrelated parents were more vigorous than self-fertilized plants because they had inherited new combinations of genes.

      h) George Shull, at the Station for Experimental Evolution in Cold Spring Harbor, New York, developed hybrid corn, which revolutionized corn output worldwide (fig. 28.9).

4. Rice.
   a) For people in Asia, rice provides 80% of their calories.
   b) The oldest species of rice is *Oryza sativa*.
   c) Rice is a very ancient plant that evolved in a tropical, semiaquatic environment.
   d) Human cultivation of rice began about 15,000 years ago in the area bordering China, Burma, and India.
   e) In the 1960s, highly productive newcomers, the semidwarf rices, became very popular and soon accounted for nearly all of China's crop and almost half of the rices of many other Asian nations.
   f) International Rice Research Institute.
      (1) Founded in 1961 in the Philippines to offset a potential disaster due to reliance on a few types of rice.
      (2) It soon became a clearinghouse for the world's rice varieties, cold storing the seeds of 12,000 natural variants by 1970 and of more than 70,000 by 1983.
      (3) Representatives of other important crops are being banked as well.
         (a) Potato cells are stored at the International Potato Center in Sturgeon Bay, Wisconsin.
         (b) Wheat cells are banked at the Kansas Agricultural Experimental Station (fig. 29.11).
      (4) Plant banks offer three priceless services to humanity.
         (a) They are a source of variants in case a major crop is felled by disease or an environmental disaster.
         (b) They allow for the return of endangered or extinct varieties to their native lands.
         (c) They may provide a supply of genetic material from which researchers can fashion useful plants in the years to come, even after the species represented in the bank have become extinct.
      (5) The decrease in genetic diversity makes a crop vulnerable to disease or a natural disaster because it does not have resistant variants.
5. Amaranth.
   a) This plant stands 8 feet tall and has broad purplish green leaves and massive seed heads (fig. 28.13).
   b) Amaranth is rich in amino acids that are poorly represented in the other major cereals (the germ

and bran together are 50% protein).

   c)   The broad leaves are rich in vitamins A and C and the B vitamins riboflavin and folic acid.

II.   Nature's Botanical Medicine Cabinet

   A.   Natural products chemists.

      1.   These chemists search through the bounty of chemicals manufactured by organisms for substances that may treat illness in people.

      2.   More familiar plants have healing properties (fig. 28.14).

      3.   Most of these people work in laboratories, trying to imitate nature by synthesizing compounds that plants normally make.

   B.   A good example of medicines derived from plants used today are the chemicals called alkaloids from the periwinkle plant and other species.

      1.   Alkaloids have helped revolutionize the treatment of some leukemias, and alkaloid narcotics derived from the opium poppy, including morphine, are excellent painkillers.

      2.   Nearly half of all prescription drugs contain chemicals manufactured by plants or bacteria, and many other drugs contain compounds that were synthesized in a laboratory but were modeled after plant-derived substances.

   C.   Malaria is an interesting example of the relationship between plants and human disease, because plants contribute both to the disease's spread and to its cure.

      1.   Agriculture often ushers in malaria by replacing dense forests with damp rice fields that are a haven for mosquitoes.

      2.   Some plant products can kill the malaria parasite.

      3.   The French chemist Pierre Joseph Pelletier extracted an active ingredient from cinchona bark, which was called quinine; this substance reigned as the standard treatment for malaria until the 1930s when the malaria parasites developed resistance to the bark-derived treatment.

      4.   New natural antimalarial drugs were sought.

LEARNING OBJECTIVES

After reading this chapter, the student should be able to answer these questions:

1.   In what ways do we use and depend upon plants?

2.   How might agriculture have arisen? Where and when did it arise?

3.   What are the major parts of a grain, and how are they used in the human diet?

4.   From what types of plants is it thought that the modern bread wheats and modern corn evolved?

5.   In what major way does the history of rice differ from that of wheat and corn?

6.   How might we use the plant amaranth?

7.   What are some examples of plant biochemicals used as medicines?

8.   How have plants helped to spread malaria, as well as to treat it?

ANSWERS TO THE TEXT QUESTIONS

1.   Cereals are the seeds of grassplants such as wheat, oats, rice, corn, and barley.

2.   Agriculture is the domestication of animals and the intentional planting and cultivation of crops.

3.   Agriculture may have arisen through the accidental dropping of seeds followed later by the observation that this increased productivity. Later it was likely noticed that increasing the water supply also increased productivity. Individual animals that would succumb to human domination and were able to interbreed increased the reproductive rates from those in the wild.

4.   Wheat: It is believed that an accidental crossing of a domesticated variety of wheat called Einkorn with a wild type of wheat resulted in nondisjunction. This allowed for the offspring to have twice as many chromosomes as either parent. The

food production of this crossbreed was much higher than the parent generation.

Corn: The present-day species of corn are believed to have arisen through the artificial selection for those individual plants that produced female structures and plump, tasty kernels. Later others, including Charles Dauscn and George Shull, crossed unrelated lines to increase the quantity and quality of kernels.

Rice: The rice species used in cultivation remained virtually unchanged until the 1960s, when semidwarf rices were produced for increased yield.

5. Environmental disasters can result in devastation of the entire species. In 1970 the southern corn leaf blight fungus destroyed 15% of the United States corn crop. In 1846, the mold *Phytophthora infestans* decimated the potato crop in Ireland.

6. Alkaloids from plants have been used in the treatment of cancer. Quinine extracted from cinchona bark was used as the standard treatment of malaria. Since 1979 a newly extracted chemical called artemisinin has been tested to replace quinine.

ANSWERS TO THE "TO THINK ABOUT" QUESTIONS

1. Although there is no correct answer, the student may wish to consider the consequences to corn crops if humans are unable to cultivate them due to a depression or catastrophe, such as earthquakes, nuclear war destruction, or traditional war destruction. Would it be advisable to keep larger supplies of corn in plant banks so massive cultivation could occur after devastation?

2. Careful studies in the greenhouse between present-day bread wheat and wild grass that is diploid show that cross-pollination is possible. This would be followed by chromosomal studies to see if nondisjunction occurred. The student should keep in mind that evidence in support of the hypothesis is only a clue to the actual events that occurred years ago.

3. Teosinte can be grown in large numbers so selection for certain individuals is possible. Various environmental stress factors could be applied over time to select for various phenotypes. Careful selection and interbreeding of individuals with traits most similar to modern-day corn would then be made. It might not be possible to get a close species through experimentation, since natural selection requires mutation, and it may not be possible to recreate the actual selection factors of environmental stress of the past.

4. The student will have his or her own ideas and opinions.

5. The pest-resistant variety sounds like a good bet considering the company is highly experienced. However, the student should remember that pests are evolving too. Having only one species, even though a good one, may result in no crop if a new pest is successful in penetrating the plant's defense mechanisms. This can be prevented by raising more than one species.

6. Since new plant varieties would be impossible without genetic stock, it might be argued that some of the cost to donating countries be defrayed.

7. The substance would need to be disguised such that it was undiscernible by appearance, taste, or texture from a control substance that was known not to have healing powers. Relief or lack of relief could be measured through interview. The two groups would then be compared.

AUDIOVISUAL MATERIALS

The Green Machine, Time Life Video, 49 min., 1978.

The Growth of Plants, Encyclopaedia Britannica Educational Corp., 20 min., 1962.

29 Plant Form and Function

KEY CONCEPTS

The primary axis of a plant consists of a shoot above ground for photosynthesis and the root for support.

A plant is made up of many tissue types specialized for functions of photosynthesis, transport of water and sap, primary and secondary growth, and protection from predators and water loss.

Plants never stop growing, unlike other organisms, because they contain meristem cells that continue to undergo cell division.

Stems provide support, food storage, and transport of nutrients and water but can be specialized for reproduction, climbing, protection, and storage.

Leaf structure is designated to maximize light absorption, and gas exchange is designed for the process of photosynthesis.

Roots anchor plants and absorb, transport, and store nutrients.

Plants evolved the ability to put on girth so they could increase their height to compete for sunlight.

KEY TERMS

abscission zone
adventitious roots
aerial roots
apical meristems
axil
bark
blade
bracts
Casparian strip
chlorenchyma
collenchyma
cork cambium
cork cells
cortex
cotyledons
cuticle
cutin
deciduous trees

dermal tissue
dicots
endodermis
epidermis
fibrous root
    system
ground tissue
growth rings
guard cells
hardwoods
heartwood
hypodermis
insect-trapping
    leaves
intercalary
    meristem
internodes
lateral meristems

leaf abscission
mesophyll
monocots
mucigel
mycorrhizae
nodes
parenchyma
pericycle
periderm
petiole
phelloderm
phloem
pith
primary body
primary growth
primary tissues
quiescent center
radicle
root cap
root hairs
roots
rosettes
sapwood
sclereids
sclerenchyma
secondary growth
sieve cells
sieve plate
sieve pores
sieve tube members

softwoods
spines
stolons
stomata
storage
    leaves
subapical
    region
suberin
succulent
taproot
    system
tendrils
thorns
tracheids
trichomes
tubers
vascular
    bundles
vascular
    cambium
vessel elements
xylem
zone of cellular
    division
zone of cellular
    elongation
zone of cellular
    maturation

CHAPTER OUTLINE

I. Primary Tissues
    A. The primary body of a plant is an axis consisting of a root and a shoot, which are extensive and increase available surface area on which the chemical reactions of life can occur.
    B. The primary tissues of the plant body are groups of cells with a common function.
    C. Meristems.
        1. Meristems are localized regions of cell division consisting of

202

undifferentiated plant tissue from which new cells arise (fig. 29.1).

2. Apical meristems are found near the tips of roots and shoots; primary growth is when these meristematic cells divide and elongate.

3. Lateral meristems grow outward, thickening the plant, which is called secondary growth.

4. Intercalary meristem is a type of dividing tissue found in grasses between mature regions of stem.

D. Ground tissue.

1. Ground tissue comprises most of the primary body of a plant and is make up of three cell types.

2. Parenchyma cells.

   a) These are the most abundant cells in plants; they are unspecialized and at maturity are alive and capable of dividing.

   b) These are usually the edible parts of the plant.

   c) These cells are important sites where photosynthesis, cellular respiration, and protein synthesis are involved.

   d) Chlorenchyma are chloroplast-containing parenchyma cells that take part in photosynthesis, imparting to leaves their green color (fig. 29.2).

3. Collenchyma cells.

   a) These cells support the growing regions of shoots.

   b) They have unevenly thickened primary cell walls that can stretch, enabling the cells to elongate.

4. Sclerenchyma cells.

   a) They have nonstretchable secondary walls that support regions of plants that are no longer growing and are usually dead at maturity (fig. 29.3).

   b) They consist of sclereids that have many shapes and occur singly or in small groups.

   c) Many sclerenchyma fibers are used to produce textiles.

E. Dermal tissue.

1. Dermal tissue covers the plant body.

2. The epidermis covers the primary plant body and is usually only one cell thick. Special features of the epidermis are the cuticle, stomata, and trichomes.

3. The cuticle.

   a) This is a covering over all but the roots of a plant that protects and keeps out water.

   b) It consists mainly of cutin, a fatty material produced by the epidermal cells.

   c) This acts as the first line of defense against predators and infectious agents.

4. Pores called stomata are where plants exchange water and gases with the environment.

5. Trichomes.

   a) These are outgrowths of epidermal cells found in almost all plants.

   b) They help deter predators: hooked-shaped trichomes may impale marauding animals, and predators may also break off the tips of trichomes, which release a sticky substance that traps the invading animals.

   c) In carnivorous plants they secrete enzymes to digest trapped animals.

   d) They often reflect light to help prevent overheating; root hairs are trichomes that appear near root tips and absorb water and minerals from the soil.

F. Vascular tissues.
   1. Vascular tissues are specialized conducting tissues.
   2. There are two kinds of vascular tissue.
      a) Xylem forms a continuous system that transports water with dissolved nutrients from roots to all parts of a plant.
         (1) There are two kinds of conducting cells that have thick walls that are essential to prevent collapse.
         (2) Water moves from cells called tracheids to other tracheid cells through thin areas in cell walls called pits.
         (3) Vessel elements are more specialized cells that evolved from tracheids.
      b) Phloem transports water and food materials, primarily dissolved sugars, throughout a plant.
         (1) Phloem transports substances under positive pressure.
         (2) The conducting cells of phloem are alive at maturity.
         (3) Their cell walls have thin areas perforated by many sieve pores, through which solutes move from cell to cell.

II. Parts of the Plant Body
   A. Stems.
      1. Nodes are areas of leaf attachment, and internodes are portions of stem between nodes; the region between a leaf stalk and stem is a leaf axil (fig. 29.1).
      2. Stem elongation occurs in the internodal regions.
      3. The epidermis surrounding a stem contains stomata and may possess protective trichomes.
      4. Vascular tissues.
         a) They are organized into vascular bundles.
         b) Food-conducting phloem forms on the outside of a bundle, whereas water-conducting xylem forms on the inside of a bundle.
         c) Vascular bundles are arranged differently in different types of plants: monocots, which have one first, or "seed," leaf, and dicots, which have two seed leaves.
            (1) Monocots such as corn have vascular bundles scattered throughout their ground tissue.
            (2) Dicots such as sunflower have a single ring of vascular bundles (fig. 29.7).
         d) A single layer of cells between the xylem and phloem remains meristematic.
      5. Cortex.
         a) Ground tissue that fills the area between the epidermis and vascular tissue in stems.
         b) In plants having concentric cylinders of xylem and phloem—pine, for example—the ground storage tissue in the center of the stem is called pith.
      6. Plants modify their stems.
         a) Stolons, or runners, are stems that grow along the soil surface.
         b) Thorns often are stems modified for protection, such as thorns on a rosebush.
         c) Succulent stems of plants such as cacti are fleshy and store large amounts of water.

d) Tendrils are shoots that support plants by coiling around objects.

B. Leaves.

1. Leaves are the primary photosynthetic organs of most plants.

2. Most leaves consist of a flattened blade and a supporting, stalklike petiole.

3. There are four basic kinds of leaves.

   a) Simple leaves have flat, undivided blades.

   b) The blades of compound leaves are divided into leaflets.

   c) Pinnate leaflets are paired along a central line.

   d) Palmate leaflets are all attached to one point at the top of the petiole.

4. Leaves can be arranged in different patterns on stems.

   a) Plants with one leaf per node have alternate, or spiral, arrangements (fig. 29.10).

   b) Plants with two leaves per node have opposite arrangements.

   c) Plants with three or more leaves per node have whorled arrangements.

5. The epidermis is usually nonphotosynthetic and has many stomata.

6. Vascular tissues.

   a) In leaves, they occur in strands called veins.

   b) Leaf veins may be net, with minor veins branching off from larger, prominent midveins, or parallel.

   c) Most dicots have netted veins, and many monocots have parallel veins.

7. The ground tissue of leaves is called mesophyll, which is parenchyma.

   a) The upper side of a leaf contains tissue called palisade mesophyll cells, which are specialized for light absorption.

   b) Located below the palisade layer are spongy mesophyll cells, which are specialized for gas exchange.

8. Specializations.

   a) tendrils are modified leaves that wrap around nearby objects.

   b) Spines of plants such as cacti are leaves modified to protect plants from predators and excessive sunlight.

   c) Bracts are floral leaves that protect developing flowers.

   d) Storage leaves are fleshy and store food.

   e) Insect-trapping leaves are found in carnivorous plants and are adapted for attracting, capturing, and digesting prey.

9. Cotyledons are embryonic leaves found in flowering plants that often store energy used for germination.

   a) Leaf abscission is the normal process by which a plant sheds its leaves.

   b) Deciduous trees shed their leaves at the end of a growing season.

   c) Leaves are shed from an abscission zone, a region at the base of the petiole.

C. Roots.

1. These provide anchorage and absorb, transport, and store water and a nutrient; they absorb oxygen from between soil particles.

2. The first root to emerge from a seed is the radicle.

3. In a taproot system, the radicle enlarges to form a major root that persists throughout the life of the plant.

4. In a fibrous root system, the radicle is short-lived, and adventitious roots, which are roots that form on stems or leaves, replace the radicle, resulting in an extensive system.
5. The root cap, a thimble-shaped structure covering the tips of roots, produces mucigel, which protects root tips from desiccation.
6. Just behind the root cap is a cluster of seemingly inactive cells called the quiescent center, which functions as a reservoir to replace damaged cells of the adjacent meristem.
   a) The region immediately behind the root cap is called the subapical region.
   b) The zone of cellular division is meristematic.
   c) The zone of cellular elongation lies behind the zone of cellular division; here cells elongate as their vacuoles fill with water, and this action pushes the root rapidly through the soil.
   d) Cells mature and differentiate in the zone of cellular maturation, also called the root hair zone, which greatly increases the surface area of the root, through which water is absorbed (fig. 29.15).
7. The epidermis surrounds all the cortex, which consists of three layers.
   a) The hypodermis.
      (1) This is the outermost, protective layer of the cortex, consisting of parenchyma cells that form a vast collecting system that absorbs water and minerals through the epidermis.
      (2) It stores these nutrients for future growth.
   b) The endodermis.
      (1) It is the innermost ring of the cortex and consists of a Casparian strip.
      (2) It contains a waxy, waterproof material called suberin, which can regulate the movement of nutrients and water into and out of the central vascular tissue of the root (fig. 29.16).
   c) The pericycle is the ring of parenchyma cells that produces branch roots.
      (1) Storage is a familiar root specialization.
      (2) Pneumatophores are specialized roots of plants growing in oxygen-poor environments that grow up into the air, and oxygen diffuses into the plant body.
      (3) Adventitious roots that form and grow in the air are called aerial roots.
   d) Many roots form mutualistic associations called mycorrhizae with beneficial fungi.
III. Secondary Growth
   A. Formation of lateral meristems that increase the girth of stems and roots is called the vascular cambium and cork cambium and produce the secondary plant body.
   B. Vascular cambium.
      1. This produces most of the secondary plant body and is a thin cylinder of meristematic tissue found in roots and stems.
      2. It produces wood during the spring and summer.
         a) Spring wood is made of large, thin-walled cells.
         b) Summer wood consists of small, thick-walled cells

and is specialized for support.

    c)   Because of the differences between spring and summer wood, wood often displays visible demarcations called growth rings.

    3.   Types of wood.

        a)   Hardwoods are woods of dicots such as oak, maple, and ash, and softwoods are woods of gymnosperms such as pine, spruce, and fir.

        b)   Hardwoods contain tracheids, vessels, and supportive fibers, whereas softwoods are 90% tracheids.

        c)   Heartwood, wood in the center of a tree, collects a plant's waste products.

        d)   Sapwood, wood located nearest the vascular cambium, transports water and dissolved nutrients within a plant.

    4.   Bark.

        a)   This includes all of the tissues outside of the vascular cambium.

        b)   The periderm portion of the bark is the outer protective covering on mature stems and roots.

C.   The cork cambium.

    1.   This is the lateral meristem that produces the periderm.

    2.   These are waxy, densely packed cells covering the surfaces of mature stems and roots; they are dead at maturity and form waterproof, insulating layers that protect plants.

    3.   Phelloderm consists of living parenchyma cells, which may be photosynthetic and store nutrients.

    4.   Ninety percent of a typical tree is secondary xylem.

LEARNING OBJECTIVES

After reading this chapter, the student should be able to answer these questions:

1.   What are the main tissues and structures of the plant body?

2.   What functions are provided by a plant's tissues and structures?

3.   How does a plant transport water and nutrients?

4.   How does a plant elongate?

5.   How does a plant increase in girth?

ANSWERS TO THE TEXT QUESTIONS

1.   Meristematic tissue allows a plant to continue putting on girth and height all of its life.

2.   The most abundant tissue in plants is the ground tissue.

3.   a.   Cutin is a fatty material produced by epidermal cells.

        b.   Mucigel is a slimy substance that protects root tips from desiccation.

        c.   Suberin is a waxy, waterproof substance in the endodermis, which helps with water regulation.

4.   The trunk of a banana plant is composed of leaves. The organization of tissues is dependent upon the structure. This is determined by taking a cutting and demonstrating its basic leaf characteristics.

5.   Leaves are classified by their basic forms, arrangement on stems, vein palmation, or organization around a node.

6.   Trichomes are cells that make up defense structures such as barbs, tips of needles, or thorns. They also protect the plant from overheating by deflecting light.

7.   a.   In monocots the vascular bundles are scattered, and in dicots the vascular tissue is in a ring.

        b.   Monocots have parallel veins and dicots have netted veins.

        c.   Most dicots develop taproot systems, and monocots form fibrous root systems.

ANSWERS TO THE "TO THINK ABOUT" QUESTIONS

1. Lateral meristem on stems and roots gives rise to branches and leaves.
2. Stomata are openings in the epidermis that allow for gas exchange. The cuticle that covers the leaf is impermeable to gases to reduce water loss.
3. Xylem cells are under negative pressure due to transpiration. Phloem transports substances under positive pressure, which forces the sap in many directions at once, like blowing up a balloon.
4. The site of attachment for the leaf contains a region called the abscission zone. In response to environmental cues, a separation layer forms here isolating the dying leaf.
5. The root cap protects the root tip by secreting mucigel and replacing destroyed cells. The cells in the zone of elongation take up water in their vacuoles, causing a change in the cell length. This pressure causes the root to move through the ground.
6. Root hairs increase the surface area for the absorption of nutrients and water just as villi do in the human intestine.
7. The zone of cellular division would be exposed, killing off meristematic tissue.

AUDIOVISUAL MATERIALS

The Green Machine, Time Life Video, 49 min., 1978.

The Growth of Plants, Encyclopaedia Britannica Educational Corp., 20 min., 1962.

Nitrogen and Living Things, Universal Education and Visual Arts, 14 min. 1961.

The Plant as an Energy Trap, CRM McGraw-Hill, 17 min., 1967.

Plant Propagation: From Seed to Tissue Culture, Harvard University, 28 min., 1977.

Plant Tropisms and Other Movements, Coronet Instructional Films, 11 min., 1965.

30 Plant Life Cycles

KEY CONCEPTS

Plants evolved from green algae.

Terrestrial plants have made adaptations to land that reduce desiccation and allow for the combination of gametes.

Plants have alternate stages to their reproductive cycles that include both sexual and asexual stages.

Flowers are reproductive structures.

Petals and sepals are involved in protection and attracting pollinators.

The female portion of the flower is composed of the ovary, which produces and protects the egg and then the seed; its stigma and style are receptacles for pollen.

The male portion of the flower is composed of pollen-producing structures called anthers.

Microspore mother cells in pollen sacs undergo meiosis and then mitosis to produce a generative nucleus and a tube nucleus.

Megaspore mother cells go through meiosis and then mitosis to produce a mature megagametophyte that contains the egg.

Pollination occurs by animals or wind.

Pollen landing on the stigma initiates growth of the pollen tube down the style.

Double fertilization results with the fusion of the egg with one sperm to form the zygote and fusion of the other sperm nuclei with two polar nuclei to form the endosperm.

The two cotyledons of dicots absorb the endosperm.

The monocot cotyledon transfers food from the endosperm to the embryo.

Seed dormancy allows the embryo to wait for environmental conditions that enhance its chance of survival.

Fruits protect the seed from desiccation and assist in seed dispersal.

Germination begins with imbibition of water.

Gymnosperms produce unprotected seeds on the scales of cones.

Ovules produce three haploid degenerative cells and one megaspore, which produces the egg.

Microsporangia or male cones produce pollen that is carried in the wind.

Asexual reproduction is much simpler than sexual reproduction and is used when environmental conditions are stable.

Vegetative propagation occurs from adventitious buds on roots, leaves, and modified stems.

KEY TERMS

alternation of
   generations
androecium
angiosperms
anthers
auxin
bryophytes
calyx
carpels
coleoptile
cones
corolla
cotyledons
dicots
double
   fertilization
embryo sac
endosperm
epicotyl
ethylene
flowers
fruit
gametophyte
generations
generative
   nucleus
gynoecium
hypocotyl
imbibition
megagametophyte
megasporangia
megaspore mother
   cell
megaspores
microgametophyte

microsporangia
microspore mother
   cells
microspores
monocots
outcrossing
ovary
ovules
petals
pistil
plant embryo
plumule
polar nuclei
pollen grains
pollen sac
pollination
radicle
root apical
   meristem
scale
seed
seed coat
sepals
shoot apical
   meristem
sporophyte
stamens
stigma
style
tracheophytes
tube nucleus
zygote

I. Tracheophytes and bryophytes are the two main groups of plants.
   A. Tracheophytes, or vascular plants, dominate on land and possess vascular tissue, which transports water and dissolved nutrients throughout a plant.
   B. Bryophytes lack vascular tissue and are smaller than vascular plants.

II. Alternation of Generations
   A. In plants a complete life cycle of an individual includes both a diploid stage and a haploid stage, which alternate within a life cycle.
      1. The diploid generation of sporophytes produces haploid spores through meiosis.
      2. Haploid spores divide mitotically to produce a multicellular haploid individual, the gametophyte.
      3. Eventually, the gametophyte produces haploid gametes—eggs and sperm—which fuse to form a zygote, and the zygote grows to become a sporophyte.
   B. In most vascular plants, gametophytes produce either eggs or sperm, but not both.
      1. Female egg-producing gametophytes are larger than the male sperm-producing gametophytes and are called megagametophytes.
      2. Male and female gametophytes arise from two different types of spores called megaspores and microspores.
      3. These spores form in two different types of structures, megasporangia and microsporangia.
   C. The gametophyte stage is a more obvious phase in bryophytes, whereas vascular plants have a reduced gametophyte phase and a dominant sporophyte phase (figs. 30.2 and 30.3).

III. Flowering Plant Life Cycle
   A. Flowering plants, or angiosperms, are the dominant group of vascular plants.
   B. Structure of the flower.
      1. Complete flowers have four types of organs, which whorl, or circle, about the end of the flower stalk (fig. 30.4).
      2. The calyx.
         a) The outermost whorl of the flower.
         b) It is made up of green, leaflike sepals that enclose and protect the inner floral parts.
      3. The corolla.
         a) Inside the calyx and is a whorl made up of petals.
         b) The petals are often large and colorful, especially when they are important in attracting pollinators to the flower.
      4. The androecium.
         a) The whorl to the inside of the corolla.
         b) It consists of the male reproductive structures.
         c) These are the stamens, which are built of stalklike filaments bearing pollen-producing oval bodies called anthers at their tips.
      5. Gynoecium.
         a) The whorl at the center of a flower.
         b) It consists of the female reproductive structures of pistil.
         c) It is formed from one or more carpels, leaflike structures enclosing the ovules.
         d) The carpels and their enclosed ovules are referred to as the ovary; the ovary gives rise to a stalklike style bearing a stigma at its tip, which receives pollen.
   C. Formation of gametes.
      1. Microspores.
         a) Structures that form in the anthers.

b) Each anther contains four microsporangia called pollen sacs, which contain microspore mother cells that divide meiotically to produce four haploid microspores.

c) Each microspore then divides mitotically, producing a haploid generative nucleus and a haploid tube nucleus.

d) A thick, resistant wall forms around each two-celled structure, forming male microgametophytes, more familiarly known as pollen grains.

2. Megaspores.

a) These form in the ovary, and each ovule within the ovary is a megasporangium containing a megaspore mother cell, which divides meiotically to produce four haploid cells.

b) Three of these cells quickly disintegrate, leaving one large haploid megaspore.

c) The megaspore undergoes three mitotic divisions, forming a female megagametophyte with eight nuclei but only seven cells—one large cell has two nuclei called polar nuclei.

d) In a mature megagametophyte, also called an embryo sac, one of the cells with a single nucleus is the egg (fig. 30.5).

D. Pollination.

1. The transfer of pollen from an anther to a receptive stigma.

2. Other angiosperms are outcrossing species in which pollen grains from one flower are carried to the stigma of another flower.

3. Self-fertilization leads to reduced genetic variability and is common in temperate regions where environmental conditions are relatively uniform.

4. Outcrossing is crucial in a changing environment because it produces a genetically variable population that can adapt to changing conditions (fig. 30.6).

5. Pollen is often assisted in traveling from plant to plant (fig. 30.7).

a) In animal-pollinated angiosperms, floral characteristics such as morphology, color, shape, and odor often attract particular animals to particular flowers.

b) Some flowers produce heat, which volatizes their aromatic molecules.

c) Pollinators.

(1) Different types of insects have characteristic floral preferences.

(2) Birds are attracted to red flowers, a color that insects cannot distinguish.

(3) Some flowers are pollinated by butterflies and moths, and these flowers are white or yellow and are heavily scented.

(4) Bats are important pollinators in the tropics.

(5) Rotting flesh smell of the carrion flowers in South Africa attracts flies.

(6) Wind-pollinated plants grow closely together; wind pollination is far less precise than animal pollination.

E. Fertilization.

1. After a pollen grain lands on a stigma, the pollen grains gives rise to a growing pollen tube.

2. The pollen grain's two sperm cells enter the pollen tube as it grows down through the tissues of the stigma and the style towards the ovary.

3. When the pollen tube reaches an ovule, it discharges its two sperm cells into the embryo sac.

4. One sperm cell fuses with the egg, forming a diploid zygote.

5. After a series of cell divisions, the zygote will become the plant embryo.

6. The second sperm cell fuses with the two polar nuclei, forming a triploid nucleus that will divide to form a nutritive tissue called endosperm.

F. Seed development.
   1. A seed is a temporarily dormant sporophyte individual surrounded by a tough protective coat.
      a) Initially, the endosperm nucleus divides more rapidly than the zygote, forming a large mass of nutritive endosperm.
      b) The developing embryo forms cotyledons, or seed leaves.
      c) Monocots (fig. 30.9).
         (1) Angiosperms with one cotyledon.
         (2) In monocots, the cotyledon does not absorb the endosperm but absorbs and transports food from the endosperm to the embryo.
      d) Dicots.
         (1) These have two cotyledon.
         (2) In many dicots, the cotyledons become thick and fleshy as they absorb the endosperm.
   2. Embryonic development.
      a) The shoot apical meristem forms at the tip of the epicotyl.
      b) The root apical meristem differentiates near the tip of the embryonic root, or radicle.
      c) The epicotyl plus its young leaves is called a plumule.
      d) The stem like region below the cotyledons is the hypocotyl.
      e) In monocots, a sheathlike structure called the coleoptile covers the plumule, and the ovary wall remains attached to a monocot seed, forming a fruit.

G. Seed dormancy.
   1. At a certain point in embryonic development, the embryo becomes dormant and is protected by a tough outer layer called the seed coat.
   2. Seed dormancy enables seeds to postpone development when the environment is unfavorable, such as during drought.

H. Fruit formation.
   1. When a pollen tube begins growing, the stigma produces large amounts of ethylene.
      a) It triggers senescence of the flower.
      b) The floral parts wither and fall to the ground, and the ovary swells and develops into fruit, which is a ripened ovary enclosing a seed (fig. 30.10).
   2. Seeds with fruits synthesize the hormone auxin, which stimulates fruit growth.

I. Fruit and seed dispersal.
   1. In addition to protecting vulnerable seeds from desiccation, fruits facilitate seed dispersal by being eaten by birds and other animals.
   2. Mammals spread seed from place to place when fruits bearing hooked spines become attached to their fur.

212

3. Wind-dispersed fruits such as dandelions and maples have wings or other structures for dispersal.

J. Seed Germination.
1. Germination usually requires water, oxygen, and a source of energy.
2. The first step is imbibition, which is the absorption of water by a seed (fig. 30.12).
3. Imbibition causes the embryo to release hormones that stimulate the breakdown of the endosperm or stored food reserves.

K. Plant development.
1. The hypocotyl, with its attached radicle, emerges first from the seed.
2. In response to gravity, the radicle grows downward and anchors the plant in the soil.
3. In most dicots, the elongating hypocotyl forms an arch that breaks through the soil and straightens in response to light, pulling the cotyledons and epicotyl out of the soil.
4. In monocots the epicotyl begins growing upward shortly after the initiation of root growth and is protected by its coleoptile; the single cotyledon remains underground.

IV. The Pine Life Cycle
A. Gymnosperms are "naked seeds" that are not protected and include the conifers.
B. Cones are the reproductive structures of pines.
1. Large, female cones bear two ovules (megasporangia) on the upper surface of each scale.
   a) Each ovule produces four haploid megaspores, three of which degenerate and one of which continues to develop into a female gametophyte.
   b) Over many months the female gametophyte undergoes mitosis, and two to six structures called archegonia form, each housing an egg that is ready to be fertilized (fig. 30.13a).
3. Small male cones have pairs of microsporangia borne on thin, delicate scales.
   a) Microsporangia produce microspores through meiosis.
   b) Each microspore eventually becomes a four-celled pollen grain (microgametophytes).

C. Pollination occurs when airborne grains drift down between the scales of female cones and adhere to drops of a sticky secretion.
1. The cone scales grow together, and the pollen tube begins growing through the ovule toward the egg.
2. The pollen grain must undergo two more cell divisions to become a mature, six-celled microgametophyte.
3. Two of six cells become active sperm cells, one of which fertilizes the egg cell.
4. Fertilization occurs about 15 months after pollination.

D. Within the ovule, the developing embryo is nourished by the haploid tissue of the megagametophyte, and the embryo becomes dormant and the ovule develops a tough, protective seed coat.

E. The gymnosperm life cycle differs from the angiosperm life cycle in several ways.
1. The reproductive structures are cones instead of flowers.
2. Ovules lie bare on female reproductive structures rather than embedded in their tissues.
3. Gymnosperms have no fruit.
4. Gymnosperms have single fertilization rather than double fertilization.
5. The haploid tissue of the female gametophyte rather than a

triploid endosperm nourishes the gymnosperm embryo.

V. Asexual Reproduction

A. Many plants reproduce by asexually forming new individuals by mitotic cell division.

B. The parent plant simply gives rise to genetically identical individuals called clones.

1. Adventitious buds form on roots of cherry, pear, apple, and black locust plants.

2. A few plants use their leaves to reproduce asexually, and the roots and shoots develop at their edges.

3. Asexual reproduction also occurs from modified stems and roots, and shoots form at intermittent nodes and eventually form new plants some distance from the parent plant.

4. Tubers are swollen regions of stems that grow below ground and produce nodes that can grow roots.

LEARNING OBJECTIVES

After reading this chapter, the student should be able to answer these questions:

1. What are the diploid and haploid phases of the life cycle of a sexually reproducing plant?

2. What are the functions of the parts of a flower?

3. How do the gametes of flowering plants come together?

4. What events follow fertilization to form and nourish the plant embryo?

5. What is a fruit, and how does it assist a plant's reproduction?

6. What factors provoke a seed to germinate?

7. How does sexual reproduction in pines differ from that in flowering plants?

8. How do plants reproduce asexually?

ANSWERS TO THE TEXT QUESTIONS

1. The sporophyte produces haploid spores through meiosis. Haploid spores divide mitotically to produce a multicellular haploid individual, the gametophyte. Eventually this generation produces haploid gametes that fuse to form a zygote.

2. The female and male portions of the plant are directly involved in reproduction. The egg-producing portion of the plant is the ovary, which gives rise to the style and the stigma. These are the receptacles for pollen. The male portion of the plant consists of pollen-bearing anthers that sit atop filaments in a whorl around the ovary. Petals and sepals of flowers are indirectly involved in reproduction. The petals are brightly colored and sometimes have UV markings to catch the eye of a pollinator. They may also give off an aroma that attracts specific pollinators. The sepals make up an outermost whorl for the protection of the other structures.

3. The male wasp likely touches the anthers and stigma.

4. The plant with only female or male structures cannot self-pollinate. Some plants physically isolate the anthers and stigma while others release pollen only after fertilization has occurred.

5. a. birds, b. beetles, c. bees.

6. Wind-pollinated species must disperse a lot of pollen since the mode of transmission is so unpredictable.

7. a. haploid—egg, sperm, megaspore, microspore; b. diploid—megasporangium, microsporangium; c. triploid—endosperm.

8. Germination requires absorption of water and a source of energy in order to grow. Others require a series of cold days before they germinate.

9. The life cycle of corn plants is like that of a spruce pine in that both have distinct alternations of generations. The conspicuous plant in each case is the sporophyte generation. The microgametophyte generation is represented by pollen and the megagametophyte generation develops from the megaspore. The two differ in that the pine has cones instead of flowers and ovules that lie naked on the female scales. The pine bears no fruit and has single fertilization rather than double. The haploid tissue of the

female gametophyte rather than a triploid endosperm nourishes the gymnosperm embryo.

ANSWERS TO THE "TO THINK ABOUT" QUESTIONS

1. A fruit is defined by biologists as the outgrowth of the ovary for the purpose of protecting the seed or for distribution.
2. If the pollinator visited other species, precious supplies of pollen would be wasted by being deposited in the wrong place.
3. By doing a chromosome study of tissues from each, you would know that the tissue with twice the number of chromosomes than the other was the diploid.
4. When environmental conditions were not conducive to getting gametes together by wind, water, or pollination, asexual reproduction would be more beneficial. However, under dramatic environmental changes, such as the depletion of nutrients in the soil, plant competition, predators, or climate changes, variation in offspring is beneficial as provided by sexual reproduction.
5. Flowering plants have alternating generations where a complete life cycle includes both a diploid and haploid stage. In humans there is no haploid stage.
6. Petals attract pollinators, which are necessary to bring pollen to the egg for fertilization.
7. The farmer wishes to have cross-pollination to get hybrid vigor.

AUDIOVISUAL MATERIALS

The Green Machine, Time Life Video, 49 min., 1978.

The Growth of Plants, Encyclopaedia Britannica Educational Corp., 20 min., 1962.

Nitrogen and Living Things, Universal Education and Visual Arts, 14 min., 1961.

The Plant as an Energy Trap, CRM McGraw-Hill, 17 min., 1967.

Plant Propagation: From Seed to Tissue Culture, Harvard University, 28 min., 1977.

31 Plant Responses to Stimuli

KEY CONCEPTS

Plants respond to their environment with the aid of hormones.

Plants move their leaves toward the sun to gather more light for photosynthesis.

Roots move deep into the ground to absorb water and nutrients, support the plant, and protect stored nutrients.

Leaves curl in response to touch to protect from damage, predators, and water.

Detection of environmental cues of light, water, nutrients, and temperature allow the plant to invest in its reproductive effort when it is most advantageous.

Circadian rhythms are programmed responses that act independently of environmental clues.

KEY TERMS

abscisic acid	gibberellins
acclimation	indoleacetic acid (IAA)
anthocyanins	nastic movements
auxins	nyctinasty
circadian rhythms	obligate photoperiodism
cytokinins	photoperiodism
day-neutral plants	phototropism
dormant	phytochrome
entrainment	positive tropism
ethylene	seismonasty
facultative	short-day plants
photoperiodism	thigmomorphogenesis
flavin	thigmotropism
geotropism	tropism

CHAPTER OUTLINE

I. Plant Hormones
   A. Plant growth is regulated by hormones, which are synthesized in one part of the organism and transported to another part, where they stimulate or inhibit growth.
   B. They are not produced in tissues specialized for hormone production,

and they do not have definite target areas.

   C. A single hormone can elicit numerous responses.
   D. Five major classes of plant hormone are known.
      1. Auxins.
         a) The first plant hormone to be identified.
         b) It stimulates cell elongation in grass seedlings and herbs by stretching their walls.
         c) The most active auxin is indoleacetic acid, which is produced in shoot tips, embryos, young leaves, flowers, fruits, and pollen.
      2. Gibberellins.
         a) Gibberellin (Ga) stimulates shoot elongation in mature regions of trees, shrubs, and a few grasses and is also found in immature seeds, apices of roots and shoots, and young leaves of flowering plants.
         b) It induces both cell division and elongation.
      3. Cytokinins.
         a) They stimulate cytokinesis (the division of the cell after the genetic material has replicated and separated) by pushing cells into mitosis.
         b) An auxin must be present before mitosis begins.
         c) They promote cell division and growth and participate in development, differentiation and senescence.
         d) Most cytokinins are found in roots and actively

developing organs such as seeds, fruits, and young leaves.
- 4. Ethylene.
  - *a)* This ripens fruit in many species.
  - *b)* Ripening is a complex process involving pigment synthesis, fruit softening, and breakdown of starches to sugars.
  - *c)* It is synthesized in all parts of flowering plants, but large amounts form in roots, the shoot apical meristem, nodes, and ripening fruits.
  - *d)* By blocking this gene, scientists can block ethylene production.
- 5. Abscisic acid (ABA).
  - *a)* This inhibits the growth-stimulating effects of many other hormones.
  - *b)* ABA is used commercially to inhibit growth of nursery plants.
- 6. Hormonal interactions.
  - *a)* Several plant hormones influence abscission, which is the shedding of leaves or fruit.
  - *b)* Senescence begins when auxin production drops in response to environmental stimuli such as injury or the shorter days of autumn.
  - *c)* During senescence, cells in the abscission zone begin producing ethylene, which produces biochemicals that digest cell walls causing the cells to separate and the leaf or fruit to drop.
  - *d)* Synthetic auxins prevent preharvest fruit drop in orchards.

II. Tropisms
- A. Tropism refers to plant growth toward or away from environmental stimuli, such as light or gravity.
- B. Phototropism—responses to unidirectional light.

1. During phototropism, cells on the shaded side of a stem elongate faster than cells on the lighted side of the stem, and the rapid elongation of cells along the shaded side of coleoptiles is controlled by auxin coming from the apex.
2. Cells in the shade elongate more rapidly than cells in the light, which curves toward the coleoptile towards the light (fig. 31.6).
3. Only blue light induces phototropism.
4. The yellow pigment flavin transports auxin to the shaded side of the stem or coleoptile.

C. Geotropism—response to gravity.
1. Charles Darwin and his son Francis also studied geotropism, which is a growth response to gravity, and they found that decapped roots grew, but not downward, in response to gravity.
2. One hypothesis is that root tissue is far more sensitive to auxin than shoot tissue, and the amount of auxin accumulating is so great that it actually inhibits cell elongation.

D. Thigmotropism—response to touch.
1. Contact with an object is detected by specialized epidermal cells, which induce differential growth of the tendril.
2. The tendril completely encircles the object.
3. Auxin and ethylene control thigmotropism.

III. Nastic Movements
A. Nondirectional plant motions are called nastic movements.
B. Seismonasty.
1. A nastic movement resulting from contact or mechanical disturbance is seismonasty.
2. This depends upon a plant's ability to transmit rapidly a stimulus from touch-sensitive cells in one part of the plant to

responding cells located elsewhere.

3. Leaf movement in the "sensitive plant" mimosa.
   a) Touching the leaf elicits a reaction that causes the cell membrane of motor cells at the base of the leaflet to become more permeable to K^+ and other ions.
   b) As ions move out of the motor cells, water moves out of the motor cells by osmosis.
   c) This loss of water shrinks the motor cells, and seismonastic movement occurs.
   d) This decreases a leaf's chances of being eaten.
   e) Sharp prickles located along a leaf axis are exposed when the leaflets close, and motor cells secret noxious substances called tannins, both of which discourage hungry animals.

4. Movements of the Venus's flytrap.
   a) This results from irreversible increases in cellular size, which can be initiated by acidifying the cell walls to pH 4.5 and below.
   b) The leafy traps are built of two lobes, each of which has three sensitive "trigger" hairs overlying motor cells.
   c) When a meandering animal touches these hairs, signals are sent to the plant's motor cells, which then initiate transport of H^+ to the walls of epidermal cells along the outer surface of the trap.
   d) Motor cells use almost one-third of their ATP to pump the H^+ that acidifies the cell walls and closes the trap.
   e) A trap will not close unless two of its trigger hairs are touched in succession or one hair is touched twice.

C. Nyctinasty
   1. The nastic response caused by daily rhythms of light and dark is known as nyctinasty, or "sleep movement."
   2. The movement of the prayer plant's leaves is in response to light and dark and occurs by changes in turgor pressure of motor cells at the base of each leaf.
      a) In the dark, K^+ moves out of cells along the upper side and into cells along the lower side of a leaf base.
      b) This ion flux moves water by osmosis into cells along the lower side of the leaf bases, swelling them; cells along the upper side lose water and shrink.

IV. Thigmomorphogenesis
   A. Thigmomorphogenesis is the response to mechanical disturbances such as rain, hail, wind, animals, and falling objects.
   B. It is controlled by ethylene.

V. Seasonal Responses of Plants to the Environment
   A. Seasonal changes affect plant responses.
   B. Flowering—a response to photoperiod.
      1. Flowering reflects seasonal changes.
      2. Garner and Allard established the term photoperiodism, which is a plant's ability to measure seasonal changes by the length of day and night.
      3. The adaptive significance is that the day length is consistent due to the position of the earth as it travels around the sun.

4. Plants are classified into one of four groups, depending upon their responses to photoperiod.

    a) Day-neutral plants do not rely on photoperiod to stimulate flowering and include roses and snapdragons.

    b) Short-day plants require light periods that are shorter than some critical length and usually flower in late summer or fall, such as asters and strawberries.

    c) Long-day plants flower when periods are longer than a critical length such as spring or early summer and include lettuce, spinach, beets, clover, and corn.

    d) Intermediate-day plants flower only when exposed to days of intermediate length, growing vegetatively at other times; they include sugar cane and purple nutsedge (fig. 31.11).

5. Geographical distribution of plants is greatly influenced by their flowering response to photoperiod.

6. Photoperiod is sensed by leaves; plants whose leaves are removed do not respond to photoperiod changes.

7. Some plants will not bloom unless they are exposed to the correct photoperiod and are said to exhibit obligate photoperiodism.

8. Some will eventually flower even without an inductive photoperiod, and this response is called facultative photoperiodism.

C. Do plants measure day or night?

1. Researchers were startled to discover that plants responded to the length of the dark period rather than the light period.

2. Plant flowering requires a specific period of uninterrupted dark rather than uninterrupted light.

3. Short-day plants are more accurately described as long-night plants.

D. Phytochrome—is a pigment controlling photoperiodism.

1. Phytochrome exists in two interconvertible forms: P_r and P_{fr}.

    a) The inactive form of phytochrome is synthesized P_r, and when P_r absorbs red light, it is converted to P_{fr}, which is the active form of phytochrome in flowering. P_{fr} promotes flowering of long-day plants and inhibits flowering of short-day plants; P_{fr} is converted to Pr when it absorbs far-red light (fig. 31.12).

    b) During the day, P_r is converted to P_{fr}, and this abundance of P_{fr} could signal the plant that it is in light.

    c) When the plant is placed in darkness, P_{fr} is slowly converted to P_r.

2. A plant can respond to a photoperiodic initiation of flowering only if it is reproductively mature.

E. Other responses influenced by photoperiod and phytochrome.

1. Seed germination is also affected by phytochrome.

    a) Red light stimulates germination, and far-red light inhibits germination.

    b) Seeds alternately exposed to red and far-red light are affected by the last exposure only.

    c) The phytochrome system can inform a seed that sunlight is nearby for photosynthesis and thus promote germination.

2. Phytochrome also controls the early growth of seedlings.

a) Seedlings grown in the dark have abnormally elongated stems, small roots and leaves, and a pale color; this condition is termed etiolated (fig. 31.14).

b) By rapidly elongating, etiolated plants reach the light before exhausting their food reserves.

c) Red light controls transformation from etiolated to normal growth.

3. Phytochrome may also help direct shoot phototropism.
   a) A gradient is established from P_r and P_{fr} across the stem.
   b) This gradient of phytochrome could bend a shoot.
   c) It also promotes stem elongation, and P_{fr} inhibits this.

4. Another function of phytochrome is to provide information about shading by overhead plants, and since P_r promotes stem elongation (as in etiolated plants), information provided by the phytochrome system can help a plant reach sunlight more rapidly.

5. Horticulturists use photoperiodism to produce flowers when they are wanted for sale.

F. Senescence.
   1. This is aging and also a seasonal response of plants.
   2. Aging occurs at different rates in different species.
   3. It is an energy-requiring process brought about by new metabolic activities.
   4. It begins during the shortening days of summer as nutrients are mobilized and proteins broken down.
   5. By the time a leaf is shed, most of its nutrients have long since been transported to the roots for storage.

6. The destruction of chlorophyll in leaves is part of senescence in which carotenoid pigments, which were previously masked by chlorophyll, become visible.

7. Senescing cells produce pigments called anthocyanins, which contribute to the color.

G. Dormancy.
   1. Before the onset of harsh environmental conditions such as cold or drought, plants often become dormant, a state of decreased metabolism.
      a) This involves structural and chemical changes, and cells synthesize sugars and amino acids, which function as antifreeze, preventing or minimizing cold damage.
      b) These changes in preparation for winter are called acclimation.
   2. Growth resumes in the spring as a response to changes in photoperiod and/or temperature.
      a) Lengthening days release dormancy.
      b) In some plants, dormancy is triggered by other factors, such as when rainfall releases the dormancy in desert plants.

VI. Circadian Rhythms
A. Some rhythmic responses in plants are not seasonal.
   1. Regular, daily rhythms are called circadian rhythms.
   2. Other plant activities that occur in daily rhythms include cell division, stomatal opening, protein synthesis, secretion of nectar, and synthesis of growth regulators.

B. Circadian rhythms in many species do not coincide with a 24-hour day.
   1. Circadian rhythms are probably controlled internally by a little-understood mechanism called a biological clock.
   2. A plant's circadian rhythm may be altered by environmental

factors, such as a change in photoperiod, regardless of the rhythm's internal control; this resynchronization of the biological clock by the environment is called entrainment.

3. Flowers of a particular species open when they are most likely to be visited by pollinators, and for some plants this timing is quite precise.

LEARNING OBJECTIVES

After reading this chapter, the student should be able to answer the following questions:

1. How do hormones regulate a plant's growth and development?
2. What is the mechanism behind a plant's growth towards or away from a particular stimulus?
3. How do carnivorous plants trap insects?
4. How are plants affected by the length of daylight?
5. How do plants respond to seasonal changes?

ANSWERS TO THE TEXT QUESTIONS

1. Plants must deal with the environmental conditions they find themselves in since they cannot move like animals. They adapt by growing. Like animals they have hormones that respond to internal and external factors.
2. Gibberellin induces both cell division and elongation in more mature tissues than those affected by auxin. It stimulates elongation after a 1-hour delay, which is much slower than auxin-induced growth.
3. Thigmotropism is movement in response to touching an object. Specialized epidermal cells in the plant's tendrils grow differentially when touched. These cells are under the control of auxin and ethylene. Seismonastic movements involve touch-sensitive cells in one part of the plant that transmit a stimulus to cells located elsewhere. This is due to alterations of the membranes, which become more permeable to K^+ and other ions.

4. In the plant Memosa, K^+ ions move out of motor cells into surrounding areas followed by water. This shrinking of motor cells causes seismonastic movements. In Venus's-flytrap, H^+ ions are transported to epidermal cells from motor cells, which causes them to expand, shutting the leaf trap. In Prayer plants K^+ ions move out of cells on the upper side of the leaf to those on the lower side in response to darkness.
5. In phototropism plants move their leaves into a position where they can absorb more sunlight. Geotropism ensures that roots move deep into the soil to get water. Thigmotropism provides structural support by the wrapping of stems around objects. Nastic responses allow plants to protect themselves from predators and water loss. Response to photoperiods gives the plant clues when it is most advantageous to engage in reproductive effort.
6. Interruption of the light cycle made no significant difference, whereas interruption of the dark cycle resulted in a decrease in flowering.
7. Auxin appears to migrate to the shady side of the stem, where it stimulates elongation of cells. Since the sunny side has a slower rate of elongation, unequal growth results, tipping the stem with its leaves toward the sun. Light coming from one direction would create a gradient of Pr and Pr across the stem. Pr promotes stem elongation, while Pr inhibits it.

ANSWERS TO THE "TO THINK ABOUT" QUESTIONS

1. Scientists observed that seemingly related phenomenon occurred. For example, they noticed in studies of phototropism that red light inhibits flowering, whereas red light followed by far-red light reversed inhibition. They predicted a pigment existed that acted differently under varying conditions. When phytochrome was isolated, it was found that one form, P_{fr}, promoted flowering of long-day plants and inhibited flowering of short-day plants.

221

2. Plant hormones or their synthetic equivalents can be used to stimulate plants to flower early or late and fruit to ripen early or late.

3. To get bananas to ripen quickly, wrap the bananas in a plastic bag with oranges, which emit the chemical ethylene.

4. It requires stimulation of two hair triggers simultaneously or one hair twice.

5. Ragweed will only flower if the photoperiod is 14 hours or less, so it begins in the fall when the days are short. Spinach, which blooms if it has 14 hours or more of light, needs the long days of summer.

6. Phytochrome is similar to rhodopsin in that it is activated in the presence of light and converts to the inactive form in the dark.

AUDIOVISUAL MATERIALS

The Green Machine, Time Life Video, 49 min., 1978.

The Growth of Plants, Encyclopaedia Britannica Educational Corp., 20 min., 1962

Nitrogen and Living Things, Universal Education and Visual Arts, 14 min., 1961.

The Plant as an Energy Trap, Factory, CRM McGraw-Hill , 17 min., 1967.

Plant Propagation: From Seed to Tissue Culture, Harvard University, 28 min., 1977.

Plant Tropisms and Other Movements, Coronet Instructional Films, 11 min., 1965.

32 Plant Biotechnology

KEY CONCEPTS

Protoplast fusion results in hybrids that have
unpredictable combinations of traits and
may be difficult to propagate.

Artificial seeds allow for the encapsulation of
insecticides and herbicides along with the
embryo.

Somaclonal variants have a great degree of
variation.

Gametoclonal variation techniques produce
homozygous plants quicker than can be
produced by inbreeding plants.

Mutant selection allows for production of plants
that are resistant to various chemicals.

New varieties of plants can be produced by
introducing new combinations of
mitochondria and chloroplasts.

Transgenic plants are generated by introducing
foreign DNA into a cell and regenerating a
plant from it.

Agricultural biotechnology can provide valuable
new variants, but we must understand how
altering these organisms will affect the
world.

KEY TERMS

artificial seed	mutant selection
biotechnology	organelle transfer
callus	protoplast fusion
chlybrid	recombinant DNA
clonal propagation	technology
clones	somaclonal
cybridization	variation
electric discharge	somatic embryo
particle	somatic hybrid
acceleration	Ti plasmid
electroporation	transgenic
explants	
gametoclonal variation	

CHAPTER OUTLINE

I. It is estimated that by 1995, 5% of
agricultural output in the United States
will come from technologies that
manipulate plants at the cellular and
subcellular levels.

II. The Challenge of Agricultural
Biotechnology

   A. Biotechnology is defined as the use
of organisms or their components to
provide goods or services.

      1. It includes fermenting, cell
manipulation, and genetic
alterations.

      2. The first products of modern
biotechnology served the
medical, microbiological,
veterinary, and agriculture
communities.

   B. Plant biotechnology has lagged behind
similar work on animals and bacteria
because the arrangement of genetic
material in plants is more complex
than in organisms.

   C. Hurdles of biotechnology.

      1. One problem is growing a
mature plant from an initial cell
that has been manipulated,
demonstrating the new trait in
the regenerated plant, and
showing that the regenerated
altered plant passes on the
characteristic to its progeny.

      2. Another is being able to
understand which combinations
of individual altered traits
contribute desired
characteristics.

III. Traditional Plant Breeding Versus
Biotechnology

   A. The steps of biotechnology.

      1. An interesting trait is identified
and bred, or engineered, into a
plant whose other characteristics
comprise a valuable package.

      2. The new variety is then tested
in several different habitats and
during different seasons.

      3. Seeds are distributed to growers.

B. Because each sex cell brings with it a different combination of the parent plant's traits, offspring from a single cross are not genetically uniform.

C. Most biotechnologies begin with somatic (body) cells.

D. The somatic cells are usually genetic replicas or clones, which assure consistency.

IV. Protoplast Fusion—The Best of Two Cells

A. Protoplast fusion is when new types of plants are created by combining cells from different species and then regenerating a mature plant hybrid from the fused cell.

B. A protoplast is simply a plant cell whose cell wall has been removed by treatment with digestive enzymes.

C. Two protoplasts may join on their own, or they can be stimulated to do so by exposure to polyethylene glycol (an antifreeze), a brief jolt of electricity, or being hit by a laser beam (figs. 32.3 and 32.4).

D. A plant regenerated from a protoplast fusion of two plant types is called a somatic hybrid.

E. Protoplast fusion is more successful when the parent cells come from closely related species.

F. Protoplast fusion has limitations.
   1. Unpredictable results: single cells, fused cells of the same species, and the sought-after cells of two different types.
   2. Not all species form protoplasts that can be coaxed to fuse.
   3. Only some of the fusion products go on to divide and develop into plants.
   4. There is no guarantee that they will have useful traits.

V. Cell Culture

A. Explants are tiny pieces of plant tissue nurtured in a dish with nutrients and plant hormones.
   1. The cells lose the special characteristics of the tissue from which they were taken and form a white lump called a callus.
   2. Certain cells of the callus grow into either a tiny plantlet with shoots and roots or a tiny embryo.
   3. An embryo grown from a callus is called a somatic embryo.

B. Embryos or plantlets from a callus.
   1. Most embryos or plantlets grown from a single callus are genetically identical.
   2. Sometimes the embryos or plantlets differ from each other because certain of the callus cells have undergone genetic change (mutation).
   3. By altering the nutrients and hormones in the callus culture new varieties may be sought.

C. Cell culture for uniformity.
   1. Somatic embryos as artificial seeds.
      a) Artificial seed can be fashioned by suspending a somatic embryo in a transplant polysaccharide gel containing nutrients and hormones and providing protection and shape with an outer biodegradable polymer coat (fig. 32.7).
      b) Biotechnologists can improve upon nature by packaging somatic embryos with pesticides, fertilizer, nitrogen-fixing bacteria, and even microscopic parasite-destroying worms.
      c) The advantage of artificial seeds for the farmer is their guarantee of a uniform crop because they are genetically identical.
      d) Artificial seed technology so far has been most successful for dicots.
   2. Clonal propagation.
      a) Clonal propagation is when cells or protoplasts are cultured in the laboratory and then grown into genetically identical plants (clones) (fig. 32.8).

b) Advantages are that it can speed growth of plants that are difficult to grow or are slow growing.

c) They are often too costly to be practical.

D. Cell culture for variety.

1. Somaclonal variation.

a) These are unusual plants arising from protoplast or cell culture, and they are derived from a single somatic cell (the one that gave rise to the callus).

b) New plant varieties arise literally once in a million by spontaneous mutation, but somaclonal variants arise much more frequently.

c) Possibly a genetic variant would not be noticeable if only one leaf cell among thousands became obvious when the cell that contains it gives rise to an entire new plant.

d) Somaclonal variation makes it possible to alter or add traits one at a time to an existing genetic background.

2. Gametoclonal variation.

a) Genetically variant plantlets that grow from callus initiated by sex cells are said to arise from gametoclonal variation (fig. 32.10).

b) Because such a callus consists of mass-produced sex cells, each cell has half the number of chromosomes found in somatic cells of that particular species.

c) A plant regenerated from such a gamete-derived callus cannot itself form gametes, and so it cannot reproduce.

d) To get around this drawback, gametoclonally derived plantlets are exposed to the drug

colchicine, which creates a polyploid.

3. Mutant selection.

a) Specific characteristics of new plant variants arise in cell cultures by exposing cells or protoplasts to noxious substances and then selecting only those cells that survive.

b) Looking for genetic variants that offer a desired characteristic is called mutant selection.

c) Seeds can be tailored to be resistant to particular herbicides.

VI. Altering Organelles

A. Chloroplasts and mitochondria are good candidates for organelle transfer because they contain their own genes, some of which confer such traits as male sterility (important in setting up crosses), herbicide resistance, and increased efficiency in obtaining and using energy.

B. Cybridization is a technique that produces a plant cell having cytoplasm derived from two cells but containing a single nucleus.

1. This is created by fusing two protoplasts, then destroying the nucleus of one with radiation.

2. In another approach, individual chloroplasts or mitochondria are isolated and encapsulated in a fatty bubble called a liposome.

3. The liposome can transport its contents across the cell membrane into a selected cell.

4. Introducing a chloroplast in this manner creates a cell called a chlybrid; sending in a mitochondrion produces a mibrid (figs. 32.12 and 32.13).

VII. Within the Nucleus—Recombinant DNA Technology

A. In recombinant DNA technology, single genes are transferred from a cell of one type of organism to a cell of another.

B. Multicellular organisms that contain a "foreign" gene in each of their cells, resulting from foreign DNA introduced at the fertilized egg stage, are called transgenic.
C. Recombinant DNA technology steps.
  1. First the researcher must identify and isolate an interesting gene.
  2. The donor DNA as well as the "vector" DNA are then cut with the same restriction enzyme so that they can attach to each other at the ends to form a recombinant molecule.
  3. The vector and its cargo gene are then sent into the plant cell.
  4. Dicots.
     a) Foreign genes are introduced into cells by the plasmid in the microorganism, which then causes a cancerlike growth called crown gall disease (figs. 32.14 and 7.10).
     b) A Ti plasmid in nature causes a cancerlike growth called crown gall disease.
  5. Genetic engineering of the monocots.
     a) This requires more creative approaches because many naturally occurring plasmids do not enter monocot cells.
     b) To use monocot protoplasts, you remove the cell wall, making the cell membrane more likely to admit foreign DNA.
     c) In electroporation, a brief jolt of electricity opens up transient holes in the cell membrane of monocots that may permit entry of foreign DNA.
     d) Genetic material can also be injected into monocot cells using microscopic needles or sent across the cell membrane within liposomes (fig. 32.15).
     e) Electric discharge particle acceleration (gene gun) shoots tiny metal particles, usually gold, that have been coated with the foreign DNA (fig. 32.16).
     f) Once foreign DNA is introduced into a target cell, it must enter the nucleus and then be replicated along with the cell's own DNA and transmitted when the cell divides.

VIII. Biotechnology Provides Different Routes to Solving a Problem
  A. In the past herbicides were devised to fit crops.
  B. Today, biotechnologists are altering crops to fit herbicides.

IX. Beyond the Laboratory—Release of Altered Plants to the Environment
  A. The first "deliberately released" genetically engineered organisms were bacteria that prevent ice crystal formation on crop plants.
  B. The concerns of agencies and the public.
    1. How long will the altered plant or bacterium survive?
    2. How quickly does it multiply?
    3. Has it been altered in a way not seen in nature?
    4. How far can the organism travel?
    5. What are the effects of the genetically engineered organism on the living and nonliving environment?
    6. Can the altered organism pass on its new characteristic—such as herbicide or disease resistance—to a weed?

LEARNING OBJECTIVES

After reading this chapter, the student should be able to answer these questions:

1. What is biotechnology?
2. What are the similarities and differences between plant biotechnology and traditional plant breeding?

226

3. How can protoplast fusion produce new plant varieties?
4. How can plant cells be grown in culture to produce identical offspring or variant offspring?
5. How can novel combinations of plant cell nuclei, cytoplasms, and organelles be developed?
6. How can recombinant DNA technology introduce new traits into plants?
7. How do dicots and monocots differ in their abilities to be manipulated by specific biotechnologies?
8. What are some of the environmental questions that must be addressed when plants altered by biotechnology are field tested?

ANSWERS TO THE TEXT QUESTIONS

1. A mature plant must be regenerated from the engineered cell and express the desired trait in the appropriate tissues at the right time in development. The plant must then pass the characteristic on to the next generation.
2. Biotechnology is similar to traditional plant breeding techniques in that the grower is still selecting for those individuals that are most useful. It still requires that a gene be identified in nature that would have a benefit. It is different in that the desired gene can be isolated and inserted into the genome of a plant without eliminating any of the other genes. In traditional methods, many generations of plants have to be produced to get the right combination of genes.
3. A plant biotechnology that is very precise is the practice of inserting a specific desired gene into a host cell and getting it to demonstrate the trait in the next generation. A technique called protoplast fusion results in the combination of genetic material from two different species. The result may not yield mature plants or those that are not useful.
4. Identify a variant with the characteristic of identical-looking flowers that are easy to pick and through clonal propagation produce genetically identical plants.

5. Isolate all variants grown on a sweet carrot plant calli. Select the sweetest and produce explants. The selection should be made for nonvariants.
6. Gametoclonal variation allows for a rapid production of homozygous plants compared with conventional techniques. The drawback is that the plants are haploid and therefore cannot reproduce. To solve this, they are treated with colchicine, which makes polyploids.
7. In protoplast fusion of monocots, cells must be taken from undifferentiated tissue, whereas in dicots, tissue can be taken from the adult plant. In recombinant DNA technology, the insertion of a gene through the cell membrane of a monocot requires the removal of the cell wall and electroporation to permit the entry of foreign DNA. Other methods include microinjection, liposomes, and "gene guns."

ANSWERS TO THE "TO THINK ABOUT" QUESTIONS

1. The major benefit of biotechnology is to increase productivity and introduce new variants that are beneficial economically or medically. The major risk is that new genes may be introduced into the environment that are detrimental.
2. Another explanation for somaclonal variation is that the process itself enhances mutation.
3. A gene for an insecticide that is effective against potato beetle could be introduced into the genetic material of the plant. Isolate a plantlet that has grown from a single callus that has undergone a mutation, making it resistant to the insect. Produce a somatic hybrid from a protoplast fusion of a potato plant and a like species that is resistant to the potato beetle.
4. a. One thing that must be known is the relative danger of the altered *Pseudomonas syringae* to other organisms. It may not only kill the fungus that causes Dutch elm disease but kill other organisms as well.
   b. Whether Strobel acted hastily or not can be debated. It would seem that

additional tests should have been conducted to determine the effect of *P. syringae* on other organisms in the environment.

5. The field test of a crop plant genetically engineered to tolerate extreme heat next to your home should cause little difficulty given your understanding of plant biotechnology.

6. Education is probably the most important answer. This can take the form of educating people in the public schools and institutions of higher learning as well as the general public through television and the press.

AUDIOVISUAL MATERIALS

Agricultural Research at the University of Wisconsin, University of Wisconsin Instructional Service Center, 25 min., 1980.

The Green Machine, Time Life Video, 49 min., 1978.

The Growth of Plants, Encyclopaedia Britannica Educational Corp., 20 min., 1962.

Nitrogen and Living Things, Universal Education and Visual Arts, 14 min., 1961.

The Plant as an Energy Trap, Factory, CRM McGraw-Hill, 17 min. 1967.

Plant Propagation: From Seed to Tissue Culture, Harvard University, 28 min., 1977.

Plant Tropisms and Other Movements, Coronet Instructional Films, 11 min., 1965.

33 Darwin's View of Evolution

KEY CONCEPTS

Evolution is a genetic change in a population.

Islands provide geographical isolation where evolutionary changes can be observed.

Microevolution accounts for the mechanisms involved in small genetic changes that occur in a population.

Macroevolution accounts for the mechanisms involved in large changes in populations that result in speciation and extinction.

Rock strata reveal fossil records of when organisms lived.

Darwin's observations on his journeys lent support to geological findings by others; he also noted differences and similarities between organisms and the correlation of organisms' characteristics to their environment.

Darwin predicted that the different varieties of finches from the Galápagos ascended from a common ancestor, a phenomenon he called "descent with modification."

Natural selection, the tendency for the better adapted of each generation to survive to reproduce, is one mechanism by which evolution occurs.

Adaptive radiation refers to when several species can diverge from a common ancestor.

Evolution acts on preexisting variants in the population.

Lamarck thought that evolution was driven by traits acquired during an individual's lifetime to suit a particular need.

KEY TERMS

adaptive	microevolution
adaptive radiation	mosaic
coevolution	catastrophism
convergent evolution	natural selection
evolution	Neptunism
extinction	principle
fossils	of superposition
macroevolution	speciation

species
sexual selection

survival of
the fittest
uniformitarianism

CHAPTER OUTLINE

I. Land Evolution (fig. 32.2)
  A. Mutual dependence of two types of organisms is termed coevolution; examples are the extinct dodo bird and the Calvaria tree, which has nearly died out.
  B. Local populations confined to isolated habitats accumulate their own distinct sets of adaptations, so that the island is home to hundreds of species seen nowhere else on the planet.

II. Evolution in Action—A Tale of Three Islands
  A. Evolution is the process by which the genetic composition of populations of organisms changes over time.
  B. In sexually reproducing organisms, when changes accumulate that prevent members of one population from successfully breeding with members of another population, two species have diverged from the ancestral group.
  C. It is likely that over time, certain individuals were better able to survive and leave offspring than other individuals, because they had inherited variations of traits that were helpful, or adaptive, in that particular environment.
  D. One way that a new species is born is from isolation afforded by islands.

III. Macroevolution and Microevolution
  A. The appearance of new species (speciation) and the disappearance of species (extinction) are large changes termed macroevolution.
  B. Microevolution includes more subtle, incremental single-trait changes that accumulate to the point that two

groups of organisms can no longer interbreed.

C. Macroevolutionary events tend to span very long periods of time; microevolutionary changes happen so rapidly that they can be seen experimentally.

D. An example of microevolution is the influenza viruses, which change their surface proteins, reflecting underlying genetic mutation.

E. Evolutionary biologists debate whether macroevolutionary changes reflect the buildup of step-by-step microevolutionary changes or if the more sweeping events involve some other, yet unknown, genetic mechanism.

F. The discovery of genes explained the mechanics of evolutionary change.

G. The idea of evolution as a gradual change in life forms actually grew out of many observations made by several individuals; these concepts were crystallized into coherent theory by Charles Darwin.

IV. The Influence of Geology—Clues to Evolution in Rock Layers

A. Principle of superposition: it had been recognized since the 1600s that lower rock layers were older than those above them.

B. Several interesting theories attempted to explain how rock layers came to be (fig. 33.4).

1. Neptunism held that a single great flood organized the earth's surface, receding to reveal the mountains, valleys, plains, and rock strata present to this day.

2. Mosaic catastrophism held that a series of great floods molded the earth's features and were responsible for extinctions of some life forms and creations of others.

3. Uniformitarianism is the continual remolding of the earth's surface.

   a) James Hutton first noted the weathering of exposed rock on the land and the gradual deposition of sediments in bodies of water.

   (1) According to Hutton, the seas receded to reveal new, uplifted sediments, as ancient mountains were eroded down, slowly contributing future sediments to the sea.

   (2) These geological changes must have taken longer than the 6,000 to 10,000 years of earth's existence according to the Bible, and the long time of earth history was inferred because of the forces of geological changes.

   (3) Hutton published his ideas in the book *Theory of the Earth*.

   b) William Smith was a surveyor who discovered that lower strata were older than strata closer to the surface.

   (1) He proposed that the positions of certain fossils within rock layers indicated the relative times of existence of the organisms they represent.

   (2) Fossils could now be put into a relative time frame.

   c) Geologist Charles Lyell, in the nineteenth century, suggested that sandstone was once sand, shale rock was once mud, and islands were once active volcanoes; he described these ideas in his three-volume Principles of Geology.

V. The Voyage of the HMS *Beagle* (fig. 33.5)

A. Lyell's ideas fit nicely with what Darwin had observed on his excursions prior to boarding the HMS *Beagle.*

1. Darwin suggested that fossils, like sediments, were arranged in a chronological sequence, with those most like present forms occupying the highest, or most recent, layers.
2. Today we know that organisms of different species can evolve similar adaptations to similar environments, a phenomenon called convergent evolution (fig. 33.6).
3. Islands were the ultimate barriers and often lacked the types of organisms that could not get to them.

B. After the voyage.
1. Darwin brought back or described 14 distinct types of finch, each different from the finches seen on the mainland (fig. 33.7).
2. He thought that the different varieties of finch on the Galápagos were descended from a single ancestral type of bird that dwelled on the mainland.
3. He called this gradual change from an ancestral type "descent with modification."
4. Thomas Malthus's "Essay on the Principle of Population" stated that the size of a human population is limited by food availability, disease, and war, and individuals who were better able to obtain those resources would survive to reproduce, contributing some individuals like themselves to the next generation (fig. 33.8).
5. This would explain the observation that more individuals were produced in a generation than survived, and over time, challenges posed by the environment would "select" these better-equipped variants, and gradually the population would change in a process called natural selection.

a) The divergence of several types from a single ancestral type.
b) An example was from Darwin's observation: because each of the islands had slightly different habitats, different varieties of finches were selected on each one.
7. Alfred Russell Wallace, author of *On the Tendency of Varieties to Depart Indefinitely From the Original Type,* had seen the principles of evolution demonstrated among the diverse life forms of South America.

VI. Natural Selection—The Mechanism Behind Darwinian Evolution
A. An individual who is successful in mating is considered to be "fit," and reproductive success is what is meant by "survival of the fittest."
B. Any trait that ensures an individual's survival can make reproduction more likely.
C. Sexual selection.
1. Some traits that directly boost reproductive success in a population are due to this type of natural selection.
2. Such traits include elaborated feathers in male birds and horns in beetles and elk (fig. 33.9).
D. Natural selection does not produce perfection, but it reflects transient adaptation to a prevailing environmental condition.
E. Natural selection can mold the same population in different directions under environmental conditions.
F. Darwin recognized that the natural variations that play a part in the accumulation of new traits and the origin of species are those that are passed on to future generations, but he did not know how they arose or were transmitted.
1. Jean-Baptiste Lamarck thought that inherited variations were acquired during an individual's

lifetime through want or need and then passed on to the next generation.

2. Darwin thought that new characteristics were acquired by an individual's somatic cells and then transmitted to the sex cells, so that they could be passed to subsequent generations (table 33.1).

LEARNING OBJECTIVES

After reading this chapter, the student should be able to answer these questions:

1. What can be learned about the evolution of species when cataloging the types of organisms that populate islands over known periods of time?
2. What natural phenomena does the science of evolution consider?
3. How do geology and geography influence evolution?
4. How did Charles Darwin's observations of the distribution of organisms throughout the world contribute to his theory of evolution?
5. What is natural selection, and how does it explain and underlie evolutionary change?

ANSWERS TO THE TEXT QUESTIONS

1. Since islands are geographically isolated from the mainland, only limited species can fly or swim to them. A small population that is isolated will have a high amount of cross-breeding. This would result in new, rare phenotypes that would provide the variation for natural selection to work upon.
2. Microevolution refers to small genetic changes that occur over a relatively short period of time and result in variation in the species. Macroevolution refers to separation of a population into more than one species or the extinction of a species.
3. James Hutton contributed the idea that the earth was much older than formerly believed. This provided a timeframe that made possible the idea that gradual genetic change in populations gave rise to new species. William Smith gave Darwin the idea

that fossils could be put in a relative timeframe by studying rock strata. Fossil evidence showed that species had changed with time. Charles Lyell revealed that the earth's surface was continually changing. If surface environments continually change then species would have to change continually to remain adapted. Thomas Malthus identified the natural tendency of populations to overreproduce and that factors such as limited food, space, and aggression kept populations in check. Darwin concluded that since not all individuals in a population lived, the best adapted would survive to pass on those genes that made them more "fit."

4. Natural selection is the differential survival and reproduction of better-adapted variants in a population. It is one mechanism by which the genetic composition of a population changes over time.
5. Traits that lead to "fitness" are physical traits that allow the species to survive and have reproductive success.
6. Since the environment is constantly changing, species must continue to change genetically in order to stay adapted. Consequently there is no perfect or end form. Because variation in a population occurs by chance, the best adapted may be far from perfect.

ANSWERS TO THE "TO THINK ABOUT" QUESTIONS

1. They may have come via water mats or natural rafts. They may have swam or evolved from flighted birds already present on the island.
2. Uniformitarianism theory best explains Darwinian evolution because it states the environment is continually changing and there is pressure on each species to change to remain adapted.
3. The theory of evolution predicts that changes in species are transient adaptations to the environment. It assumes all species arose from common ancestors and that there are no perfect forms.

4. Darwin predicted that genetic variation in a population was the material upon which natural selection worked.

5. Though there is no right answer, the student should ask these questions: Is human behavior subject to the same laws of evolution as other traits? If so, does that necessarily justify not giving assistance to others?

AUDIOVISUAL MATERIALS

Evolution, National Film Board of Canada, 11 min., 1971.

Evolution—The Four Billion Year Legacy, Films Incorporated Education, 29 min., 1980.

Evolution and the Origin of Life, McGraw-Hill Training Systems, 35 min., 1972.

Evolution by DNA—Changing the Blueprint, Cinema Guild, 22 min., 1972.

The Evolution of Vascular Plants—The Ferns, Encyclopaedia Britannica Educational Corp., 16 min., 1962.

Natural Selection—Evolution at Work, BBC TV Media Guild, 24 min., 1978.

34 The Forces of Evolutionary Change

KEY CONCEPTS

On a small scale evolution affects the frequency of genes in a population.

Microevolution can be measured by applying the Hardy-Weinberg formula.

Hardy-Weinberg equilibrium is never met in large populations because of mutation, migration, natural selection, and nonrandom mating; in small populations it is never met because of genetic drift, which includes the founder effect and bottlenecks.

Harmful genes are introduced in a population by mutation and are maintained in the heterozygote condition.

Natural selection may be towards better-adapted individuals or towards two extreme expressions of a trait or the intermediate between two extreme phenotypes.

Occasionally a harmful allele is maintained because the heterozygote has some advantage.

Macroevolution refers to speciation and extinction.

Speciation depends upon natural selection and biogeography.

Once speciation occurs, populations stay separate due to both premating and postmating mechanisms.

Extinction results when species cannot adapt to a changing environment.

Specific mass extinctions have been linked to oxygenation of the atmosphere, formation of large continents, meteor impacts, and the hunting behavior of humans.

KEY TERMS

allopatric	disruptive selection
allopolyploid	ecological isolation
allopatric speciation	founder effect
autopolyploid	gene pool
balanced polymorphism	genetic drift
behavioral isolation	genetic load
biogeography	Hardy-Weinberg
directional selection	equilibrium

hypothesis	population
impact theory	population
industrial melanism	bottleneck
keystone herbivore	punctuated
hypothesis	equilibrium
paleontologists	stabilizing
plate tectonics	selection
Pleistocene	sympatric species
overkill	temporal isolation
polyploidy	tetraploid

CHAPTER OUTLINE

I. On a small scale, evolution reflects the changing representations of particular traits among groups of individuals, and on a large scale, evolution is the formation and the extinction of species.

II. Evolution After Darwin—The Genetics of Populations
    A. Genes influence evolution at the population level.
    B. A population is a broad term for any group of interbreeding organisms.
    C. Gene pool.
        1. All the genes in a population.
        2. The proportion of different alleles of each gene determines the characteristics of the population.
    D. Microevolution.
        1. This occurs when the frequency of an allele in a population changes.
        2. A gene's frequency can be altered when new alleles are introduced into a population by mutation; when individuals migrate between populations; when an environmental condition is more easily tolerated by those who have a particular phenotype (natural selection); and when genes are eliminated because individuals with certain genotypes do not reproduce.

234

III. When Gene Frequencies Stay Constant—Hardy-Weinberg Equilibrium
   A. The phenotype and genotype frequencies could be determined by applying a simple algebraic expression, the binomial expansion $p^2 + 2pq + q^2 = 1.0$, where p^2 represents homozygous dominant individuals, q^2 represents homozygous recessive individuals, and $2pq$ represents heterozygotes (fig. 34.2).
   B. The Hard-Weinberg equation is analogous to a monohybrid cross.
   C. This equation can reveal single-gene frequency changes that underlie evolution.
   D. If the proportion of genotypes remains the same from generation to generation, as indicated by the equation, then evolution is not occurring for that gene and this situation is called the Hardy-Weinberg equilibrium.
IV. When Gene Frequencies Change
   A. Migration and nonrandom mating.
      1. Many people choose mates who have particular characteristics, and therefore mating is not random.
      2. This is quite common in populations of all species.
   B. Genetic drift.
      1. A phenomenon where gene frequencies can change when a small group is separated from a larger population.
      2. By chance, the small group may not represent the whole.
      3. An example of genetic drift in human populations is the founder effect.
         a) This occurs when small groups of people leave their homes to found new settlements.
         b) An example is type B blood seen in the Asian population, from whom the Indians descended.
         c) The founder effect can also increase the proportion of an allele in a population.
         d) An example is the Afrikaner population of South Africa, who descended from a small group of Dutch immigrants.
      4. Genetic drift can occur when members of a small community choose to mate only among themselves, resulting in inbreeding.
         a) The frequencies of some genotypes are different among the Dunkers than among their neighbors who are not part of their society and different from people living today in the part of Germany that the Dunkers left between 1719 and 1729 to settle in the New World (table 34.1 and fig. 34.4).
         b) It also results from population bottleneck, a situation in which many members of a population die, and the numbers are then restored by mating among a few individuals; the new population has a much more restricted gene pool than the large ancestral population.
   C. Mutation.
      1. A major source of genetic variation; the changing of one allele into another.
      2. The spontaneous mutation rate varies for different genes and in different organisms.
      3. Genetic load is the collection of deleterious alleles in a population.
         a) The frequency of a mutant allele is maintained in a population in heterozygotes and by reintroduction by further mutation.
         b) Because of heterozygosity and mutation, all populations have some

alleles that would be harmful if homozygous.

4. Each person probably has four or five such deleterious recessive alleles, and when mating occurs among blood relatives, the chance of conceiving a homozygous recessive individual occurs as soon as two people mate who carry the same mutant allele, inherited from a shared ancestor.

D. Natural selection.
   1. The predominant color of moth populations also varies with the degree of industrialization in a particular geographic region.
   2. Natural selection is also evident in the declining virulence of certain human infectious diseases over many years, such as the spread of tuberculosis.

E. Types of natural selection.
   1. Industrial melanism and the decline in severity of tuberculosis illustrate directional selection, in which a previously "normal" characteristic of a population alters in response to a changing environment as the number of better-adapted individuals increases.
   2. Disruptive selection.
      a) Two extreme expressions of a trait are the most fit, such as a population of marine snails that live among tan rocks.
      b) The snails are either white and camouflaged while near the barnacles or tan and hidden while on the bare rock.
   3. Stabilizing selection.
      a) When extreme phenotypes are less adaptive.
      b) An intermediate phenotype has greater survival and reproductive success, for example, human body weight.

F. Balanced polymorphism—the sickle cell story.

1. This allows a genetic disease to remain in a population even though the illness clearly diminishes the fitness of affected individuals.
2. The disease persists because carriers have some advantage over those who have two copies of the normal allele.
3. Balanced polymorphism maintains sickle cell disease, an autosomal recessive disorder that causes anemia, joint pain, a swollen spleen, and frequent, severe infections; carriers of sickle cell disease are resistant to malaria.
4. Those people who inherited one copy of the sickle cell allele were better able to fight off the fever and chills of malaria.
5. Gradually, the frequency of the sickle cell allele in East Africa rose, from 0.1% to a spectacular 45% in 35 generations.

V. How Species Arise
A. Charles Darwin saw the evolution of species as a gradual series of adaptations molded by a changing environment selecting certain naturally occurring variants.
B. Speciation depends upon natural selection and the physical distribution of organisms, or biogeography.
   1. The original geographic isolation may have, by chance, created two subgroups with different allele frequencies.
   2. Two populations that are geographically isolated from one another are said to be allopatric, and formation of a new species initiated by geographic isolation is called allopatric speciation (fig. 34.9).
   3. For a new species to form, geographic isolation must be followed by reproductive isolation.
   4. Two closely related groups of organisms that occupy the same region but cannot reproduce with each other are called

sympatric species and may arise following geographical isolation or by itself.

C. Premating reproductive isolation.
1. In ecological isolation, members of two populations prefer to mate in different habitats.
2. In temporal isolation, they have different mating seasons.
3. In behavioral isolation, the organisms perform different repertoires of courtship cues as the male and female approach one another.

D. Postmating reproductive isolation.
1. Reproductive isolation due to chromosome incompatibility can occur among individuals of the same species if something creates subgroups with different organizations of genetic material.
2. Polyploidy.
   a) Instantaneous reproductive isolation also results from polyploidy.
   b) This can occur when meiosis fails, producing, for example, diploid sex cells in a diploid individual.
   c) If diploid sex cells in a plant self-fertilize, a tetraploid individual results, having four sets of chromosomes.
   d) When chromosome sets derive from the same species, the organism is called an autopolyploid.
   e) Polyploids can also form when gametes from two different species fuse, creating a hybrid from which an allopolyploid may develop.
   f) Polyploidy is rarely seen in animals, because the disruption in sex chromosome constitution usually leads to sterility; one exception is the grey tree frog *Hyla versicolor*.

VI. How Species Become Extinct
A. Extinction reflects the inability of organisms to adapt to an environmental challenge.
B. What causes mass extinctions?
1. Earth history has been marked by at least a dozen periods of mass extinctions (table 34.2).
2. Paleontologists, who study evidence of past life, can find clues in the earth's sediments to the catastrophic events that heralded the disappearances of many species over relatively short expanses of time.
3. Two general hypotheses have emerged in recent years.
   a) The impact theory.
      (1) A meteor or comet may have hit earth, sending debris skyward.
         (a) This dust and debris may have blocked the sun, resulting in a decline in photosynthesis.
         (b) With the decline in plant life, first herbivores and then carnivores would have died.
      (2) An extraterrestrial object landing in the ocean would be equally devastating.
         (a) Oxygen-poor waters would be shot upward in the turbulence.
         (b) Upper-dwelling organisms adapted to the oxygen carried in their water surroundings would die of oxygen starvation.

237

(3) Evidence for the impact theory includes centimeter-thin layers of earth that are rich in iridium, an element rare here but common in meteors (fig. 34.12).

b) The geological theory of plate tectonics.
   (1) The restlessness of the planet's rocks may explain some mass extinctions.
   (2) The plates of the earth's surface continually drift away from oceanic ridges.
   (3) The environmental changes thrust upon organisms must have been profound.
   (4) The shifting of continents often altered shorelines, diminishing shallow sea areas packed with life.

C. Mass extinctions through geological time.
   1. Oxygenation.
      a) This drastic change in the atmosphere happened as cyanobacteria evolved.
      b) Oxygen they released was toxic to the reigning anaerobic species, many of which died out as the more efficient energy users flourished.
   2. Formation of Gondwana.
      a) Life in the seas was severely disrupted as a huge continent, called Gondwana, formed and covered the south pole.
      b) This caused an ice age, resulting in the death of many fish species and nearly 75% of all marine invertebrates.

3. Formation of Pangeae.
   a) The Permian period of 240 million years ago saw the greatest mass extinction.
   b) The cause is thought to be the fusion of landmasses into a single gigantic continent, Pangeae.
4. Just 40 million years after the Permian extinction, 75% of marine invertebrate species vanished, perhaps due to a meteor impact.
5. Sixty-five million years ago, when 75% of all plant and animal species, including the great dinosaurs, disappeared, so did the plankton, microscopic food for many larger marine dwellers; this suggests a series of impacts rather than a single cataclysmic event.
6. Pleistocene overkill hypothesis.
   a) Humans from Asia crossed a land bridge over the Bering Strait to a corridor east of the Canadian Rockies, traveling down through North America; along the way they encountered 50 to 100 million large herbivores, which they hunted to extinction.
   b) The keystone herbivore hypothesis suggests that the demise of the large herbivores led to overgrowth of vegetation that these animals ate, and this vegetation change threatened the existence of smaller herbivores (fig. 34.13).

LEARNING OBJECTIVES

After reading this chapter, the student should be able to answer these questions:

1. What is a population?
2. What is a gene pool?

238

3. Under what conditions does evolutionary change not occur?
4. How can algebra be used to describe whether or not microevolution is occurring?
5. How can migration, nonrandom mating, mutation, and natural selection alter gene frequencies?
6. How do some genetic diseases remain in populations, even though they decrease reproductive success?
7. How do species arise?
8. How do species become extinct?

ANSWERS TO THE TEXT QUESTIONS

1. a. 40%; b. 60%; c. 48; d. 16; e. 36.
2. Predation under these conditions results in a decreased chance of survival for the green fraggles. Since the gray variety can flourish, the frequency of the gray-green will increase in the population.
3. Those individuals who survived tuberculosis were those with a lesser virulent infection than those populations of bacteria that killed their host. Hence the less virulent strain became more prevalent.
4. Stabilizing selection is demonstrated when the average phenotype is selected.
5. Since a chimera is an individual who consists of tissue of more than one genotype, the organism will likely pass on either the goat or the sheep chromosomes, depending on which cell line becomes the germ line. Therefore it will have no evolutionary effect.
6. In rock strata that is devoid of fossils, the element iridium which is not usually found on earth but often found in comets, would suggest an extraterrestrial cause of extinction.

ANSWERS TO THE "TO THINK ABOUT" QUESTIONS

1. Each student will have his or her own opinion but may want to reconsider the scientific method before answering. In science, many hypotheses are often offered to explain a particular phenomenon. Each is tested to determine which is most likely to be true. Even a well-supported theory is only a tentative explanation and may continually undergo scrutiny to see if it is still thought to be true.
2. Three characteristics of the human population in the United States that violate the Hardy-Weinberg equilibrium are nonrandom mating, migration, and mutation.
3. Yes, if they can mate and produce fertile, viable offspring.
4. In order for gene therapy to have an effect on evolution it must be applied to the reproductive cells in gonads so that the next generation has a lower frequency of the defective genes.
5. a. The underground population may have undergone natural selection to make them better adapted to that environment, while the smaller group above ground may have responded to genetic bottleneck.
   b. This is also an example of bottleneck.
   c. Natural selection through an artificial process would result in a selected small population that would then change on the new planet due to founder effect.

AUDIOVISUAL MATERIALS

Evolution, National Film Board of Canada, 11 min., 1971.
Evolution—The Four Billion Year Legacy, Films Incorporated Education, 29 min., 1980.
Evolution and the Origin of Life, McGraw-Hill Training Systems, 35 min., 1972.
Evolution by DNA—Changing the Blueprint, Cinema Guild, 22 min., 1972.
The Evolution of Vascular Plants—The Ferns, Encyclopaedia Britannica Educational Corp., 16 min., 1962.
Natural Selection—Evolution at Work, BBC TV Media Guild, 24 min., 1978.

35 Evidence for Evolution

KEY CONCEPTS

Evidence for evolution comes from past life as well as clues in present-day organisms.

Fossil records are left when minerals replace parts of an organism.

In order to leave fossils, organisms had to be buried rapidly to avoid oxygen.

Fossils are dated by the rock layer where they are found, isotope ratios, altered crystal structures, and amino acid isomer ratios.

Homologous structures reflect ancestry, whereas analogous structures do not.

Vestigial organs may be left over from ancestors, where they served a function.

Evolutionary trees are constructed by evidence from chromosome bands and gene and protein sequences.

Humans are most closely related to chimpanzees.

Life's history is organized into a geological time scale according to eras, periods, and epochs.

The Precambrian era accounts for five-sixths of the earth's existence and included forerunners of bacteria, algae, fungi, and animals.

The Cambrian period saw an explosion of life forms.

The Ordovician period marked the appearance of the first vertebrates.

Life came to the land during the Silurian period.

Bony fishes evolved during the Devonian period.

During the Carboniferous period, amphibians and reptiles first appeared.

During the Triassic period of the Mesozoic era, thecodonts shared the gymnosperm forests with therapsids the ancestors of modern mammals.

Dinosaurs reigned during the Jurassic period and angiosperms debuted.

By the end of the Mesozoic era, many species were extinct.

The Cenozoic era was dominated by mammals, first by marsupials and monotremes and then by placental mammals.

Formation of land bridges allowed for migration to the American continents.

The common ancestor of the gibbons, apes, and humans, the *Aegyptopithecus,* lived about 30 to 40 million years ago.

The first hominoid was *Dryopithecus,* followed by *Ramapithecus,* and then *Australopithecus.*

Homo habilis followed *Australopithecus* and gave rise to *Homo erectus.*

Homo erectus gave rise to *Homo sapiens,* who were followed by Neanderthals and then Cro-Magnons.

Modern humans appeared about 40,000 years ago.

KEY TERMS

allosaurs	graptolites
amino acid	half-life
racemization	hominoids
amniote egg	homologous
analogous	ichthyosaurs
apatosaurs	Jurassic period
archaeopteryx	maiasaurs
brachiopods	marsupials
Cambrian period	molecular
Carboniferous	evolution
period	monotremes
cotylosaurs	nautioloids
Cretaceous period	Neanderthals
Cro-Magnons	Ordovician
Crossoptergians	period
Devonian period	paleontologists
diplodocus	pelycosaurs
DNA hybridization	periods
electron-spin	Permian period
resonance	phylogenies
epochs	placental mammals
eras	Precambrian era
eurypterids	psilophytes
evolutionary tree	radiometric dating
diagrams	Ramapithecus
fossil	relative date
geological	Silurian period
time scale	stegosaurs

thecodonts	Triassic period
therapsids	trilobites
thermoluminescence	vestigial

CHAPTER OUTLINE

I. Paleontologists use fossils to reconstruct evolutionary relationships between species, which are called lineages, or phylogenies.

II. In the biological science of evolution, the pictures painted by these different types of evidence replace the experimental portion of the scientific method.

III. Fossils
   A. How fossils form.
      1. Many fossils are hard parts of organisms that have been replaced by minerals.
      2. The living matter of trees can be replaced with minerals, producing petrified wood.
      3. Evidence of our recent relatives often consists of teeth.
      4. Bones are fossilized when minerals replace cells.
      5. When sudden catastrophes rapidly bury organisms in an oxygen-poor environment, tissue damage is minimal.
         a) Even soft-bodied life forms can leave exquisitely detailed portraits of their anatomies.
         b) A block-long section of the Canadian Rockies called the Burgess shale, for example, houses an incredibly varied collection of 530-million-year-old soft invertebrates.
      6. A perfectly preserved baby mammoth was found in the ice of the Arctic circle in Siberia.
      7. Insects and frogs are found trapped in sticky tree resin tar pits.
      8. Microscopes can aid in identifying fossils.
         a) Colonies of microscopic organisms are sometimes preserved.
         b) Dinosaur footprints and worm borings reveal how the animals that made them traveled.
      9. Much evidence was destroyed by the formidable forces of nature, and a species may have lived for such short periods of time or dwelled in such restricted areas that their fossils, if they left any, have never been discovered.
   B. Determining the age of a fossil.
      1. A relative date is based on how far beneath the surface the rock layer lies in which the fossil is found.
      2. Radiometric dating involves using natural radioactivity such as a "clock."
         a) A date obtained in this way is absolute because it is expressed in a number of years, although it is not exact.
         b) Isotopes of certain elements are naturally unstable, causing them to emit radiation.
         c) Each radioactive isotope decays to its alternate form at a characteristic and unalterable rate, called its half-life, or the time it takes for half of the isotopes in a sample of the original element to decay into the second form.
         d) Fossils up to 40,000 years old are radiometrically dated by measuring the proportion of carbon 12 to the rarer carbon 14.
         e) When an organism dies, however, its intake of carbon 14 stops, and from then on, carbon 14 decays to the more stable carbon 12, with a half-life of 5,710 years.
         f) Carbon dates extend back to 40,000 years ago, but potassium-argon dates do not begin reliably until 300,000 years ago.

241

3. Electron-spin resonance and thermoluminescence.
   a) These techniques measure tiny holes made in crystals over time due to exposure to ionizing radiation.
   b) Each method counts the holes differently and works on samples up to 1 million years old.
4. Amino acid racemization.
   a) This chronicles the rate at which amino acids in biological matter alter to mirror-image chemical structures called isomers.
   b) It is used for dating eggs and shells up to 100,000 years old.
   c) The fact that electron-spin resonance and thermoluminescence values are perturbed by natural deposits of radioactive substances, and that amino acid racemization is affected by temperature and moisture, means that results must be corrected for these influences.

IV. Comparing Structures in Modern Species
   A. Comparative anatomy.
      1. The more similar two living species are, the more recently they diverged in their lineage from a common ancestor.
      2. Similarly built structures with the same general function that are inherited from a common ancestor, such as the vertebrate skeleton, are termed homologous.
      3. Such structures that are similar in function, but not in architecture, are termed analogous.
   B. Vestigial organs.
      1. A structure that seems not to have a function in an organism, yet resembles a functional organ in another type of organism.
      2. Human vestigial organs include wisdom teeth and the appendix.

C. Comparative embryology.
   1. Darwin suggested that the striking similarity between vertebrate embryos reflects adaptations to their similar environments.
   2. For example, embryos float in a watery bubble either in the mother's uterus or in an egg.

V. Molecular Evolution
   A. The greater the molecular similarities between two modern species, the closer their evolutionary relationship.
   B. Even more direct evidence is found in preserved DNA from extinct species, which can be amplified and sequenced and compared to DNA from its descendants.
   C. Comparing chromosomes.
      1. The number, shape, and banding patterns of stained chromosomes can be compared as a measure of relatedness.
      2. Human chromosome banding patterns match closest those of chimpanzees, then gorillas, and then orangutans.
   D. Comparing protein sequences.
      1. The fact that all species utilize the same genetic code to build proteins argues for a common ancestry to all life on earth, and in addition, many different types of organisms use the same proteins, with only a slight variation in amino acid sequence.
      2. The amino acid sequences of dozens of proteins have now been deciphered in many species to track lineages, and quite often the results are consistent with fossil or anatomical evidence.
   E. Comparing DNA sequences.
      1. DNA hybridization.
         a) Examining genomic similarity relies on complementary base pairing.
         b) DNA double helices from two species are unwound and mixed together.

c) The rate at which hybrid DNA double helices reform—that is, DNA molecules containing one helix from each species—is a direct measure of how similar they are in sequence. The faster the DNA from two species forms hybrids, the more closely related the two species are.

2. The rate of base change in DNA is used as a "molecular clock," just as the rate of radioactive decay of an isotope is used as a natural clock in radiometric dating.

3. Evolutionary tree diagrams indicate on a time scale the "branching points" when two species diverged from a common ancestor.

4. Mitochondrial DNA provides an even better molecular clock, because it mutates about 5 to 10 times faster than nuclear DNA.

VI. The History of Life on Earth
A. The Geological time scale.
1. It is divided into major eras of biological and geological activity lasting vast stretches of time.
2. The division is into periods within these eras, and finally epochs within some of the periods.

B. Precambrian life.
1. The fossils resemble more recent fossils and modern prokaryotes (bacteria and cyanobacteria).
2. A billion years ago algae and fungi left fossils.
3. Between 700 and 600 million years ago, a profound biological change took place—multicellular life appeared.
4. Unicellular organisms may have formed colonies and then undergone a division of labor, with certain cells specializing in certain functions eventually to form a coordinated multicellular organism.

C. The paleozoic era.
1. Cambrian period (600 million years ago).
a) Ancestors of all modern animal phyla debuted.
b) There were abundant soft-bodied organisms, including algae, sponges, jellyfish, and worms.
c) The hard-bodied organisms included trilobites, nautiloids, scorpionlike eurypterids, and brachiopods.
2. The Ordovician period (500 to 435 million years ago).
a) Organisms consisted of graptolites, so-named because their fossilized remains resemble pencil markings.
b) The first vertebrates to leave fossil were jawless fishes from this period.
3. Silurian period (435 to 395 million years ago).
a) Organisms first ventured onto the land.
b) They were odd-looking plants called psilophytes.
4. The Devonian period (395 to 345 million years ago).
a) This was the "Age of Fishes."
b) As well as fishes with skeletons built of cartilage or bone, corals were abundant.
c) The Crossoptergians, or lobe-finned fishes, lived in the Devonian period and were probably ancestral to amphibians because they could gulp air and use its oxygen: fleshy fins allowed them to haul themselves along the land that separated shallow pools.
5. Carboniferous period (345 million years ago).
a) This was the "Age of Amphibians," the

243

descendants of the lobe-finned fishes that were prominent on the land.

    b) Fossil evidence of these early reptiles has not been found.

    c) When plants died, they were buried beneath the swamps to form, over the millennia, coal beds, and so the Carboniferous period is also called the "Coal Age."

6. Permian period (275 to 225 million years ago).

    a) Half the families of vertebrates and more than 90% of the species dwelling in the shallow seas became extinct, but reptiles were becoming more prevalent.

    b) The reptile introduced a new biological structure, the amniote egg, in which an embryo could develop completely, without the need to be laid in water.

    c) Cotylosaurs were early Permian reptiles that gave rise to the dinosaurs, as well as to modern reptiles, birds, and mammals; they coexisted with their immediate descendants, the pelycosaurs, or sailed lizards.

D. The Mesozoic era.

1. Triassic period (225 million years ago).

    a) This was the "Age of Reptiles."

    b) The thecodonts were small ancestors of the great dinosaurs that flourished.

    c) Other animals called therapsids were reptiles, but they held their limbs and heads in a position more like those of mammals, and their teeth were more mammalian than reptilian.

    d) At the close of the Triassic period, about 190 to 185

million years ago, the numbers of both theocodonts and therapsis were dwindling as much larger animals began to infiltrate a wide range of habitats.

2. Jurassic period (185 to 135 million years ago).

    a) The dinosaurs had invaded nearly all habitats, from the ichthyosaurs in the seas, to archaeopteryx in the air, to the familiar apatosaurs (brontosaurs), stegosaurs, diplodocus, and allosaurs that dwell in the second half of the period.

    b) The first flowering plants (angiosperms) appeared, but forests still consisted mostly of tall ferns and conifers, club mosses, and horsetails.

3. Cretaceous period (135 to 65 million years ago).

    a) Angiosperms spread in spectacular diversity, and the number of dinosaur species declined.

    b) Triceratops were widespread.

    c) the reign of the great dinosaurs ended about 65 million years ago.

    d) the demise of the dinosaurs opened up habitats for many other species, including the primates that eventually gave rise to our own species.

E. The Cenozoic era.

1. At the start of the Cenozoic era 65 million years ago, many of the mammals were marsupials (pouched mammals) or egg-laying monotremes, ancestors of the platypus.

2. A new type of animal that evolved was the placental mammal, whose young remain within the female's body for a

relatively long time, where they are nurtured by a specialized organ, the placenta.

3. Then about 2 to 3 million years ago, the Bering land bridge rose connecting Asia to North America, and many mammals, including our ancestors, probably journeyed from what is now the Soviet Union to Alaska and southward through what is now Canada and the continental United States.

F. The evolution of humans.

1. Our species *Homo sapiens* (the wise human), probably first appeared during the Pleistocene epoch, about 200,000 years ago.

2. The ancestral primates were rodentlike insectivores.

3. *Aegyptopithecus.*

   a) This is a monkeylike animal about the size of a cat that appeared about 30 to 40 million years ago.

   b) These animals are possible ancestors of gibbons, apes, and humans.

4. Hominoids.

   a) These are animals ancestral only to apes and humans and appeared in Africa about 20 to 30 million years ago.

   b) They could swing and walk farther than *Aegyptopithecus.*

5. *Ramapithecus* are apelike animals that evolved in Africa about 15 million years ago.

6. *Australopithecus.*

   a) Hominoid and hominid (ancestral to humans only) fossils from 4 to 8 million years ago are scarce; this is thought to be a time when the stooped, large-brained ape gradually became the upright, smaller-brained ape-human.

   b) By 4 million years ago, an animal known as

Australopithecus ventured forth.

   c) These fossils have been found with those of animals that grazed, indicating that this ape-human had left the forest.

   d) These animals stood about 4 to 5 feet tall and had a brain about the size of a gorilla's but teeth that were very much lie those of humans.

   e) Four species of *Australopithecus* existed by 3.6 million years ago, and a form called *A. afarensis,* represented by the famous fossil "Lucy," may have been our direct ancestor (fig. 35.19b).

7. *Homo habilis.*

   a) By 2 million years ago, Australopithecus coexisted with *Homo habilis.*

   b) "Habilis" means handy, and this primate is the first for whom evidence of extensive tool use exists.

   c) They coexisted with and were followed by *H. erectus,* who left fossil evidence of cooperation, social organization, and tool use, including the use of fire.

8. *H. erectus.*

   a) Some paleontologists believe that a species intermediate between *H. erectus* and *H. sapiens,* with a big brain and robust build, lived from 50,000 to 30,000 years ago.

   b) About 75,000 years ago, Europe and Asia were home to the Neanderthals, members of *Homo sapiens.*

   c) Neanderthals mysteriously vanished about 40,000 years ago, just as the light-weight, finer-boned, and

245

less hair Cro-Magnons, known for their intricate cave art, appeared.

LEARNING OBJECTIVES

After reading this chapter, the student should be able to answer these questions:

1. What is a fossil?
2. Why are fossils important to the study of evolution?
3. How do fossils form?
4. What methods are used to determine the age of a fossil?
5. Why is the fossil record incomplete?
6. How can comparative anatomy, embryology, and biochemistry provide clues to evolutionary relationships between species?
7. What were some of the major events in life history that we know about?
8. What sorts of animals living within the past 40 million years were probably our direct ancestors?

ANSWERS TO THE TEXT QUESTIONS

1. Fossils are remnants of prehistoric organisms that are formed when hard parts are replaced with minerals.
2a. Much of the fossil evidence has been destroyed.
2b. Some species only lived for short periods of time.
2c. Some species have never been discovered.
3a. Relative Date is based on how far beneath the surface the rock layer lies in which the fossil is found. Since different rock strata form at different rates, assigning fossils to particular layers cannot offer a precise date.
3b. Radiometric Dating. A date is obtained by measuring the decay of natural radioactive isotopes. Many sedimentary rocks contain isotopes. It is difficult to determine whether the isotope came from organisms or was natural to the rock.
3c. Amino Acid Racemization - This technique measures the rate at which amino acids in biological matter alter to mirror-image

chemical structures called isomers. Dating is impeded by temperature and moisture.
3d. Electron-spin resonance and thermoluminescence are physical techniques that measure tiny holes made in crystals over time due to ionizing radiation. Each method counts the holes differently and works on samples up to one million years old. Both methods are perturbed by natural deposits of radioactive substances.
4. The closer the percentage in nitrogen base sequence in DNA and amino acid sequence in protein, the more closely related are two species.
5. Vestigial organs are structures in modern day organisms which have no apparent function and are thought to be remnants left over from the past. Similar phases in embryology in different kinds of organisms suggest relatedness. Organisms that have homologous parts are thought to be closely related.
6. Homology refers to parts that have similar structure and function. Analogy refers to structures which have similar function but different origin.
7. The Carboniferous period is also known as the "Coal Age" because this is when the plants which would eventually be compressed to make coal, flourished on Earth.

ANSWERS TO THE "TO THINK ABOUT" QUESTIONS

1. As new techniques are developed, different ways of gathering scientific information becomes possible. Often what was once believed to be true is altered with more experimentation.
2. a. It is assumed that the closer the layer is to the surface, the more recent the organism lived.
   b. It is assumed that carbon 14 in fossils is decaying to the more stable carbon 12.
   c. It is assumed that first, mitochondrial DNA mutates faster than nuclear DNA, and second, that the more similar the mDNA between two species, the more recently they have descended from a common ancestor.

3. Land appears to move by tectonic plates that, when sliding past each other, elevate landmasses to mountains that were once ocean floors.

4. It was during this time that it is thought that the large-brained ape became the upright, smaller-brained, ape-human.

5.
   a. The fleshy fins allowed the lobe-finned fish to crawl on land.
   b. The amniote egg allowed animals to spend more of their life cycles on land, free from competition with animals confined to the oceans.
   c. The placenta made possible the carrying of offspring within the mother until birth, which enhanced survival.
   d. The ability to walk upright allowed primates to increase their foraging range by expanding their habitat to include grasslands.

6. Present-day conditions that would allow for fossilization include glacier formation, sedimentation in river basins, and floods. Oil spills may duplicate tar pit conditions in both terrestrial and aquatic environments. Land slides and volcanic ash continue to trap plant and animal populations.

AUDIOVISUAL MATERIALS

Evolution, National Film Board of Canada, 11 min., 1971.

Evolution—The Four Billion Year Legacy, Films Incorporated Education, 29 min., 1980.

Evolution and the Origin of Life, McGraw-Hill Training Systems, 35 min., 1972.

Evolution by DNA—Changing the Blueprint, Cinema Guild, 22 min., 1972.

The Evolution of Vascular Plants—The Ferns, Encyclopaedia Britannica Educational Corp., 16 min., 1962.

Natural Selection—Evolution at Work, BBC TV Media Guild, 24 min., 1978.

36 The Behavior of Individuals

KEY CONCEPTS

Behavior is controlled by environment and genes.

Closed behavior programs are mostly genetically controlled.

Open behavior programs are flexible and amenable to environmental change.

Innate behaviors are important when immediate success is required for survival.

Learning is constrained by genetics.

Ethology studies how behaviors are adaptive and follow rules of natural selection.

Fixed action patterns are innate, stereotyped behavior triggered by specific releasers.

A complex behavior may result from sequences of releasers.

Habituation is when an animal no longer responds to an unimportant stimuli.

In classical conditioning, an animal associates a known response to an unconditioned stimulus.

In operant conditioning, an animal voluntarily behaves to receive a reward.

Imprinting is a rapid form of learning that occurs during a critical time in development and does not require a reward.

Insight learning is when prior learning is applied to a new situation.

Latent learning uses prior experiences without a reward.

KEY TERMS

biofeedback	fixed action
biological clocks	pattern (FAP)
circadian	habituation
circalunadian	imprinting
circannual	innate
classical conditioning	insight learning
closed behavior	insomnia
program	learning
conditioned stimulus	map sense
critical period	melatonin
entrained	narcolepsy
ethology	negative
extinction	reinforcement

open behavior	sleep apnea
program	supernormal
operant	releaser
conditioning	suprachiasmatic
pineal gland	nuclei (SCN)
positive	unconditioned
reinforcement	stimulus
releaser	
seasonal affected	
disorder (SAD)	

CHAPTER OUTLINE

I. Behavior Is Shaped by Genes and Experience
   A. All behaviors are determined by heredity (nature) as well as the environment (nurture), each contributing to different degrees for particular responses.
   B. Behavior is limited by the anatomical systems for detecting stimuli, integrating the input, and responding.
   C. Learning can be very important in mastering certain very complex behaviors.
   D. Closed and open behavior programs.
      1. A closed behavior program is rigid and not easily modified by experience.
      2. An open behavior program is flexible and easily altered by learning.
      3. A primarily innate (and therefore mostly genetically determined) behavior occurs when it is critical that the action be performed correctly on the very first trial, such as escape behaviors.
   E. Genes may influence learning.
      1. Animals may be unable to be trained.
      2. The sparrow can only learn the song of its own species, and the bird innately recognizes its

species' song, perfecting the performance through experience.

3. In baby male zebra finches, one part of the brain loses half of its neurons, and in another region, the number of neurons increases by 50% during the 70-day period when songs are learned.

II. Innate Behavior

A. Ethology examines how natural selection shapes behavior to enable an animal to survive.

B. Fixed action patterns (FAP).

1. This is an innate, stereotyped behavior.

2. An example is a dog digging on the floor as if trying to bury a bone.

3. A FAP is performed nearly identically by every member of a species and is modified very little, if at all, by the environment.

4. The egg rolling response of a female greylag goose is a FAP.

5. Once a FAP is initiated, the action continues until completion, even in the absence of appropriate feedback.

C. Releasers.

1. The specific factor that triggers a FAP is called a releaser or sign stimulus.

2. The animal must select and respond to only the few key stimuli that are reliable cues to the situation requiring response.

3. Releasers are important in human parent-infant interactions, and many species retain juvenile characteristics of their ancestors, a phenomenon called neoteny.

4. Many releasers are auditory, for example, a mosquito is attracted to his mate by the buzz of her wings.

5. Other releasers are tactile, such as during the mating ritual of the stickleback fish: the female enters the nest, stimulating the male to thrust his snout against

her rump in a series of quick, rhythmic trembling movements.

6. Pheromones.

a) Airborne chemical releasers secreted by animals that influence the behavior of other members of the same species.

b) The sex attractants of some insects, crustaceans, fishes, and salamanders are pheromones (fig. 19.21).

7. A supernormal releaser.

a) A model that exaggerates a releaser and elicits a stronger response than the natural object.

b) For example, oystercatcher birds and the herring gull prefer to sit on large eggs.

D. Chain reactions.

1. A complex behavior, sometimes called a chain reaction, is often built from simpler units joined by a sequence of releasers; this sequence of behaviors is always performed in the same order because the completion of each step is the releaser for the next.

2. Chain reactions can also result from an exchange of releasers between two individuals, such as the mating behavior of stickleback fish.

E. Behavior is adaptive.

1. Because behavior has a genetic basis, it is shaped by natural selection and is therefore adaptive.

2. The ethologist Niko Tinbergen's studies of parental care in two bird species illustrate this point.

a) He hypothesized that white pieces of shell make the cryptic intact eggs more noticeable to predators such as crows.

b) The parent gull decreased the chance of a predator noticing unhatched eggs by removing broken eggshells.

c) For an oystercatcher who has no neighbors, it is better to remove the eggshells immediately so that predators are less likely to notice the unhatched eggs.

3. A behavior amenable to a more precise "cost-benefit" analysis is foraging, because calories used to gather food can be compared to calories gained in eating it.

   a) An example is crows that gather whelks (large marine snails) at low tide and drop the whelk over a rock so they can eat the meat inside the smashed shell.

   b) The greater the height from which the whelk is dropped, the better the chances of breaking it.

F. How genes can determine behavior.

1. One way to demonstrate that genes influence behavior is that individuals who perform a particular behavior in distinctly different ways are mated, and then the behavior of the hybrid offspring is observed.

   a) The genetic basis of nest-building behavior in two closely related species of lovebirds was shown by such hybridization studies.

   b) When offered sheets of paper, both species tear them into strips, and then the peach-faced lovebird tucks the materials in its rump feathers, while Fisher's lovebird transports them in its beak.

   c) Hybrids between the two species are befuddled.

2. Some behaviors can be associated with the protein product of a particular gene.

   a) For example, normal fruit flies given a shock in the presence of a certain odor learn to avoid that odor.

   b) Also, mutant fruit flies called "dunce" forget the association so quickly that they continually receive shocks; this is attributed to a gene that reduces the breakdown of cyclic AMP.

G. A genetic link to behavior has been identified in a species of snails called *Aplysia.*

1. The egg-laying behavior and physiological changes in Aplysia are brought about by proteins in the bag cells, two clusters of nerve cell in the abdomen.

2. A single gene codes for one large protein, which is cut to yield the several smaller proteins that oversee egg laying.

III. Learning

A. Defined as a change in behavior that results from experience.

B. Habituation.

1. In the simplest form of learning, habituation, an animal learns not to respond to certain irrelevant stimuli.

2. When a stimulus is presented many times without any consequences, the animal usually decreases its response, perhaps to the point where it is eliminated completely.

3. Habituation provides protection, since organisms learn not to escape from innocuous, common stimuli.

4. An example is seabirds, who learn not to show aggression toward their neighbors and retain aggression toward other birds that are encountered less frequently.

C. Classical conditioning.

1. When an animal learns to respond in a familiar way to a new stimulus.

2. A new association between stimulus and response is formed when the new, or conditioned, stimulus is presented repeatedly immediately before a familiar, or

unconditioned, stimulus that normally triggers the response.

3. After the stimuli have been paired in a series of trials, the new stimulus presented first is enough to elicit the response.

4. The most familiar example of this is that of dogs associating the sound of a bell with the presence of food.

D. Operant conditioning.

1. A trial-and-error type of learning in which an animal voluntarily repeats any behavior that meets with success, such as a reward (positive reinforcement) or avoiding a painful stimulus (negative reinforcement).

2. In animal training, operant conditioning first reinforces any behavior that vaguely resembles what the trainer wishes and then restricts the reward to better and better approximations.

3. This shapes human behavior in many ways, including the basis of a technology called biofeedback.

E. Imprinting.

1. A type of learning that occurs quickly, during a limited time (called the critical period) in an animal's life, and is performed usually without obvious reinforcement.

2. For example, young chicks, goslings, or ducklings are imprinted to follow the first moving object they see, which in the wild is usually their mother.

3. Mammal mothers usually will accept and nurse any young that they smelled during the critical period and reject any youngsters they do not recognize.

F. Insight learning.

1. Also called reasoning, this is the ability to apply prior learning to a new situation without observable trial-and-error activity.

2. For example, in one experiment, captive chimpanzees stacked

boxes to climb and reach a banana hanging from the ceiling.

G. Latent learning.

1. This takes place without any obvious reward or punishment and is not apparent until sometime after the learning experience.

2. Even without reinforcement, a rat that has been allowed to run freely through a maze will master its twists and turns more quickly than a rat with no previous experience in the maze.

IV. Biological Clocks

A. Many behaviors and physiological processes cycle.

1. It is often adaptive for an organism to perform a specific behavior at a particular time relative to predictable environmental changes.

2. Animals use inner timekeepers called biological clocks to sense when it is time to act if no reliable environmental cues are present.

3. In plants, biological clocks are crucial to photosynthesis, growth, and flowering (chapter 35).

B. A biological clock runs without environmental clues.

1. The most extraordinary feature of biological rhythms is that they are not caused by the regular day/night cycles in light and temperature.

2. However, the period length of a rhythm (the interval from one peak to the next) usually deviates slightly from the precise 24-hour frequency it displays in nature.

3. Rhythms that are no longer precisely 24 hours under constant conditions are termed circadian (about a day).

C. Disrupted clocks affect how we feel.

1. Until the circadian rhythms adapt to , or are entrained to, the new lighting regime,

physiological cycles are no longer synchronous with one another or with the environment.

2. This internal disharmony causes a range of physical and psychological disturbances, including insomnia, constipation, depression, irritability, and decreased intellectual function.

D. Physical basis of biological clocks.
1. It is known that multicellular animals seem to have several clocks operating in sync.
2. Three "master clocks" have been identified—the retina or other light-sensing structure; the pineal gland, which sits atop the brain stem; and two clusters of 10,000 neurons each in a region of the brain's hypothalamus called the suprachiasmatic nuclei (SCN).
   a) The clock that sets the pace varies among species.
   b) In mammals the SCN is the major clock.
   c) The hormone melatonin is thought to be involved in biological timing.
   d) From the results of crosses made between normal hamsters who exercised every 24 hours and a mutant strain that exercised every 20 hours, the behavior appears to be genetic.

E. Rhythms with other period lengths.
1. Tidal rhythms.
   a) On most of the coats of the world, the tide comes in twice during this interval, one each 12.4 hours.
   b) Many of the activities of life match these tidal rhythms.
   c) Tidal rhythms persist in the laboratory with a period length that is only approximately the same as it was in nature and are therefore termed circalunadian (about a lunar day).

2. Other rhythms have a monthly frequency, such as the reproductive activity of certain marine annelids who release gametes into the water synchronously at a particular time of the year.
3. Many rhythms have a period length of a year and are termed circannual, such as the hibernation of the ground squirrel.

V. Orientation and Navigation
A. Birds migrate north from the tropics using the positions of the sun or stars, subtle shifts in the earth's magnetic field, sights, sounds, and the direction of winds and the aromas they carry.
B. The simplest type of navigation relies on recognizing landmarks in a random search.
C. The compass sense.
1. Most migrating species use environmental cues such as a compass to orient them in one direction.
2. The sun is a compass.
   a) This was demonstrated when birds were placed in a cage with food in boxes positioned along the periphery.
   b) Trained birds could find food as long as the sun was visible but became disoriented when their view of the sun was blocked.
3. Some birds migrate at night, using the position of the stars as a compass.
   a) To show that nocturnal migrants use the stars as a compass, a bird was caged in a structure made of blotting paper rolled into a funnel with an ink pad on the floor and a ceiling of wire mesh.
   b) When the bird touched bottom, its feet were stained with the ink, so it

left telltale footprints wherever else it alighted; in the morning, the number of footprints at each compass direction were counted.
  - c) Most of the footprints were in the half of the funnel corresponding to the direction in which the birds would migrate.
  4. Evidence from pigeons suggests that the earth's magnetic field, which runs in a generally north-south direction, can serve as a compass.
    - a) Bar magnets placed on a pigeon's wings impair its ability to find its way home on cloudy days.
    - b) Further evidence of the role of magnetic cues in orientation is shown by giving pigeons hats made of small Helmholtz coils; when an electric current runs through the coil, a magnetic field is created that appears to disrupt a bird's flight pattern.
- D. Homing.
  1. The most complex navigational skill is the ability to home—that is, to return to a given spot after being displaced to an unfamiliar location using no environmental cues.
  2. Homing requires both a compass sense for direction and a map sense telling the animal where it is relative to home.
    - a) It may depend on regular variations in the strength of the earth's magnetic field, which is about twice as strong at the poles as it is at the equator.
    - b) An animal very sensitive to the strength of a magnetic field would know how far north or south it is.

LEARNING OBJECTIVES

After reading this chapter, the student should be able to answer these questions:

1. How do biologists study behavior?
2. How do we know the extent to which a behavior is controlled by heredity or the environment?
3. Why is behavior adaptive?
4. What are the different forms of learning?
5. How do biological rhythms influence behavior?
6. What environmental cues do animals use to navigate?

ANSWERS TO THE TEXT QUESTIONS

1. The mockingbird's song.
   Observation: Bachelor males sing more than mated males; mated males sing certain songs at different stages of their life cycles.
   Hypothesis: If songs attract mates then mated and bachelor males should sing differently.
   Experiment: Measure the direction of singing by mated and unmated males.
   Conclusion: The unmated males search far and wide, while mated males sing in one direction.
2. Egg-rolling by the greylag goose.
   Observation: Brooding females retrieve eggs that are outside the nest.
   Hypothesis: Egg-laying in greylag geese is a fixed action pattern (FAP).
   Experiment: Remove the egg during retrieval and measure a change in behavior.
   Conclusion: Once the action of the egg retrieving is initiated, it continues in the absence of appropriate feedback.
3. Aggression in the male stickleback.
   Observation: Males attack rival males, which are red.
   Hypothesis: The color red signals aggression in male sticklebacks.
   Experiment: Red and nonred models of sticklebacks are shown to male sticklebacks.
   The red are attacked while the nonred models are not.

Conclusion: The color red is a releasing factor for male sticklebacks.

4. Mating behavior in the male mosquito.
   Observation: Male mosquitos are attracted to female mosquitos.
   Hypothesis: The buzz of the female wings attracts the male mosquito.
   Experiment: Measure the response of male mosquitos to a model that has the same frequency of vibrations as the female mosquito.
   Conclusion: The auditory signal of buzzing wings is the releasing factor for the mating behavior of the male mosquito.

5. Nesting behaviors of the oystercatcher and blackheaded gull.
   Observation: The oystercatcher and blackheaded gull sit on the largest eggs available in the nest.
   Hypothesis: Oystercatchers and blackheaded gulls use size of eggs to determine which eggs are their own offspring.
   Experiment: Measure preferences of oystercatchers and bullheaded gulls for large eggs of a different species over their own egg.
   Conclusion: Size is a supernormal releaser factor in oystercatchers and blackheaded gulls.

6. Crows dropping whelks.
   Observation: Crows take only the largest whelks and drop them from a height of 5 feet.
   Hypothesis: Crows drop whelks from the minimum distance that results in opening.
   Experiment: Measure the distance at which whelks still open when dropped and compare to actual average height of drops.
   Conclusion: Reduce energy output by dropping whelks from the minimum height required to break the shell.

7. Nest building in lovebirds.
   Observation: The peach-faced lovebird tucks the materials in its rump feathers while the Gisher's lovebird transports them in its beak.
   Hypothesis: Nest-building in lovebirds is genetic.
   Experiment: Measure behavior of hybrids.
   Conclusion: Nest building is genetic because hybrid exhibits befuddled behavior.

8. Observation: A person feels hungry when passing a donut store.
   Hypothesis: Odors of food elicit the hunger response.
   Experiment: Ask people in the vicinity of the donut shop if they feel hungry. Ask people in the area of a dry cleaners at the same time of day if they feel hungry.
   Conclusion: Odor of food triggers hunger.

9. Ducklings follow the first thing they see after birth.
   Observation: Ducklings follow their mother after hatching.
   Hypothesis: Ducklings imprint on the image of their mother.
   Experiment: Prevent the ducklings from seeing their mother at hatching and substitute a surrogate mother to see if they will follow it.
   Conclusion: Ducklings imprint the image of their mother at hatching.

10. Maze running by rats demonstrates latent learning.
    Observation: Rats run in a maze without previous experience.
    Hypothesis: Rats will learn to run a maze without being given a reward.
    Experiment: Place rats in a maze until they learn how to find their way out. Test the rats' ability to run the maze at a later time.
    Conclusion: Rats learn to run mazes with prior experience but without being rewarded.

11. A mouse's sleep/wake cycle is not exactly 24 hours.
    Observation: Sleep rhythm in mice in nature is a 24-hour cycle.
    Hypothesis: Sleep is approximately the same amount of time regardless of light or temperature.
    Experiment: Measure awakening time in laboratory in constant light or darkness.
    Conclusion: Rhythms deviate under artificial conditions according to an internal biological clock.

12. The suprachiasmatic nuclei are the master biological clocks in hamsters.
    Observation: Hamsters vary genetically in the intervals of training on an exercise wheel.

Hypothesis: The suprachiasmatic nuclei (SCN) are the master biological clocks in hamsters.

Experiment: Destroy the SCN in wild types, heterozygotes, and super short hamsters. Measure differences in training intervals compared to animals that have the SCN intact.

Conclusion: The suprachiasmatic nuclei (SCN) are the master biological clocks in hamsters.

13. Pigeons use the earth's magnetic field as a compass.

Observation: Pigeons can find their way home.

Hypothesis: Pigeons have the ability to detect magnetic fields and use them to aid in homing.

Experiment: Place small magnets on the head of pigeons to alter their ability to detect the earth's magnetic field. Test the animal's ability to find their way home.

Conclusion: Pigeons have the ability to detect magnetic fields and use them to aid in homing.

ANSWERS TO THE "TO THINK ABOUT" QUESTIONS

1. Two possibilities exist: the broken horse is exhibiting innate characteristics that were suppressed by operant conditioning; the new behavior is operant conditioning in response to the other animals, with access to mates and food resources acting as positive reinforcement.

2. Since the culture is quite unrelated to American culture, it suggests that the behavior is innate to humans.

3. If infants responded to smiling by mimicking, a blind infant would not smile, so it must be innate.

4. This may be due to imprinting, which is the association with sensory stimuli in the environment during a critical period of time.

5. Lack of sleep makes it difficult to do boring tasks such as studying, and little information is retained. Also, this would make it more difficult to take the test.

6. Birds often use familiar landmarks to migrate. If the landscape has changed through building, low light conditions might confuse them.

7. In order to eliminate the factors of light, sound, and pressure, the experimenter would have to construct a chamber where atmospheric pressure was maintained. Also the bird would have to be blindfolded and hearing would be blocked in some fashion.

8. The mechanisms involved in imprinting behavior would be helpful to biologists in altering spawning areas in the river. Also, a study of their ability to negotiate obstacles may give insight into building fish ladders.

AUDIOVISUAL MATERIALS

Animals in Autumn and Winter, Encyclopaedia Britannica Educational Corp., 11 min., 1973.

Animal Behavior—A First Film, Phoenix/BFA Film and Videos, 11 min., 1973.

Animal Behavior—The Mechanism of Imprinting, Coronet Instructional Films, 15 min., 1972.

Animals in Spring and Summer, Encyclopaedia Britannica Educational Corp., 11 min., 1973.

Behavioral Biology, Biomed Coronet, 15 min. 1981.

Behavior of Waterfowl, Nebraska Education TV Council for Higher Education, 30 min., 1970.

37 Social Behavior

KEY CONCEPTS

A eusocial colony exhibits communication, cooperative care of the young, overlapping generations, and division of labor.

Communication is at the root of behaviors requiring cooperation among individuals, such as foraging, defense, aggression, and mating.

The type of communication depends upon the environment.

Animal signals may be chemical, auditory, tactile, or visual.

Establishing territories and dominance hierarchies minimize aggression.

Courtship rituals temper aggression so that individuals can mate with members of the opposite sex.

KEY TERMS

altruism
confusion effect
courtship rituals
dilution effect
dominance hierarchy
eusocial
fountain effect
inclusive fitness
monogamy

polygamy
polygyny
round dance
school
sexually dimorphic
territory
threat postures
waggle dance

CHAPTER OUTLINE

I. Biological Societies
   A. A eusocial grouping exhibits four characteristics:
      1. Communication among its members.
      2. Cooperative care of the young.
      3. Overlapping generations.
      4. Division of labor.
   B. So efficient and predictable are biological societies that they are often called "superorganisms."

C. When large groups of individuals perform only a few tasks, errors are minimized.
   1. Several factors influence a social insect's fate, including nutrition, the temperature of the nest, and pheromones.
   2. An important determinant of an insect's job is age.
D. The insect society is a series of temporal castes—groups whose roles change with time.
E. The proportion of an insect society's population that carries out a specific task is constant, even though individual members proceed through the temporal castes.
F. The best-studied eusocieties are those of ants, termites, and some species of bees and wasps.
G. Inside a beehive.
   1. Social hierarchy of a honeybee hive is a highly regimented, efficient living machine.
   2. At the summit of the organizational ladder is the queen, a large, specialized female whose major function is to keep the hive populated.
      a) Royal jelly is a substance that places the eggs on a developmental pathway toward eventual queendom.
      b) Only one queen reigns per hive.
      c) If the queen leaves, workers chemically detect her absence and hasten the development of the young potential queens.
   3. Male drones fertilize the eggs, and then males develop from unfertilized eggs; they develop through the characteristic insect larval and pupal stages into female workers.

4. A worker's existence is more complex, including several stages and specializations, including making and maintaining the wax cells of the hive, acting as scouts or recruits, and collecting food.

H. A tent caterpillar community.
   1. Other animal societies may interact for only one or a few activities.
      a) Tent caterpillars and their treetop silken communities are centers for exchanging information about food locations.
      b) Contingents of 200 caterpillars leave to explore new food sources.
      c) Should they find food they drag their hind ends through the original markings and the path is now a "recruitment trail."
   2. Social behaviors such as communication and cooperation may be shown by groupings as small as an individual family such as the beetle *Necrophorus*.

II. Advantages of Group Living
A. Groups of animals, from the dozen or so lions in a pride to the intricate insect societies of thousands, enjoy food, protection, and care of the young more easily than more solitary types.
B. Creating more favorable surroundings.
   1. By forming groups, animals may change their environment to their advantage to conserve moisture or heat, such as during the winter, when pigs generally sleep in heaps.
   2. A group can physically alter its surroundings.
      a) Water fleas cannot survive in alkaline water, but when large numbers of them congregate, the carbon dioxide produced by their respiration decreases the alkalinity of the water to acceptable levels.
      b) The maintenance of prairie dogs' burrows depends on the animals' constant chewing of the vegetation—otherwise their homes would easily be overgrown.

C. Better defense.
   1. A predator can more easily fell a lone prey than pick one from a group, where a few alert individuals warn the others.
   2. Group defense.
      a) When faced with a school, a predator often suffers a confusion effect and is unable to decide which fish to attack.
      b) Ostriches use the dilution effect of passive grouping to protect their young as shown when the minor hens must deposit their eggs in the nest of a bonded pair, who sit on them, with their own eggs, for 6 weeks; if not all of the eggs are theirs, chances are that some of those lost to predators will not be theirs.
   3. Sometimes the best defense is a good offense, with help from others, and many prey species engage in mobbing behavior, where adults harry a predator that is frequently larger than any individual of the group.
      a) An example is redwing blackbirds, who make hit-and-run attacks against owls.
      b) Also, baboons and chimpanzees use a similar strategy against leopards.
   4. Forming a circle with the most formidable parts of the body facing outward is a common defense strategy, for example, adult musk oxen form a ring

257

with their heads pointed outward toward the attacker.
- D. Enhanced reproductive success.
    1. Failure to reproduce is evolutionary suicide.
    2. An increase in the number of animals in an area may not only raise the chance of finding a mate, but it may also trigger physiological changes in individuals that are necessary for successful reproduction.
    3. An example is a pig's ovaries, which may not develop normally unless she has a chance to hear and smell a boar.
    4. The sights, sounds, and scents of other courting individuals appear to enhance synchronized breeding, such as the spotted salamanders, who provide an extreme example of explicitly timed group mating.
- E. Improved foraging efficiency learning.
    1. If food is plentiful in some places and scarce elsewhere, it helps enormously to have others around to locate good feeding sites.
    2. Some birds living in groups establish food stores available to all members to tide them over during periods of food scarcity.
    3. Spiderwebs vividly illustrate the advantages of group living in securing a meal.
- F. Learning.
    1. Sociality may enhance learning and the passage of tradition.
    2. Many animals acquire information by watching others.
    3. Certain birds learn to break into milk bottles to steal sips of cream.

III. Disadvantages of Group Living
- A. Individuals may suffer as a result of group living, although the group is not endangered.
- B. Group living is disadvantageous when members have to compete for scarce resources.

- C. Colonial living also makes it easy for infections to spread, such as in honeybees, whose numbers are dwindling die to mite infections.

IV. Social Structure and the Situation
- A. The social structure of a species is influenced both by the animals' physiology and their surroundings.
- B. When related species occupy similar habitats, yet behave differently, the cause of the diversity may be difficult to identify.
    1. Food availability and mating practices are key to the differences in aggression between the two species of monkey.
    2. The spotty food supply fosters aggression among the brown capuchins.
    3. The dominant male protects his offspring because he is their father.
    4. With abundant, easy-to-eat-food, aggression is rare among the white-fronted capuchins.
    5. However, the males do not bother with the juveniles, and perhaps this is because each adult cannot know which offspring he has fathered.

V. Communication Fosters Group Cohesion
- A. Chemical messages.
    1. The best way to send a message depends upon the environment.
    2. Sound travels far under water, and the songs of whales can be heard hundreds of miles away.
- B. Auditory messages: elephants live in matriarchal family herds with a cacophony of rumbles, trumpets, and screams.
- C. Tactile messages.
    1. Karl von Frisch, in 1910, showed that bees not only see color (except red), but they can also detect ultraviolet and polarized light, which humans cannot see.
    2. Von Frisch eventually deciphered the "dances" that bees use to communicate the location of food sources.

a) The scout dances in tight little circles on the face of the comb, and this round dance incites the others to search for food close to the hive.

b) The longer the dance, the sweeter the food.

c) A waggle dance, a path similar to a figure eight traced on a honeycomb, signifies food that is farther from the hive.

d) The orientation of the straight part of the run indicates the direction of the food source.

e) To specify other directions, the dancer changes the angle of the waggle run relative to gravity so it matches the angle between the sun and the food source.

VI. Altruism
   A. A behavior that harms the individual performing it but helps another organism, for example, a worker bee defends its hive by stinging and intruder.
   B. How could the gene(s) behind altruistic behavior persist in a population if those possessing them are more likely to die or not produce offspring than the more selfish members of the group?
      1. Kin selection is one explanation for altruism.
         a) Inclusive fitness considers both personal reproductive success as well as that of relatives that share some of one's genes.
         b) The more genes in common, the greater the chace that the gene(s) favoring altruism will be present in the altruist's relative.
         c) Natural selection works through differing reproductive successes of individuals.
         d) Since closer relatives share more genes, the degree of altruism should be proportional to the number of shared genes.
         e) Therefore, in parent-offspring or sibling relationships, with the individuals sharing 50% of their genes, an altruistic act should at least double the reproductive success of the surviving relative.
      2. Helpers who are unrelated to the young they help raise may derive some benefit from their unselfish behavior, such as a home.

VII. Aggression
   A. Aggressive behavior is often displayed when members of the same species compete for resources, such as mates, food, shelter, and nesting sites.
   B. Territoriality.
      1. One way that animals distribute resources with a minimum of aggression is to defend an area of their habitat against invasion.
      2. This area, or territory, may be defended against all members of the species or only those of the same sex.
      3. A territory may be fixed or moving.
      4. Some territories are held by an individual; others are defended by a group.
      5. For example, group territories are maintained by wolf packs, hamadryas baboons, and pack rats.
      6. Some animals defend territories only during the breeding season, but others hold them year round.
      7. Territorial markers.
         a) Once boundaries are set, they are generally respected without contest.
         b) An animal becomes less aggressive when it crosses

the boundary to leave its territory.

c) Sometimes a territory is marked by visual advertisement, involving display of threat postures.

d) Territories may be indicated by a strong scent produced in the resident's glands, such as badgers, martins, mongeese; ungulates may have hoof glands, and rabbits have both chin and anal glands.

e) Some animals vocally proclaim ownership of a territory.

8. Traversing territories.

a) When animals ignore territorial markers, threat behavior follows.

b) A threat display indicates an animal's strength or endurance.

c) At any time during a battle, one combatant can call off the fight with a display of submission, without gloating retaliation from the victor.

d) Another response to a threat is diversion, and the "loser" behaves in a way unrelated to the situation to distract the aggressor.

e) When competition does lead to combat, it is more like a tournament with a set of rules than a bloody fight to the death.

9. Dominance hierarchies.

a) Another way to distribute resources with a minimal amount of aggression is by forming a dominance hierarchy, a social ranking of each group member relative to every other adult of the same sex.

b) One of the simplest dominance relationships is a linear hierarchy, in which one animal is dominant over all the rest.

c) A second animal is dominant over all but the first, and a third is dominant over all but the first two, and so on.

d) The dominance arrangement is set up during the initial encounters of the member's group, when there may be many threats and an occasional fight.

e) Subordinates indicate submission by lowering their heads and tails and flattening their ears.

VIII. Courtship

A. Courtship ritual are stereotyped, elaborate, and conspicuous behaviors that overcome aggressive tendencies long enough for mating to occur.

B. Males also behave to reduce aggression in the female as in the case of a male wolf spider who cautiously approaches a female and, keeping a safe distance, signals its intent to mate by waving specially adorned appendages.

C. Courtship displays are specific, preventing the costly error of mating with the wrong species.

D. Mating systems.

1. In monogamy, a permanent male-female pair is established.

2. In polygamy, a member of one sex associates with several members of the opposite sex.

a) The most common polygamous arrangement is polygyny, in which one male mates with several females.

b) the sexes of polygynous species such as lions are usually very different in appearance and are said to be highly sexually dimorphic.

E. Primate mating behavior.

1. Primate sexual behavior varies even within species.

2. Every mating system observed in nonhuman primates has also been practiced in humans.

3. However, the fact that behavior in many other species is partially genetically determined—from simple fixed action patterns to complex social interactions—suggests that the same is true for us.

LEARNING OBJECTIVES

After reading this chapter, the student should be able to answer these questions:

1. What are the characteristics of an animal society?
2. What are the advantages and disadvantages of group living?
3. How do animals communicate?
4. Why does altruistic behavior persist?
5. How do animals minimize aggression?
6. How do animals behave during courtship and mating?

ANSWERS TO THE TEXT QUESTIONS

1. Honeybees communicate among themselves by pheromones and through dances that tell the location of food sources. Cooperative care of the young is accomplished by female workers. When a queen is absent, workers hasten the development of young potential queens. Several generations coexist, with each age group performing a different function for the hive. Division of labor consists of a queen, whose major function is to keep the hive populated; drones, which breed with the queen; and workers, who do the work of the hive. The workers are further subdivided into caretakers, undertakers, and guards.

2. Group living offers advantages such as alteration of the environment to conditions more favorable for the group, better defense, enhanced reproductive success, improved foraging efficiency, and increased learning.

3. Brown capuchin dominant males protect the infants in the troop because they are most likely their own offspring or kin. In the white-fronted capuchin, mating is not restricted and so the offspring can belong to anyone. Since there is no advantage in helping the young of someone else, they may be avoided. The case of the ostrich hen sitting on the eggs of another's young is not kin selection since the other hens are not necessarily related.

4. An individual who pounds his or her fists on a conference table or stands up and shouts is exhibiting threat behavior. Appeasement is shown by those who attempt to calm him or her with a word or a hand on the sleeve. Others may look down or away to avoid the eyes of the angry person. A diversion might be created by suggesting a 5-minute break. Displacement might happen when a new topic becomes the focus.

5. The dog is protecting his territory by displaying a threat posture.

6. Courtship rituals are stereotyped, elaborate, and conspicuous behaviors that overcome aggressive tendencies long enough for mating to occur. They also help identify members of the same species.

ANSWERS TO THE "TO THINK ABOUT" QUESTIONS

1. She would look for the four characteristics: communication among members, cooperative care of the young, overlapping generations, and division of labor.

2. The biologist may be unaware of disturbances in usual social behavior by the introduction of the Plexiglas and alterations in temperature and lighting. A video camera system set up in a natural site may reduce these variables.

3. The limitation of human perception may bias our judgment of the behaviors of other animals.

4. Since proposed behavior genes could be under environmental control, the knowledge of their presence should be used with caution. They could give insight into how to modify negative behaviors in some individuals.

5. Human aggression, if expressed under the same circumstances as other animals, is a result of scarce resources and competition for mates. Overpopulation is likely to result in increased aggression.

AUDIOVISUAL MATERIALS

Animal Camouflage, Phoenix Films and Videos, 17 min., 1977.

Animal Communities and Groups, Coronet Instructional Films, 11 min., 1963.

Animal Imposters, "NOVA" Series, Time Life Films, 59 min., 1980.

Animal Migration, Phoenix Films and Videos, 12 min. 1977.

Animal Navigation, Phoenix Films and Videos, 24 min., 1970.

Animal Predators and the Balance of Nature, Life Science Series, 10 min., 1966.

38 Populations

KEY TERMS

CHAPTER OUTLINE

I. Human Population Growth
    A. The growth of the human population
        has not been a steady march; it has
        accelerated in recent years thanks to
        medical advances and our ability to
        manipulate the environment.
    B. In July 1986, the 5-billion mark was
        reached.
II. Population Dynamics
    A. A population is a group of
        interbreeding organisms.
    B. Births and immigration increase
        population size; deaths and emigration
        decrease it.
    C. Population growth is measured by
        rates as well as by absolute numbers.
        1. A rate is the number of events
            (such as births or deaths)
            divided by the average number
            of individuals in the population
            during a specific time period.
        2. Population growth is usually
            expressed as the change in the
            number of individuals during a
            set interval of time.
    D. The larger the population, the faster
        it grows.

III. Biotic Potential and Exponential Growth
  A. Under ideal conditions, populations would grow explosively.
  B. Growth resulting from repeated doubling of numbers (1, 2, 4, 8, 16, ...) is mathematically described as exponential, and plotting the number of individuals in a population over time as it grows exponentially yields a characteristic J-shaped curve.
  C. Unrestricted by the environment and with each member producing as many offspring as possible, the population would grow at its maximum intrinsic rate of increase.
  D. Biotic potential.
    1. The maximum number of offspring an individual is physiologically capable of producing.
    2. This limit is influenced by such factors as the number of offspring produced at one time, the length of time an individual remains fertile, and how early in life reproduction begins.
  E. The population grows faster when reproduction begins at an earlier age, simply because the rate at which new individuals are added is accelerated.
  F. The biotic potential for even the slowest breeding species is incredibly high, and few organisms ever attain it.
IV. A Population's Age Structure (table 38.2)
  A. If most members are young, the population will grow as the females enter their reproductive years.
  B. Assuming that the birth rate remains the same, a population consisting of mostly elderly individuals will decline as the members die.
V. Regulation of Population Size
  A. Environmental resistance refers to all the factors that reduce birth rate or increase death rate.
  B. Environmental resistance and the carrying capacity.
    1. Environmental resistance may include limited food, pollution that impairs health and ability to reproduce, natural or human-caused environmental disasters, and predation.
    2. The maximum number of individuals that can be supported by the environment for a indefinite time period is called the carrying capacity.
  C. "Boom or bust" cycles.
    1. Because of environmental resistance, exponential growth can be maintained only for a limited time before the population overshoots the carrying capacity; then numbers plummet or the population size stabilizes at the carrying capacity.
    2. A population crash, sometimes called a "boom or bust" cycle, is commonly seen in species that colonize a new area with abundant food.
    3. Population crashes are unusual; more often population growth slows as the carrying capacity is approached in response to environmental resistance.
    4. This leveling off produces a characteristic S-shaped curve or logistic growth curve.
  D. Density independent factors.
    1. These kill a certain percentage of the population regardless of its size.
    2. Natural disasters and severe weather conditions are typical density independent regulating factors.
  E. Density dependent factors.
    1. These factors have a greater effect as the size of a population increases.
    2. Behavioral responses to crowding include cessation of mating, poor parental care, and increased aggressive behavior.
    3. Physical responses to crowding include increased rate of spontaneous abortion, delayed maturation, and hormonal changes.
    4. Competition for food and space is a commonly encountered

density dependent factor in population growth.

   a)   In scramble competition, the individuals compete directly for the limited resource, which may suffice for the individual but may be counterproductive to population growth.

   b)   In contest competition, animals compete for social dominance or possession of a territory, factors that guarantee the winners an adequate supply of the limited resource.

   c)   According to the principle of competitive exclusion, coexistence of two competing species will not continue indefinitely; one species will replace the other in the area of overlap.

   d)   Laboratory experiments confirm that competition may indeed lead to local extinctions.

   e)   Two species of the protozoan *Paramecium* thrive when cultured separately, however, if these species are grown together, only one survives.

F.   Habitat and niche.

   1.   Habitat is the place where a species lives.

   2.   Niche is the ways in which it interacts with the environment.

   3.   Although division of a habitat into a variety of niches can allow different species to live together, competition may still narrow their niche so that the population size is restricted.

   4.   A species' fundamental niche includes all the places and ways in which it could possibly live.

   5.   A species' realized niche, the one it actually fits, is smaller than its fundamental niche due to competition.

VI.   Predator-Prey Interactions

   A.   Prey populations are often maintained by high reproduction rates that compensate for the loss to predation.

   B.   Prey populations are also maintained by natural selection, because predators tend to eliminate the weakest prey.

   C.   The primary cause of death of predators is starvation due to high metabolisms.

   D.   Reproduction among predators is intimately tied to the density of their prey populations.

   E.   Predators may hunt alone or in packs.

   F.   The type of prey usually depends simply on what is available.

   G.   Animals evolve adaptations that help them more readily obtain prey or avoid becoming prey, such as the leaping gait.

VII.   Human Population Growth Revisited—The Future

   A.   Population dynamics are complicated by numerous factors—migration, disease, predator-prey relationships, and the destruction of habitats by humans.

   B.   A too-large population is detrimental to its members, and, in terms of global numbers, the human population is increasing.

   C.   John Calhoun is the researcher who tracked overcrowded mouse populations.

      1.   He estimates that a point of no return—when a population self-destructs due to sheer size—occurs 2.5 to 4 generations after a population reaches twice its optimal density, based on observations of mice.

      2.   The global human population reached twice its optimal levels in 1975, and with a generation time of 27 years, this formula predicts the point of no return—the maximal population size after which overcrowding-related deaths begin to

occur—will be reached between the years 2042 to 2083.

3. However, population predictions are notorious for being changed.

D. We can conserve or find alternate resources, use agriculture to grow crops on barren land and aquaculture to farm the water, and even control our reproduction.

LEARNING OBJECTIVES

After reading this chapter, the student should be able to answer these questions:

1. What physical and behavioral responses do organisms have to overpopulation?
2. What factors influence the rate of population growth?
3. Under what conditions would a population grow exponentially?
4. What factors prevent organisms from producing as many offspring as they theoretically could?
5. How do populations respond to environmental constraints on population growth?
6. How can populations of different species coexist?
7. How do populations of predators and their prey interact?
8. Can we—and should we—control human population growth?

ANSWERS TO THE TEXT QUESTIONS

1. When food and water become scarce along with crowding, behavior changes dramatically. Juvenile behavior continues into adulthood and mating behavior gradually ceases in experiments on animals. There is also poor maternal care and increased aggression, which may also apply to human populations.
2. Scarcity of food and clean water may limit human population growth. Increasing pollution and spread of diseases as well as environmental disasters can affect the rate of growth. Contraception and having children at a later age in life would also decrease the rate of growth. However, increasing medical technology and the

ability to alter our environment for raising food crops can actually accelerate population growth.

3. This is a density dependent factor since the number of hiding places is finite. As the population of frogs increases, more of them will be at risk of predation. This will have a greater effect in limiting the population.
4. Populations will grow faster when reproduction begins at an earlier age. Since there are more young people of reproductive age in Pakistan, the population will grow at a faster rate.
5. Both organisms and cells grow at an exponential rate when conditions for growth are ideal. This would be at the maximum intrinsic rate of increase. When limited food, waste accumulation, and crowding become factors in the population, the growth rate slows. The population will then level off at the limits of the carrying capacity of the environment.
6. A niche will fall within a specific habitat. The habitat is the place in which an organism lives, and the niche is the way in which the organism interacts with the environment. The habitat may be a pond or a desert, and the niche includes physical factors such as temperature, humidity, and light, as well as biological factors such as growth rate, metabolic requirements, and behavior. A fundamental niche includes all the places and ways in which the species could possibly live. The realized niche is smaller, since another species with similar, but not identical, requirements will compete for a portion of the fundamental niche, partitioning it into smaller areas.

ANSWERS TO THE "TO THINK ABOUT" QUESTIONS

1. The different species of coral have similar but not identical requirements, therefore the fundamental niche within the habitat is partitioned into smaller realized niches for each species. The different species reproduce at different times of the year so their metabolic requirements will vary, allowing each species to survive.

2. The poorer families have more children at an earlier age, thus their population is growing at a faster rate as observed in the Third World countries in table 38.2. The percentage of the population less than 15 years old is very high, which will increase the annual rate of increase in the future. The resources available are limited and must be shared by more and more individuals, which could lead to a population crash from poor nutrition, famine, or disease. The time spent in trying to survive increases, and the opportunity for education and other opportunities decline.

3. Restrictions in population growth are essential to maintain the population within the carrying capacity of the environment. Pushing the age of child rearing to the mid-thirties could reduce the rate of growth and still allow for more than one child. Our ability to alter the environment for increased food production will raise the carrying capacity, but population growth would then have to be restricted in the future at a higher level.

4. The explosive rate of growth is supported by advances in food production and medical technology. As the population gets very large the effects of environmental resistance such as hurricanes and tornadoes have a small relative effect on the total population numbers.

5. Statistics on violent crime from high-density areas such as the inner cities could be compared to small populations in rural areas.

AUDIOVISUAL MATERIALS

Animal Populations—Nature's Checks and Balances, Encyclopaedia Britannica Educational Corp., 22 min., 1983.

Population Ecology, CBS Corp., 19 min., 1963.

Population Dynamics, University Washington Films, 30 min., 1974.

Population Time Bomb, Access Network, 28 min., 1977.

Population Story: Collision with the Future, Encyclopaedia Britannica Educational Corp., 23 min., 1984.

39 Ecosystems

KEY CONCEPTS

Ecology is the study of the interrelationships between organisms and their biotic and abiotic environments.

An individual has a niche and is part of a population, community, ecosystem, biome, and biosphere.

Primary producers use solar or geothermal energy to synthesize nutrients, which are eaten by secondary producers, including primary consumers at the second trophic level, secondary consumers at the third trophic level, and tertiary consumers at the fourth trophic level.

At each level 90% of the available energy is lost.

Some chemicals, particularly those that are soluble in fat, are concentrated as they ascend a food chain, becoming more and more toxic.

Elements are continually cycled between organisms and the physical environment.

Plants use solar energy and atmospheric CO_2 to manufacture carbohydrates, which pass through food webs and are recycled in respiration, excretion, and decay.

Nitrogen is fixed and converted to usable nitrates by bacteria and also returned to the atmosphere by bacteria.

Phosphorus cycles from nonliving reservoirs to the living.

Succession is a gradual, directional change in the communities of an ecosystem.

A biome's characteristics are determined by the types of plants, latitude and altitude, and moisture.

Life is most diverse in the tropical rain forest, where nutrients cycle rapidly.

In lentic systems water stands still, such as in lakes and ponds. Oxygen and minerals in a lake are distributed unevenly and are moved by winds and seasonal temperature changes.

The thermocline lies between a warm and cool layer.

Aging lakes accumulate nutrients, becoming eutrophic, and are murky.

Rivers and streams are lotic systems, where life is adapted to current.

In an estuary, a river meets an ocean, life is diverse, and salinity and water level fluctuate greatly.

Estuaries flow into intertidal zones, which are home to organisms that can hold fast to rocks, burrow in the sand, or resist damage by the tide.

The ocean's bottom benthic zone includes the abyssal zone, where light does not reach, and a lit pelagic zone, where life abounds.

Upwelling of lower, nutrient-rich water layers is important in building food web bases.

KEY TERMS

abyssal zone	lentic system
benthic zone	limnetic zone
biogeochemical	littoral zone
cycles	lotic systems
biome	net primary
biomagnification	production
biomass	neritic zone
biosphere	nitrifying bacteria
climax community	oceanic zone
community	oligotrophic
decomposers	pelagic zone
denitrifying	permafrost
bacteria	pioneer species
ecological	primary consumers
succession	primary producers
ecology	primary succession
ecosystem	profundal zone
energy pyramid	pyramid of biomass
estuary	pyramid of numbers
eutrophic	savanna
fall turnover	secondary consumers
food chain	secondary production
food webs	secondary succession
gross primary	scavengers
production	spring turnover
intertidal zone	stable isotope tracing

taiga
tertiary consumers
thermal
    stratification
thermocline
trophic level

tropical rain forests
tundra
upwelling
vertical
    stratification

CHAPTER OUTLINE

I. An Organism's Place in the World
    A. Ecology is the study of the relationships between organisms and their environments.
    B. An ecosystem is a unit of interaction among organisms and between organisms and their physical environment.
        1. It consists of all living things within a defined area and may be as large as the whole earth or only a small part of it—a forest or even a single rotting log on the forest floor.
        2. They always include living (biotic) and nonliving (abiotic) components, and they are relatively self-contained, with more materials cycling within than entering or leaving.
        3. An ecosystem is terrestrial if it is on land and aquatic if it is in water.
        4. They are supported by energy, which is ultimately derived from the sun (solar energy) or from the earth's interior (geothermal energy).
        5. Ecology deals with the homes of organisms.
            a) An ecological home is a niche, which includes both the species' relationship to other organisms and its physical habitat.
            b) A species is successful in an ecosystem if its niche supplies food and protection from predators and the abiotic environment.
        6. Species coexist within ecosystems by a precise balancing of their requirements.

7. A population consists of members of a single species in a particular area, such as the bark beetles in a single egg gallery of the fallen tree.
8. An ecological community includes all of the organisms, possibly several species, in a given area.
9. Living communities are not distributed evenly on the planet.
    a) Biological diversity is highest at and around the equator, possibly due to the relatively unchanging climate there.
    b) Toward the poles, biological diversity declines as climatic fluctuations increase, creating harsher, seasonal conditions.
    c) In general, similar types of biological communities appear at corresponding latitudes, because they have similar climates.
10. Ecosystems also interact with one another.
11. A group of interacting terrestrial ecosystems characterized by a dominant collection of plant species, or a group of interacting aquatic ecosystems with similar salinities, forms a larger ecological unit, the biome, which includes grasslands, deserts, and forests.
12. The entire planet can be viewed as one huge, interacting ecosystem, the biosphere.

II. Energy Flow Through an Ecosystem
    A. Energy flows through an ecosystem in one direction, beginning with solar energy that is converted to chemical energy by photosynthetic organisms.
    B. This stored energy is then passed through a food chain, a series of organisms in which one eats another.
    C. The trophic level, or feeding level, to which an organism belongs describes its position in the food chain.

D. Little energy is used for growth and development but rather energy is lost to the environment as heat, or used to power metabolism, including cellular respiration.

E. The total amount of energy converted to chemical energy by photosynthesis in a certain amount of time in a given area is called gross primary production.

F. Net primary production is what is left over for growth and reproduction.

G. From each trophic level to the next, only 10% of the potential energy in the bonds of its molecules fuels growth and development of organisms at the next trophic level.

H. For example, one tenth of the energy eaten in seeds ends up as a bird.

I. This inefficiency of energy transfer is why food chains rarely extend beyond four trophic levels.

J. Trophic levels.
1. The first link in the food chain and the base of the trophic structure is formed by primary producers.
   a) These are organisms able to use inorganic materials and energy to produce all the organic material they require.
   b) Almost all primary producers photosynthesize and include plants on land and algae in aquatic ecosystems; a few primary producers are chemosynthetic organisms.
2. The energy stored in the tissues of herbivores (plant eaters) and carnivores (meat eaters) is called secondary production, and at the second trophic level are the primary consumers or herbivores.
3. On the third level are secondary consumers, animals that eat herbivores.
4. Carnivores that eat other carnivores form a fourth trophic level and are called tertiary consumers; these animals expend a great deal of energy in capturing their prey.
5. Scavengers such as vultures eat the leftovers of another's meal, and decomposers, such as certain fungi, bacteria, insects, and worms, break down dead organisms and feces, liberating minerals held in organic molecules for use by other organisms.
6. Although energy passes through a series of organisms, chains of "who-eats-whom" often interconnect, forming complex food webs (fig. 39.1).
7. Webs form when an organism functions at more than one trophic level.

K. Stable isotope tracing—deciphering food webs.
1. The method of stable isotope tracing, borrowed from geology, provides a way to decipher who eats whom without actually being on the scene of a meal.
2. The technique analyzes the proportions of certain isotopes in tissue samples, and the basis of the approach is that primary producers each have a characteristic ratio of carbon-13 to carbon-12.

L. Ecological pyramids.
1. If each trophic level is represented by a bar whose length is directly proportional to the number of kilocalories available from food for growth and development, the bars representing the different trophic levels, stacked in order, form a steep-sided energy pyramid.
2. A pyramid of numbers shows the number of organisms at each trophic level, and the shape of this type of pyramid depends largely on the size of the producers.
3. A pyramid of biomass takes into account size or weight.

a) Biomass is the total dry weight of organisms in an area.

b) Many pyramids of biomass are wide at the bottom and narrow at the top, because energy that can be converted to biomass is lost with each transfer between trophic levels.

c) The biomass of the producers (phytoplankton) may be smaller than that of the herbivores (zooplankton) because biomass is measured at one time.

III. Concentration of Chemicals—Biomagnification

A. Some substances, particularly those not normally found in ecosystems, become more concentrated in organisms at higher trophic levels.

B. Such biomagnification occurs when the chemical is passed along to the next consumer rather than being metabolized and excreted.

C. DDT.

1. Almost all of the DDT taken in by an organism remains in its body, lingering in fatty tissue because it is soluble in fat.

2. The chemical is passed on to those who eat the organism, and at the next trophic level, the process is repeated.

3. The organisms at the top of the energy pyramid are most severely affected.

D. Mercury.

1. Biomagnification occurred in Minimata, Japan, when fish were contaminated with mercury.

2. Nearly 200 infants were born, as a result, with "minimata disease," grossly deformed and mentally retarded.

3. By marijuana smoking, 75% to 85% of the mercury is absorbed into the bloodstream, causing the classic symptoms of mercury toxicity—irritability, paranoia, forgetfulness, tremors, and narrowed vision.

IV. Biochemical Cycles

A. The recycling of chemicals essential to life involves pathways called biogeochemical cycles.

B. Water, carbon, hydrogen, nitrogen, oxygen, phosphorus, sulfur, potassium, sodium, calcium, chlorine, iron, and cobalt all pass between the atmosphere, the earth's crust, water, and organisms.

C. The carbon cycle.

1. In the carbon cycle, photosynthetic organisms capture the sun's energy and use it to produce carbohydrates, using atmospheric carbon dioxide (CO_2).

2. Both plants and animals release carbon to the atmosphere through respiration, and animals return carbon to the soil in excrement.

3. When an organism dies, bacterial, insect, and fungal decomposers break down its organic compounds to release carbon to the soil, atmosphere, or aquatic surroundings.

D. The nitrogen cycle.

1. Organisms depend on nitrogen-fixing bacteria that convert atmospheric nitrogen into ammonia (NH_3).

2. Lightning also fixes a small amount of nitrogen.

3. Nitrifying bacteria convert ammonia from dead organisms to nitrites, and then other bacteria convert nitrites to nitrates (NO_3).

4. Nitrogen is returned to the atmosphere by denitrifying bacteria that convert ammonia, nitrites, and nitrates to nitrogen gas.

5. Rhizobium bacteria live in swellings (nodules) on the roots of legumes such as beans, peas, and clover.

E. The phosphorus cycle.
   1. Phosphorus is a vital component of genetic material, ATP, and the phospholipids of membranes.
   2. As rain falls over reservoirs, phosphorus is released as phosphate that can be used by organisms to form living tissues.
   3. Decomposers eventually return phosphorus to the soil.

V. Succession
   A. The process of change in the community is called ecological succession, and it is gradual and directional.
   B. Eventually, a climax community, one that remains fairly constant if the land and climate are undisturbed, is established.
   C. Primary succession.
      1. Primary succession occurs in an area where no community previously existed.
      2. The first species to invade, called pioneer species, are hardy organisms, such as lichens and mosses, that are able to get a foot hold on smooth rock.
   D. Secondary succession.
      1. This occurs in areas where a community is disturbed, such as by a river changing course and flooding an area or fire destroying a forest.
      2. Although the sequence of changes is often similar to that of primary succession, the rate of change is often faster in secondary succession because the soil may already be formed.
      3. An example of "old field" succession is when the land is no longer farmed, fast-growing pioneer grass and weed species, such as black mustard, wild carrot, and dandelion, move in, flowed by slower-growing, taller goldenrod and perennial grasses.
      4. In a few years, pioneer trees such as pin cherries and aspens arrive, but these are eventually replaced by pine and oak, and finally, a century or so later, the climax community of beech and maple is again well developed.

VI. Terrestrial Biomes
   A. The types of plants in a biome depend upon altitude and latitude.
   B. Biomes within these categories are distinguished by their degrees of moisture.
   C. Tropical rain forest.
      1. These are found where the climate is almost constantly warm and moist and rainfall is typically between 79 and 157 inches per year; the ground is perpetually soggy.
      2. From the air, a tropical rain forest is a solid, endless canopy of green built of treetops 50 to 200 feet above the forest floor.
      3. Plants beneath the canopy compete for sunlight, forming layers of different types of organisms called vertical stratification.
      4. The ground of the tropical rain forest looks rather bare, shaded as it is by the overwhelming canopy, but it too is home to countless shade-adapted species.
      5. The lush vegetation feeds a variety of herbivores.
      6. Cycling of nutrients, including minerals, is quite rapid in the tropical rain forest, where the daily torrential rains rapidly wash away nutrients released in the soil from dead and decaying organisms.
      7. Termites recycle nutrients with the help of protozoans in their guts, which break down the cellulose in plants and release its atoms back to the environment.
      8. The soil is not very fertile because nutrients are deposited and reused so quickly.
      9. Essential in a tree's recycling program are fine roots that permeate the upper 3 inches of

ground and grow up to and attach to dead leaves; fungus on rootlets speeds decomposition, and nutrients released from the dead leaves are quickly absorbed.

D. Temperate deciduous forest.
1. Deciduous hardwood trees that lose their leaves during the winter thrive where the growing season lasts at least 120 days, annual rainfall ranges from 28 to 59 inches, and the soil is rich.
2. Vertical stratification to compete for light is seen here too, although diversity is less than in the tropical rain forest.
3. Trees in the deciduous forest often have ball-shaped tops, which maximize light absorption.
4. The decomposers in temperate forests—nematodes, earthworms, and fungi—break down the leaf litter to create rich soil; herbivores include the cottontail rabbit, whitetail deer, and grouse, and the carnivores include red fox and raccoon.

E. Temperate coniferous forest.
1. Coniferous trees are commonly called evergreens.
2. Fire recycles nutrients and selects resistant species.
3. The thick bark on certain pine trees resists flames, and some pinecones liberate their seeds only after exposure to the extreme temperature of a forest fire.

F. Taiga.
1. North of the temperate zone is the taiga, also known as the northern coniferous forest or boreal forest.
2. The cold, snowy mountainous taiga is home to few hardy, well-adapted trees such as spruce, pine, fir, and hemlock, with a few deciduous trees near lakes, streams, and rivers.
3. Little grows beneath the tall conifers, with decaying needles making the soil too acidic for

many plants to grow and the needles blocking the sun.
4. The shape of needles minimizes surface area on which water is lost, as does their thick, waxy outer surfaces.
5. Common decomposers in the taiga are fungi; familiar herbivores include squirrels, grouse, warblers, and moose, and carnivores include lynx and wolves.

G. Grasslands.
1. Grasslands have 10 to 30 inches of rainfall annually.
2. These biomes have a variety of names, including savanna in the tropics, prairies in the temperate zone, and the arctic tundra.
3. The African savanna.
   a) This is a tropical grassland.
   b) Grazing the land are zebras, gazelles, antelopes, impala, giraffes, and wildebeests, and stalking the grazers are lions, cheetahs, tigers, and hyenas.
4. The North America prairie.
   a) This is temperate grassland.
   b) The height of the grass reflects local moisture.
   c) Root systems are extensive, with some species sending roots 6 feet underground to reach water; the mat of roots holds soil together, preventing it from blowing away during a drought.
   d) Unlike trees, grasses survive damage by fire because their roots are so deep.

H. Tundra.
1. Temperatures range from an average of -26 F in the winter to a range of 40 F to 70 F in the summer.
2. A part of the ground called the permafrost remains frozen all year.

3. The colonizers are the lichens, which prepare the soil to support woody shrubs and short grasses.
4. Tundra plants tend to be low and flat, a shape that lets the wind blow over them.
5. Animals include the caribou, musk ox, reindeer, lemmings, snowy owls, foxes, wolverines.

I. Desert.
1. Deserts receive fewer than 8 inches of rainfall per year.
2. They are home to fewer species than the other tropical biomes.
3. Annual plants grow quickly, with their seeds germinating only after a soaking rain.
4. Cactuses.
   a) A lack of leaves minimizes water loss.
   b) The stem expands and stores water and is guarded by spines.
   c) The root system is shallow but widespread.
   d) Many desert plants produce chemicals that inhibit the growth of neighboring plants.
5. Desert animals cope with water scarcity.
   a) Body coverings such as an exoskeleton in a scorpion or the leathery skin of reptiles, reduce dehydration.
   b) Small mammals excrete a concentrated urine.
   c) Few animals are exposed to the midday sun.
6. The driest and largest desert in the world is Africa's great Sahara.

VII. Freshwater Biomes
A. Lakes and ponds.
1. Light penetrates to differing degrees in the regions of a lake, and these determine the types of plants that live in a particular area.
   a) Littoral zone.
      (1) This is the shallow region along the shore, where light can penetrate to the bottom with sufficient intensity to allow photosynthesis.
      (2) Most species are rooted to the bottom yet stick out of the water (fig. 39.16).
      (3) This is the richest area of a lake or pond.
      (4) Producers include cyanobacteria, green algae, diatoms, and cattails.
      (5) Animals include damselflies, dragonfly nymphs, crayfish, rotifers, flatworms, hydra, snails, snakes, turtles, frogs, and fish.
   b) Limnetic zone.
      (1) This is the layer of open water that is penetrated by light.
      (2) It is inhabited by phytoplankton, zooplankton, and fish.
   c) Profundal zone.
      (1) It is the deep region beneath the limnetic zone where light does not penetrate.
      (2) The organisms here rely on the rain of organic material from above and include scavengers and decomposers.
2. Oxygen and nutrients in a lake are distributed unevenly.
   a) Oxygen concentration is greatest in the upper layers.
   b) Phosphates and nitrates are in the lower layers due to decaying organisms.

c) Thermal stratification.
(1) This is the seasonal redistribution of nutrients.
(2) The upper layers are heated in the summer while the lower layers remain cold.
(3) The thermocline refers to the middle layer.
(4) Fall turnover occurs when the wind mixes the upper and lower layers.
(5) During the spring as the upper layers warm, cold water drops, bringing nutrients with it.

3. Lakes age.
a) Young lakes are called oligotrophic and have few foods in them.
b) As the lake ages, organic material begins to fill it in and is called eutrophic.
c) Pollution speeds up eutrophication.

B. Rivers and streams.
1. They depend heavily on the land, both for the water that fills their banks and the materials that nourish the life within.
2. Many rivers flood each year, swelling with melt water and spring runoff and spreading nutrient-rich silt onto the land.
3. Oxygen in flowing water is usually abundant because air and water mix in turbulent areas.
4. Dumping organic material such as sludge into a river or stream can create a problem because its decomposition uses oxygen needed by river residents.
5. Slow-moving rivers support more diverse life, including crayfish, snails, bass, and catfish.

VIII. Marine Biomes
A. The coast.
1. At the margin of the land, where the freshwater of a river a meets the salty ocean, is an estuary.
2. Life in an estuary must be able to cope with a range of chemical physical conditions.
a) The salinity of the water fluctuates dramatically.
b) The estuary is alternately exposed to drying air and then flooded.
3. Estuaries house very productive ecosystems, its rocks slippery with algae, its shores lush with vegetation, and the water abounding with plankton.
4. Estuaries are nurseries for many sea animals.
5. Bordering estuaries.
a) They are rocky or sandy areas of the intertidal zone.
b) Organisms include large marine algae, mussels, sea anemones, sea urchins, snails, and starfish.

B. The ocean.
1. The oceanic zone is the area beyond the continental shelf where there are deep open seas.
2. The neritic zone is the area along the coast.
3. The benthic zone is the bottom of the ocean and is home to crabs, starfish, coral reefs, and tropical fish.
a) The area where light never reaches below 6,500 feet is the abyssal zone.
b) The water above the ocean floor is the pelagic zone.
4. Nutrient-rich bottom layers are moved upwards by a process called upswelling.

LEARNING OBJECTIVES

After reading this chapter, the student should be able to answer these questions:

1. At what levels can an organism's place in the world be described?
2. How does energy flow through an ecosystem?
3. How do organisms interact to form food webs?
4. How are some chemicals concentrated as they proceed up food chains?
5. How do carbon, nitrogen, and phosphorus cycle through an ecosystem?
6. How do ecosystems naturally recover from damage?
7. What are the types of large-scale environments (biomes)?
8. How are plants and animals adapted to their environments?
9. What are the different challenges posed to organisms in a stream, lake, estuary, and ocean?

ANSWERS TO THE TEXT QUESTIONS

1. Krill directly or indirectly support nearly all life in Antarctica. Farming krill would directly affect the mammal, fish, and bird populations by reducing their primary food resource. In addition, the removal of krill could alter the plankton population on which the krill feed and the flow of organic material from the ocean floor. The Antarctica ecosystem is intimately connected to other world ecosystems and its disturbance may have far-reaching consequences.
2. Plant (primary producer) butterfly (herbivore, primary consumer) dragonfly (carnivore, secondary consumer) bullfrog (carnivore, tertiary consumer) snake (carnivore, tertiary consumer) hawk (carnivore, tertiary consumer).
3. Only 10% of the potential energy in the chemical bonds is available from one trophic level to the next. Therefore, a nation of hungry citizens would derive only 1% of the available energy if they raised animals and ate them as opposed to 10% if

they raised food crops and eliminated one trophic level.
4. For each trophic level only 10% of the energy available can be used for growth and development. Most (90%) of the energy taken in as food is lost as heat or to power metabolism, such as cellular respiration. Additionally, not all parts of the organism eaten can be utilized and are eliminated as waste.
5. Solar energy is captured by photosynthetic plants (primary producers) and is used to convert water, carbon dioxide, and minerals into organic molecules such as sugars, proteins, and fats. The energy in these organic molecules, and the molecules themselves as building blocks, are used by herbivores, which in turn are consumed by carnivores. The breakdown products of metabolism and the products of decomposition are returned to the ecosystem for use by other organisms. Therefore, energy flows through the ecosystem and the primary molecules are recycled, passing from one trophic level to another and to the physical environment.
6. Tropical rain forests support diverse and abundant life with poor soil because of a constant, warm, and moist climate and the rapid cycling of nutrients and minerals. The poor soil is actually an indicator of the efficiency by which organisms of the rain forest extract available nutrients. The nutrients and minerals are unable to accumulate in the soil as they do in other biomes.
7. Grasslands recover from fire quickly because the seeds of these fast-growing pioneer species are already present. Their deep roots are protected from fires by the soil so they are able to rapidly regrow.
8. In estuaries the chemical and physical conditions constantly change as the fresh water of a river meets the ocean. Tides alter the salinity of the water and repeatedly cover the land, then expose it to the air. The advantage of daily delivery of nutrients is used by the large number of organisms that live there. The intertidal zone is also alternately exposed and covered by water. The organism must be

able to resist drying out as well as attach firmly to the substrate to resist the pull of the tide and the pounding of waves.

ANSWERS TO THE "TO THINK ABOUT" QUESTIONS

1. a.

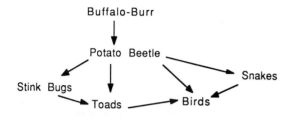

   b. The evolutionary process would be natural selection for a DDT-resistant strain of potato beetle.

   c. Attempt to identify and isolate a natural parasite that kills the potato beetle larva, such as one of the wasps. This species could be introduced into the environment and selectively kill the potato beetle without harming other insects or higher organisms.

2. One advantage would be that the introduced species could quickly replace the destroyed organisms. The difficulty may be that appropriate food sources would not be available for the introduced species and they would not survive and that existing animals would overgraze the newly planted species. Also, the feeding of deer or other large mammals may maintain an abnormally high population that could not be sustained over time until the natural habitat became reestablished.

3. Natural selection is apparent in succession by the orderly appearance of new species that are adapted to the environment as the habitat changes with time. In vertical stratification plants use different strategies to capture sunlight. Some grow very tall, where others are broad and flattened, maximizing sun exposure. Epiphytes grow on the trees at varying heights, providing resources for a large variety of organisms.

4. Fire immediately destroys habitats and individuals, so is devastating in the long term. It may be positive for some species long term in that it can remove climax communities, thus allowing secondary succession to occur. Pioneer species readily move in and thrive where they couldn't before because of competition.

5. Under controlled conditions, some interactions that are subtle in nature may be more observable since factors introduced in a changing environment are eliminated. However, because of this intervention, interpretation of data from an artificial environment should be made with caution.

6. Students will have their own definitions. One definition of an environmentalist is a person interested in protecting the environment from destruction by humans.

AUDIOVISUAL MATERIALS

Animal Habitats, Phoenix Films and Videos, 11 min., 1956.

Ecological Biology, Coronet Instructional Films, 17 min., 1981.

Ecosystems, Access, 30 min.

40 Environmental Concerns

KEY CONCEPTS

High in the atmosphere, sulfur and nitrogen oxides join with water to form sulfuric and nitric acids, which fall as acid precipitation.

Acidic water adversely affects many aquatic species.

In coniferous forests, acid rain releases aluminum into soil, which displaces calcium and magnesium, stunting growth.

Excess nitrogen spurs growth.

Carbon dioxide and other gases may cause global warming by trapping infrared radiation near the earth's surface.

Much carbon dioxide is generated from agriculture, deforestation, jets, and burning fossil fuels.

A warmer climate would alter plant distributions, which would alter animal distributions.

Ultraviolet radiation can harm or kill cells; much of this radiation from the sun is shielded from the earth's surface by ozone, but the ozone layer is being depleted by CFC compounds.

Widely used CFCs react with ultraviolet radiation in the upper atmosphere to produce chlorine compounds that hamper ozone formation.

Ozone holes have formed over the poles.

The lushness of the tropical rain forest has fooled many people trying to farm the land; the poor soil quickly turns to dust when species not adapted to the rapid nutrient cycling of the area are introduced.

The great Sahara Desert is spreading southward; the present drought probably began naturally but has been perpetuated by poor agricultural practices.

As the vast United States was settled, the native temperate forest was replaced with crops, railroads, and buildings.

Human destruction of a lake's shores hastens eutrophication by dumping sediments and debris, destroying ecotones, and adding nutrients that encourage algal growth.

Dumping sewage into the Chesapeake Bay depleted the oxygen, which encouraged growth of some algae while killing phytoplankton.

As food webs toppled, many animal species have declined or vanished from the area.

The vastness of the oceans does not protect them from pollution.

The dumping of sewage, fertilizer, motor oil, garbage, and toxic chemicals or their release from sediments upsets food webs.

Nitrogen and phosphorus trigger algal blooms, which deplete oxygen and release toxins.

Infectious diseases spread, and plastic refuse kills many animals.

KEY TERMS

acid precipitation   ecotones
chlorofluorocarbon   greenhouse effect
    (CFC)   ozone layer

CHAPTER OUTLINE

I.   The Human Influence Is Everywhere
      A.   On the North Slope in Alaska, lichens that form the bases of food webs cannot weather the sulfur emission from an oil company's machines.
      B.   At the South Pole is a gaping hole in the atmosphere's protective ozone layer opened up by chemical pollutants.
      C.   Humans may be causing more harm than good to ecosystems.
II.   The Air
      A.   Acid precipitation.
           1.   Causes.
                a)   Smokestacks in the Midwest that burn coal spew sulfur and nitrogen oxides (SO_2 and NO_2) high into the atmosphere.
                b)   Gasoline and diesel fuel in internal combustion engines and waste from heavy

metal smelters contribute noxious gases.

    c) In the atmosphere, these oxides join with moisture to fall at some distance as sulfuric nitric acids (H_2SO_4 and HNO_3), forming acid rain, snow, fog, and dew.

    d) Burning fossil fuels has made the average rainfall in the eastern United States 25 to 60 times more acidic than normal.

2. Acid precipitation is also a problem in the upper Midwest, the Pacific Northwest, the Rockies, Canada, Scandinavia, Europe, the Orient, and the Soviet Union.

3. A land polluted by acid rain has a pH in the 3.0 to 5.0 range.

    a) Fish eggs die or hatch to yield deformed offspring.

    b) Amphibian eggs do not hatch at all.

    c) The shells of crustaceans do not harden.

    d) Bacteria and plankton die, and plants are replaced with lake-clogging mosses, fungi, and algae.

    e) Over time, succession leads to the establishment of species that can live in increasingly acidic environments.

4. Study of the effects of acid rain.

    a) Canada's experimentally acidified lake shows that life continues, but diversity—the backbone of evolutionary success—plummets.

    b) The experimental lake shows that acid precipitation alters food webs and biogeochemical cycles but does not seem to affect primary production, nutrient levels, and decomposition rates, at least in the years

that the area has been monitored.

    c) Most encouraging is the demonstration of natural defenses against acid precipitation.

       (1) Calcium compounds in the earth show a buffering effect.

       (2) Sphagnum peat mosses ringing lakes soak up acidic rain and buffer it.

       (3) Bacteria deep within lakes break down nitrogen and sulfur compounds.

    d) In a coniferous forest high in the mountains of Fichtelgebirge, West Germany, the thinning trees and yellowed needles attest to the effect of acid precipitation.

B. The greenhouse effect.

1. CO_2 warms the air near the earth's surface by allowing solar radiation of short wavelengths in but not releasing the longer wavelength (infrared) heat to which the energy is converted.

2. The resulting elevation in surface temperature is called the greenhouse effect, because the CO_2 blocks the escape of heat much in the same way that the glass panes of a greenhouse retain heat.

3. The total amount of atmospheric CO_2 has increased by 15% to 30% since 1896.

4. Today, fossil fuels, crops, and to a lesser extent the burning of tropical rain forests, nuclear detonations, and supersonic jets send 5 million tons of CO_2 into the atmosphere each year.

5. The consequences of the greenhouse effect.

    a) The more dramatic predictions foresee sea levels rising and lowland regions flooding as polar ice caps melt.

b) Plant communities, over time, will change.

c) Coniferous forests of spruce and cedar will gradually be dominated by deciduous species, such as red oak, sugar maple, and yellow birch.

d) As trees vanish, the animals that live in them will have to adapt to the new tree species—or disappear too.

e) Melting of the microbe-lined labyrinths in the Antarctic ice could ultimately lead to starvation of large mammals.

f) Warming will alter when animals mate and produce young, affecting the organisms that are adapted to feeding on them according to a familiar schedule.

6. It is not clear that the predicated increase in global temperature is actually occurring.

a) Perhaps the rise in CO_2 which would increase temperature, is offset by an increase in dust from pollution, which would reflect more of the sun's energy and lower temperature.

b) Oceans may absorb much of the heat.

c) Average temperature did climb from the mid-1800s to the mid-1900s, and natural variation in temperature would be expected to be only 0.4° F.

C. Destruction of the ozone layer.

1. Ultraviolet (UV) radiation can damage or kill cells.

2. It can cause a painful sunburn and also induce mutations that could develop into cancer; mutations in germ cells could lead to infertility, spontaneous abortion, or birth defects.

3. If the amount of UV light reaching the earth increases it could kill plant cells, and this would disturb food webs at their bases.

4. Life is partially shielded from UV radiation by a layer of ozone (O_3), which forms high in the atmosphere when oxygen (O_2) reacts with high-energy ultraviolet light.

5. Ozone formation is slowed or stopped by nitrogen oxides or chlorine—which we send into the atmosphere in the form of chlorofluorocarbon (CFC) compounds, which contain carbon, chlorine, and fluorine.

6. CFC's are used in refrigerants such as freon, and until 1978 they were used as propellants in aerosol cans and are still used as fire retardants, to clean electronic and computer parts, to soften pillows, and to produce foamed plastics.

7. Eventually, they rise in the upper atmosphere, where they react with UV light to produce the chlorine compounds that slow the rate of ozone formation; slowly, but noticeably, the earth's protective ozone layer is being destroyed.

III. The Land

A. The shrinking tropical rain forest.

1. "Slash and burn" agriculture is rapidly destroying the South American and African tropical rain forests to provide food for increasing human populations.

2. Once this bountiful land is used for crops, timber, or grazing, the nutrients quickly wash away and the soil hardens into a cementlike crust, incapable of supporting plants.

3. When trees die, food webs topple.

4. If this rate of destruction continues, it is estimated that the tropical rain forest, with the millions of species in and beneath its lush canopy, may be gone by the year 2000.
B. The encroaching desert.
   1. Each year the Sahara Desert moves southward by nearly 4 miles (6 kilometers).
   2. Topsoil becomes so dry that it cannot hold seeds long enough for them to germinate; wells are dry and there are no birds, no basking reptiles, and no succulent cacti.
   3. Although the current drought probably started naturally, short-sighted human agricultural practices may be sustaining it, transforming what was once semidesert to the bleakest, most lifeless land on the planet.
C. the vanishing temperate forest today, due to agricultural and lumbering practices, less than 1% of the original temperate forest of the southeastern United States prevails.
IV. The waters
A. A lake in danger—Tahoe.
   1. Lake Tahoe, although only in the early stages of human-induced eutrophication, has already been profoundly altered.
   2. As roads, houses, hotels, and casinos rose, nutrients flowed into the lake, causing the algae and weeds to bloom into a murky tangle; pollution came from sewage, from runoff from fertilized golf courses and lawns, and from the air.
   3. Altogether, 75% of the marshes and 50% of the meadows have been disturbed.
   4. Marshes and meadows are bridges between ecosystems called ecotones, and they are important in maintaining interactions between ecosystems.

B. An endangered estuary—the Chesapeake
   1. Humans have mistreated estuaries, draining them and filling them in to build houses and dumping garbage and other pollutants in these waters.
   2. The Chesapeake Bay touches six states and houses more than 200 species of fishes and 75 species of birds.
   3. Beneath the once-beautiful waters, the food web base is in upheaval, triggered by sewage, silt, heavy metals, pesticides and oil pouring in over the past 35 years.
   4. The pollution has sharply reduced the oxygen content; dinoflagellate and green algae populations are on the rise, while the numbers of diatoms are falling, and as a result of the change in available food, the fish and crustacean populations are declining.
   5. Oysters harvested are on the decline.
   6. Cleanup efforts may help return the Chesapeake's lost diversity.
      a) New sewage facilities are being built, and the old ones repaired.
      b) A goal is to reduce the phosphorus and nitrogen flowing into the estuary by 40%.
C. The oceans.
   1. Contamination with medical waste.
      a) Along the eastern seaboard soiled bandages, sutures, lumps of solid waste, syringes, and even vials of blood that tested positive for hepatitis B and AIDS washed up on beaches.
      b) Elsewhere, dead and dying sea animals littered shores, victims of lack of oxygen or unusual new infectious diseases; some had sores or

tumors, caused by chemical pollutants in the water.

    c) San Francisco Bay is similarly polluted.

2. Chemical pollutants.

    a) Ocean fouling is a cascade of events, and pollutants pour in from several sources—animal droppings, human sewage, motor oil, fertilizer from large farms and millions of lawns, plus industrial waste, oil spills, and garbage.

    b) Rivers pick up pollutants, primarily those containing nitrogen and phosphorus, and deliver them to marshes and wetlands.

    c) The nitrogen and phosphorus trigger algae to bloom, and they rapidly displace other plants.

    d) The top millimeters of water are also where petroleum-based pollutants concentrate, choking out life from above.

3. Plastic.

    a) More than 2 million seabirds and 100,000 marine mammals have been entrapped and choked by carelessly dumped plastic, particularly the rings that hold six packs of beverages together.

    b) Many birds build their nests with plastic, and the nestlings eat it and die.

4. Oceans have natural defenses.

    a) The sheer volume of the oceans dilutes some pollutants, and some may naturally be degraded by microorganisms.

    b) Nearly everyone is more aware of what is discarded and poured down the drain.

V. Epilogue—Thoughts on the Resiliency of Life

  A. Life on earth has had many millions of years to adapt, diversify, and occupy nearly every part of the planet's surface, from tropical rain forest treetops to minute crevices in Antarctic ice.

  B. Life has prevailed through all manner of localized challenges—from natural events such as the torrential rains and high winds of hurricanes, the eruption of Mount St. Helens and the great Yellowstone fires, to the garbage heaps, oil-slicked beaches, and tarnished countryside.

  C. Individual organisms may perish but the plasticity built into genes ensures that in most cases, some individual will inherit what in one environment is a quirk, but in another, salvation.

  D. Life does not end; it changes.

  E. Life may be resilient in an overall sense, yet it is fragile in its interrelationships; intricate food webs easily collapse if a single member species declines or vanishes.

  F. Do nothing to harm life—and do whatever you can to preserve its precious diversity, for in diversity lies resiliency, and the future of life on earth.

LEARNING OBJECTIVES

After reading this chapter, the student should be able to answer these questions:

1. What are some environmental problems caused by humans?
2. How does air pollution cause acid precipitation?
3. How can acid precipitation affect lakes and forests?
4. How can buildup of carbon dioxide in the atmosphere cause global warming?
5. How do certain chemicals deplete the ozone layer?
6. Why is destroying the tropical rain forest ecologically unwise?
7. Why is the African desert expanding?
8. How do human activities pollute lakes, estuaries, and oceans?

ANSWERS TO THE TEXT QUESTIONS

1. a. The experimental lake outside of Winnipeg, Manitoba, has been deliberately acidified to measure changes in ecosystems as well as to determine natural resistance to changing pH. It has shown that diversity decreases although life continues. It showed that natural nutrient levels and decomposition rates remain for the most part unchanged. Calcium compounds in the earth acted as natural buffer. This model shows what can be expected if acid rain is not prevented.

    b. Simulations using greenhouses and artificial atmospheres show that with increased atmospheric CO_2 and 3 to 5 C increases in temperature, plants use less CO_2 from soil. This may result in increased plant productivity, but caterpillars may also increase. Computer simulations predict changes in plant communities. If we can predict changes to ecosystems caused by human intervention, then we can either choose to stop or make plans for survival in an altered environment. In the case of either the *Valdez* oil spill or Chernobyl, knowing the likely results of either disaster ahead of time and making plans for cleanup could have saved many lives. For example, better techniques for cleaning oil off of marine animals with low stress could be developed in the laboratory. Radiation exposure experiments would have predicted the outcome of farm animal offspring after the Chernobyl incident.

2. a. Natural carbonate in soil acts as a buffer for acid rain.

    b. Marshes and meadows surrounding lakes provide barriers to nutrient runoff.

    c. Oceans absorb heat, which may prevent the air from warming significantly. Also increased evaporation and present pollution may help block UV light.

    d. Left alone, some pollutants in the ocean would sink and be deposited in the sediments on the bottom. Also, microorganisms may naturally metabolize hydrocarbons.

3. In a natural ecosystem the recycling of nutrients limits growth at any particular trophic level. If producers are suddenly supplied with high levels of nutrients, a population explosion occurs that upsets the balance of the ecosystem because the higher trophic levels are not able to consume the producers. The excess die and drop to the bottom, where bacteria decompose them and deplenish the oxygen levels needed for the consumers.

4. The gradual increase of the poison sodium cyanide in the water may exceed natural mechanisms that neutralize it and result in depletion of life. Diversity would likely decrease, leaving only organisms that are capable of surviving the chemical. If the chemical is retained in tissues, biomagnification could result in fish with high levels of sodium cyanide, which may affect humans who eat them.

5. Estuaries provide a link between fresh water and salt water with fragile ecosystems that are key to the survival of other ecosystems.

ANSWERS TO THE "TO THINK ABOUT" QUESTIONS

1. Sediments that are undisturbed apparently seal over and have little effect on the life that exists on them. Dredging the Hudson River would likely stir up PCBs and pollute all layers of the waterway, which would affect life until they settled out of the water.

2. It might be argued that any life that was saved was worth the massive effort put out. However, the question should also be asked whether some life would have survived better without human assistance since stress due to capture resulted in so many deaths.

3. Though there is no agreement even among biologists in the answer to this question students should consider the following questions: Will pollution of the biosphere

continue at present levels or perhaps increase as world populations increase? Have all the major problems with pollutants been identified? Will technology be able to develop countermeasures that are effective in cleaning up pollutants?

4. The problem with cultivation of the rain forest to prevent starvation is shortsighted since after a few years, the soil is depleted of nutrients, making it uninhabitable for crops or natural vegetation. Since it takes a long time for forests to come back, other alternatives should be considered. Education to bring population rates more in alignment with natural resources would reduce starvation. Methods such as composting to increase the nutrient levels of present fields may decrease the need for slash-and-burn agriculture.

5. Upwelling brings nutrients from the bottom to the top layers of the ocean. This results in an increase in photosynthesis, which provides the base for food webs. A great decrease in upwelling due to pollutants would reduce the producer level and higher trophic levels would therefore decrease. The quantity and diversity of marine life would greatly decrease.

6. Depletion of the ozone layer if it continues will have a devastating effect on the entire world. Higher levels of UV light will increase the risk of mutation for all people. Since east Africans are likely to be as affected, they might be interested in getting other countries to stop production of CFC's.

AUDIOVISUAL MATERIALS

Air Pollution, Phoenix Film and Video, 10 min., 1968.

Acid Rain—Requiem or Recovery, NBC, 27 min., 1982.

Pollution Solution, NASA, U.S. Naval Academy, 24 min., 1976.

Pollution is a Matter of Choice, NBC, 53 min., 1970.

Test Item File

Chapter 1

1. A placebo is used in an experiment to
   a. trick the person into believing the experiment worked.
   b. ensure that only a single factor is responsible for the effect.
   c. double the sample size.
   d. eliminate the need to duplicate the study by other investigators.

   Answer: B

2. A hypothesis is simply
   a. a test.
   b. an initial observation.
   c. an educated guess.
   d. a double-blind procedure.

   Answer: C

3. Significant conclusions should be based on no fewer than _____ samples.
   a. 10
   b. 20
   c. 30
   d. 50

   Answer: C

4. When neither the researchers nor the subjects know who has been given the substance being evaluated, the experiment is said to
   a. be based on a placebo.
   b. be based on the hypothesis.
   c. be double-blind.
   d. have a double variable.

   Answer: C

5. The most serious problem with using animal models to answer human questions is that
   a. animals have rights too.
   b. there is no correlation.
   c. application is limited.
   d. they are uncooperative.

   Answer: C

6. Epidemiological studies
   a. exclude all variables except one.
   b. are done in the laboratory under controlled conditions.
   c. collect data on real-life situations.
   d. collect data on large populations of laboratory animals.

   Answer: C

7. Penicillin was discovered by
   a. Jim Schlatter.
   b. Barbara McClintock.
   c. Ed Jenner.
   d. Alexander Flemming.

   Answer: D

8. Which statement reflects an observation?
   a. The number of reported cases of breast cancer has increased.
   b. Some studies comparing the rates of breast cancer in women on the pill with those who used other means of contraception showed no correlation.
   c. Estrogen is the primary hormone used in birth-control pills.
   d. Oral contraceptives have no effect on the risk of breast cancer.

   Answer: A

9. Which statement gives a hypothesis?
   a. The number of reported cases of breast cancer has increased.
   b. Some studies comparing the rates of breast cancer in women on the pill with those who used other means of contraception showed no correlation.
   c. Estrogen is the primary hormone used in birth-control pills.
   d. Oral contraceptives increase the risk of breast cancer.

   Answer: D

10. You turn on a light switch and observe that the light did not come on. Which is the most appropriate hypothesis?
    a. There is a power outage in the area.
    b. The bulb is burned out.
    c. The circuit breaker went off.
    d. The light switch is defective.

    Answer: B

11. Which of the following hypotheses cannot be tested by the scientific method?
    a. Mary uses the phone every day after school.
    b. Heart disease is genetically linked.
    c. The flowers growing on the west face of a mountain are more beautiful than those on the east face.
    d. Bob has the knowledge to pass the written exam for his driver's license.

    Answer: C

12. Scientists have shown a higher incidence of allergies in children that were bottle-fed than those that were breast-fed. Which is the best conclusion?
   a. Cow's milk causes allergies in humans.
   b. Breast-feeding prevents allergies in humans.
   c. Allergies are unrelated to the source of milk for infants.
   d. Some individuals are less prone to allergies if breast-fed.

   Answer: D

13. Which of the following statements demonstrates a cause-and-effect relationship rather than a correlation of variables?
   a. The frequency of human lung cancer is highest in urban areas.
   b. People are more likely to get a cold in the winter than the summer.
   c. Algal growth increases downstream from sewage treatment plants.
   d. Bacterial growth is inhibited in culture media by antibiotics.

   Answer: D

14. Which of the following is an example of an epidemiological study?
   a. The study of the occurrence of multiple sclerosis in citizens of Washington state.
   b. The incidence of AIDS in monkeys.
   c. The use of mouse cells to study human aging.
   d. The effects of acid rain on forests.

   Answer: A

15. In early studies of AIDS, researchers thought that the HIV virus infected a white blood cell called the T cell. Which of the following experiments support this hypothesis?
   a. Test for the presence of the AIDS virus in other white blood cells.
   b. Test for the presence of the AIDS virus in the human population.
   c. Test for the presence of the AIDS virus in nonhuman populations.
   d. Test for the presence of the AIDS virus in human T cells.

   Answer: D

16. We wish to test the hypothesis that cancer can be caused by PCBs. Which of the following experiments eliminates both environmental and genetic factors?
   a. Determine the frequency of cancer among people exposed to PCBs.
   b. Expose mice to PCBs and determine the rate of cancer compared with their genetic clones raised under identical conditions.
   c. Compare cancer rates in identical human twins where one individual was exposed to PCBs and the other was not.
   d. Expose unrelated mice to PCBs and determine the rate of cancer compared with controls raised under identical conditions.

   Answer: B

17. Which of the following is a statement of scientific observation?
   a. Sedimentary material was deposited over a long geological period of time.
   b. Mutations are the mechanisms of evolution.
   c. Organic compounds of living organisms were synthesized from nonliving compounds in the laboratory by Dr. Urey.
   d. The oldest probable fossils discovered to date are 3.5 billion years old.

   Answer: C

18. Which of the following statements is an observation and not a hypothesis?
   a. Canadian geese fly south in the winter.
   b. Black bears hibernate in the winter to avoid cold temperatures.
   c. Stress increases the risk of heart attack.
   d. Coal burning has caused an increase in acid rain.

   Answer: A

19. Which of the following statements contains evidence for the stated conclusion that bright lights cause damage to the eyes of premature infants.
   a. Premature infants under 2 pounds often develop damaged retinas.
   b. Premature infants are often exposed to bright lights in nurseries.
   c. Nineteen of 21 premature infants exposed to bright lights had retinal damage.
   d. Twenty-one of 39 premature infants exposed to dim lights had retinal damage.

   Answer: C

20. Which of the following discoveries was made by accident?
   a. Schlatter's discovery of aspartame sweetener.
   b. Jenner's discovery of the smallpox vaccine.
   c. Mendel's discovery of inherited characteristics in peas.
   d. McClintock's discovery of jumping genes.

   Answer: A

Chapter 2

1. The relationship of one organism living on or within another organism is called
   a. mutualism.
   b. parasitism.
   c. commensalism.
   d. symbiosis.

   Answer: D

2. The relationship by which one organism living within or on another organism derives benefit from the relationship but also causes harm to its host is an example of
   a. mutualism.
   b. parasitism.
   c. commensalism.
   d. symbiosis.

   Answer: B

3. In the human colon colonies of coliform bacteria thrive that extract a chemical called vitamin K from undigested food. The human, who is unable to accomplish this feat on his or her own, benefits from the bacteria. On the other hand, the bacteria receives nutrients and a protective environment. This relationship is called
   a. mutualism.
   b. parasitism.
   c. symbiosis.
   d. commensalism.

   Answer: A

4. Owls use tall trees as perches for sighting prey. The tree benefits the owl but is in no way harmed or helped by the presence of the owl. This is an example of
   a. mutualism.
   b. parasitism.
   c. symbiosis.
   d. commensalism.

   Answer: D

5. The organisms found living today represent
   a. all the kinds of organisms that have ever existed.
   b. the most diversity in the history of the earth.
   c. less diversity than in earlier periods of time.
   d. a reduction in the numbers of individuals.

   Answer: C

6. The first living organisms were most likely related to present-day
   a. amoebas.
   b. cyanobacteria.
   c. bacteria.
   d. eukaryotes.

   Answer: C

7. Which of the following environments is believed most likely for the origin of life?
   a. volcanos
   b. crevices of rocks
   c. oceans
   d. steam clouds

   Answer: C

8. Multicellular organisms arose from
   a. prokaryotes.
   b. eukaryotes.
   c. plasmodia.
   d. mycelia.

   Answer: B

9. The age of the earth is _____ years.
   a. 5,000
   b. 1 billion
   c. 3 billion
   d. 4 billion

   Answer: D

10. The beginning of the age of bacteria began about _____ years ago.
    a. 5,000
    b. 1 billion
    c. 3 billion
    d. 4 billion

    Answer: D

11. Single-celled organisms that contain organelles are
    a. prokaryotes.
    b. eukaryotes.
    c. cyanobacteria.
    d. bacteria.

    Answer: B

12. Life appeared on land _____ years ago.
    a. 300 million
    b. 1 billion
    c. 3 billion
    d. 4 billion

    Answer: A

13. The ability of animals to live and reproduce solely on land was first dependent upon
    a. lungs.
    b. watertight skin.
    c. watertight skin and eggs.
    d. limbs designed for walking and running.

    Answer: C

14. Individuals best adapted to their environment are most likely to
    a. survive and produce offspring.
    b. reproduce in vast numbers.
    c. accumulate new traits.
    d. pass new traits to their offspring.

    Answer: A

15. Select the correct order of the appearance of life forms on earth from the earliest to the most recent.
    a. prokaryotes, eukaryotes, soft-bodied sea animals, plants, insects, amphibians, reptiles, birds, mammals
    b. eukaryotes, prokaryotes, plants, soft-bodied sea animals, insects, birds, amphibians, reptiles, mammals
    c. eukaryotes, prokaryotes, soft-bodied sea animals, birds, insects, plants, amphibians, reptiles, mammals
    d. soft-bodied sea animals, prokaryotes, eukaryotes, plants, insects, birds, reptiles, amphibians, mammals

    Answer: A

16. The definition of a species is not perfectly applicable to all organisms because
    a. they are not all reproductively isolated.
    b. plants reproduce differently than animals.
    c. many reproduce asexually.
    d. individuals are not exactly the same.

    Answer: C

17. Physical structures that look similar and have the same function in different species
    a. are proof that the organisms have a common ancestor.
    b. indicate they may have developed the structure in response to a shared environmental challenge.
    c. imply that they have the same genetic code.
    d. indicate that the organisms live in the same environment.

    Answer: B

18. The relatedness of different species can be determined by the sequence of amino acids of common proteins because
    a. these sequences do not change over time.
    b. changes in the sequence reflect changes in genetic material.
    c. they represent nongenetic changes.
    d. these sequences change rapidly in individuals.

    Answer: B

19. The subjectivity in determining which characteristics should be used to determine descent from a common ancestor has been minimized by
    a. the observation that gene and protein sequences do not vary consistently within a species.
    b. the observation that gene and protein sequences do not vary consistently between species.
    c. using only boney material present in the fossil record.
    d. the observation that gene and protein sequences vary consistently within each individual.

    Answer: A

20. The most inclusive level in the Linnaean system of biological classification is the
    a. kingdom.
    b. family.
    c. genus.
    d. species.

    Answer: A

21. In the five-kingdom taxonomic system used today, a
    a. genus is a subdivision of a family.
    b. class is a subdivision of an order.
    c. phylum is a subdivision of a class.
    d. genus is a subdivision of a species.

    Answer: A

22. The genus *Homo* is characterized by
    a. mammary glands and being warm-blooded.
    b. flexible toes and fingers, excellent vision, and a complex brain.
    c. large brain, lightweight jaw, large thumbs, and short arms.
    d. highly developed organ systems and a three-layered embryo.

    Answer: C

23. Prokaryotic cells
    a. have a nucleus.
    b. have a system of organelles.
    c. are organized in the form of "bags within a bag".
    d. have their genetic material free in the cytoplasm.

    Answer: D

24. Bacteria reproduce by
    a. fusion.
    b. budding.
    c. binary fission.
    d. meiosis.

    Answer: C

25. Bacteria that live without oxygen are called
    a. autotrophic.
    b. aerobic.
    c. facultative.
    d. anaerobic.

    Answer: D

26. Organisms that convert inorganic molecules into organic molecules are
    a. heterotrophs.
    b. autotrophs.
    c. aerobic.
    d. anaerobic.

    Answer: B

27. Cyanobacteria are photosynthetic and introduce _____ into the atmosphere.
    a. free hydrogen
    b. free oxygen
    c. water vapor
    d. ammonia

    Answer: B

28. Protista are all
    a. unicellular.
    b. photosynthetic.
    c. eukaryotic.
    d. parasitic.

    Answer: C

29. Fungi are distinguished from plants by
   a. having a cell wall.
   b. being heterotrophic.
   c. reproducing asexually.
   d. being eukaryotic.

   Answer: B

30. Except for yeasts, the basic structural unit of a fungus is the
   a. mycelium.
   b. mycorrhiza.
   c. basidia.
   d. hypha.

   Answer: D

31. Lichens are formed from the combination of
   a. bacteria and fungi.
   b. cyanobacteria and fungi.
   c. green algae and fungi.
   d. yeasts and fungi.

   Answer: C

32. Plants are distinguished from the animal kingdom by
   a. sexual form of reproduction.
   b. being eukaryotic.
   c. having cell walls.
   d. having specialized tissues and organs.

   Answer: C

33. Nonvascular plants are called
   a. bryophytes.
   b. tracheophytes.
   c. gymnosperms.
   d. angiosperms.

   Answer: A

34. Xylem
   a. conducts sugars from leaves or stems to growing shoots and fruits.
   b. transports water and minerals from roots to leaves.
   c. forms structural tubes to support the roots.
   d. conducts sugars from the roots to the leaves.

   Answer: B

35. Flowering plants are called
   a. gymnosperms.
   b. bryophytes.
   c. angiosperms.
   d. zygomycetes.

   Answer: C

36. The mesozoans and parazoans have a(n) _____ level of organization.
   a. cellular
   b. tissue
   c. organ
   d. organ system

   Answer: A

37. Which of the following statements does not apply to the animal kingdom?
   a. They have a nervous system.
   b. They are autotrophic.
   c. They are multicellular eukaryotes.
   d. They have cell membranes.

   Answer: B

38. The presence of a coelom is an advantage because it
   a. is the third germ layer.
   b. contains the open circulatory system.
   c. develops into the digestive tract.
   d. permits the development of internal organs.

   Answer: D

39. A three-layered embryo, bilateral symmetry, a coelom, a prominent head region, and a notochord are characteristics of the phyla
   a. Vertebrata.
   b. Chordata.
   c. Agnatha.
   d. Mammalia.

   Answer: B

40. The two families of primates, Pongidae and Hominidae, are separated taxonomically by
   a. the ability to grasp objects by Hominidae.
   b. more erect posture in Pongidae.
   c. reduced eyebrow ridges in Pongidae.
   d. higher intelligence in Hominidae.

   Answer: D

Chapter 3

1. Living things have in common with nonliving things the characteristic of
   a. metabolizing organic molecules.
   b. responding to external stimuli.
   c. being composed of energy and matter.
   d. growing and reproducing.

   Answer: C

2. All organisms identified to date have nucleic acids as their genetic material except _____, which is made of only protein.
   a. a virus
   b. a viroid
   c. bacteria
   d. a prion

   Answer: D

3. If an atom of phosphorus loses a neutron, it is considered to be a(n)
   a. ion.
   b. molecule.
   c. isotope.
   d. different element.

   Answer: C

4. The atomic number of hydrogen is 1 and helium is 2. These numbers identify the two atoms as being different in their number of
   a. protons.
   b. neutrons.
   c. electrons.
   d. positrons.

   Answer: A

5. The number 8 in the formula $8C_6H_{12}O_6$ represents the number of
   a. hydrogen atoms.
   b. carbon atoms.
   c. $C_6H_{12}O_6$ molecules.
   d. carbon molecules.

   Answer: C

6. The atomic number of calcium is 20. How many electrons are found in the calcium ion, Ca^{++}?
   a. 18
   b. 19
   c. 21
   d. 22

   Answer: A

7. Vomit has a pH of 2.0 and is therefore considered to be
   a. basic.
   b. neutral.
   c. a salt.
   d. acidic.

   Answer: D

8. The electrons that make up a covalent bond will likely be found associated with
   a. one atom of the bonding pair.
   b. both atoms of the bonding pair.
   c. hydrogens atoms.
   d. a single atom.

   Answer: B

9. An example of an organic molecule is
   a. methane (CH_4).
   b. water (H_2O).
   c. oxygen (O_2).
   d. ammonia (NH_4).

   Answer: A

10. A molecule that has been reduced has
    a. given up an electron.
    b. gained an electron.
    c. gained a proton.
    d. given up a proton.

    Answer: B

11. An element found in low abundance in living systems is
    a. hydrogen (H).
    b. nitrogen (N).
    c. oxygen (O).
    d. calcium (Ca).

    Answer: D

12. The tendency for water to be absorbed by dry substances is described as
    a. cohesiveness.
    b. adhesion.
    c. inhibition.
    d. libation.

    Answer: C

13. The property of water that prevents the rapid rise in body temperature on a hot day is its high
    a. surface tension.
    b. heat capacity.
    c. heat of vaporization.
    d. density at high temperatures.

    Answer: C

14. Ice floats on water because it
    a. is more dense.
    b. expands at freezing.
    c. has a high heat of fusion.
    d. has a high heat capacity.

    Answer: B

15. All carbohydrates have a common formula of
    a. $C_3H_{14}O_6$.
    b. $C_3H_4O_3$.
    c. $C_3H_9O_6$.
    d. $C_3H_6O_3$.

    Answer: D

16. Hydrolysis is exemplified by
    a. starch + water = simple sugars.
    b. amino acids = protein + water.
    c. HCl + NaOH = NaCl + H_2O.
    d. NaCl + water = Na^+ + Cl^- + water.

    Answer: A

17. Two molecules of glucose joined together are referred to as
    a. a monosaccharide.
    b. a disaccharide.
    c. a polysaccharide.
    d. fructose.

    Answer: B

18. The storage form of carbohydrates in humans is
    a. starch.
    b. glycogen.
    c. lactose.
    d. cellulose.

    Answer: B

19. The carbohydrate that gives rigidity to plant cells and is known as dietary fiber is
   a. starch.
   b. glycogen.
   c. lactose.
   d. cellulose.

   Answer: D

20. The basic unit of fat is
   a. the amino acid.
   b. glycerol.
   c. the nucleotide.
   d. glucose.

   Answer: B

21. Adipose cells in humans store
   a. carbohydrates.
   b. fats.
   c. proteins.
   d. nucleic acids.

   Answer: B

22. Saturated fats have a high concentration of
   a. carbon.
   b. oxygen.
   c. nitrogen.
   d. hydrogen.

   Answer: D

23. An element found in proteins that is not present in fats or carbohydrates is
   a. iron.
   b. oxygen.
   c. hydrogen.
   d. nitrogen.

   Answer: D

24. The functional group that distinguishes one amino acid from another is the
   a. R group.
   b. hydroxyl group.
   c. carboxyl group.
   d. peptide.

   Answer: A

25. The function of a protein is determined by its
   a. linear structure.
   b. three-dimensional shape.
   c. length.
   d. molecular weight.

   Answer: B

26. Without enzymes life could not exist because they
   a. provide structure in the cell.
   b. speed up the rate of chemical reactions.
   c. are an energy source.
   d. can be used as communication between two cells.

   Answer: B

27. The specificity of an enzyme is dependent upon its
   a. active site.
   b. half-life.
   c. complexity.
   d. ability to leave the reaction in the same form.

   Answer: A

28. DNA consists of all of the following molecules except
   a. sugar.
   b. protein.
   c. phosphate groups.
   d. nitrogen bases.

   Answer: B

29. Nucleic acids are unique among organic molecules in their ability to serve as
   a. hormones.
   b. an energy source.
   c. genetic material.
   d. enzymes.

   Answer: C

30. Vitamins and minerals are necessary for life since they facilitate the functioning of
   a. enzymes.
   b. genes.
   c. hormones.
   d. carriers.

   Answer: A

31. Which component of DNA serves as the genetic code?
   a. phosphate
   b. deoxyriboses sugar
   c. nitrogen bases
   d. ribose sugar

   Answer: C

32. This molecule is insoluble in water and contains a very high amount of calories.
   a. carbohydrates
   b. fats
   c. proteins
   d. nucleic acids

   Answer: C

33. This molecule gives up H^+ when warmed in water.
   a. carbohydrates
   b. fats
   c. protein
   d. nucleic acids

   Answer: D

34. This molecule contains sulfur.
   a. carbohydrates
   b. fats
   c. proteins
   d. nucleic acids

   Answer: C

35. This molecule contains phosphorus.
    a. carbohydrates
    b. fats
    c. proteins
    d. nucleic acids

Answer: D

36. The theory of spontaneous generation predicts that
    a. beetles and wasps sprout from cow dung.
    b. one cell gives rise to another.
    c. only life gives rise to life.
    d. mold will grow on bread.

Answer: A

37. What was the hypothesis in Francesco Redi's experiment on spontaneous generation?
    a. He filled two jars with meat, leaving one jar open and one jar covered.
    b. The covered jar produced no flies.
    c. Flies came from flies, not decaying meat.
    d. He saw flies in the open jar and maggots soon appeared.

Answer: C

38. Which of the following observations support the theory that all life descended from a common ancestor?
    a. Simple organic molecules have been found in space.
    b. All observed life forms use the same genetic code.
    c. Organic molecules have been synthesized in the laboratory.
    d. All observed life forms have eukaryotic cells.

Answer: B

39. Which of the following molecules is believed to have been absent in the earth's early atmosphere?
    a. oxygen (O_2)
    b. ammonia (NH_3)
    c. water vapor (H_2O)
    d. methane (CH_4)

Answer: A

40. Select the correct order of events in chemical evolution.
    a. energy available, production of small monomers, polymerization, replication, appearance of protocells
    b. production of small monomers, polymerization, replication, energy available, appearance of protocells
    c. energy available, polymerization, production of small monomers, replication, appearance of protocells
    d. energy available, production of small monomers, replication, polymerization, appearance of protocells

Answer: A

Chapter 4

1. The group of organisms that is unable to grow, respond to stimuli, make energy conversions, or reproduce independently of other organisms is the
    a. cyanobacteria.
    b. protista.
    c. viruses.
    d. bacteria.

Answer: C

2. The organisms that represent the simplest form of life are the
    a. bacteria.
    b. protista.
    c. viruses.
    d. algae.

Answer: C

3. Viruses always contain
    a. protein.
    b. nucleic acid.
    c. fats.
    d. carbohydrates.

Answer: B

4. The action of viral DNA in the host cell is to
    a. form viral chromosomes.
    b. replicate viral DNA.
    c. replicate viral ribosomes.
    d. take over metabolic processes of energy conversion.

Answer: B

5. The "cell," a term that we use today to describe the fundamental unit of life, was first described by
    a. Robert Hooke.
    b. Matthias J. Schleiden.
    c. Rudolph Virchow.
    d. Johann Janssen.

Answer: A

6. Bacteria and protozoa were discovered by
    a. Theodor Schwann.
    b. Zacharius Janssen.
    c. Robert Hooke.
    d. Anton van Leeuwenhoek.

Answer: D

7. An example of a prokaryote is a
   a. mushroom.
   b. bacterium.
   c. human.
   d. pine tree.

   Answer: B

8. Members of the kingdom _____ have no cell wall.
   a. Monera
   b. Fungi
   c. Animalia
   d. Plantae

   Answer: A

9. Organelles were not visible until the advent of
   a. the compound microscope.
   b. the light microscope.
   c. a high-quality lens.
   d. the electron microscope.

   Answer: A

10. The genetic material of prokaryotes differs from that of eukaryotes because it is
    a. made up of nucleic acids.
    b. inside a nuclear membrane.
    c. naked.
    d. responsible for protein synthesis.

    Answer: C

11. Which of the following has the highest surface area/volume ratio?
    a. a straw
    b. a cigarette
    c. a grain silo
    d. a 1-foot diameter pipe

    Answer: A

12. A cell that increases its surface area/volume ratio will
    a. swell.
    b. divide into two daughter cells.
    c. decrease its surface area while maintaining constant volume.
    d. increase its volume while maintaining a constant surface area.

    Answer: B

13. The jellylike fluid surrounding organelles is called
    a. cytoplasm.
    b. nucleoplasm.
    c. protoplasm.
    d. endoplasm.

    Answer: A

14. The site of polymerization of amino acids into proteins in a eukaryotic cell is the
    a. ribosome.
    b. nucleus.
    c. nucleolus.
    d. smooth endoplasmic reticulum.

    Answer: A

15. The smooth endoplasmic reticulum is the site of production of
    a. proteins.
    b. lipids.
    c. messenger RNA.
    d. glycoproteins.

    Answer: B

16. A very large substance can enter a cell through
    a. active transport.
    b. diffusion.
    c. exocytosis.
    d. endocytosis.

    Answer: D

17. The organelle that converts energy from one form to another in the cell is the
    a. nucleus.
    b. ribosome.
    c. mitochondria.
    d. peroxisome.

    Answer: C

18. Plant cells differ from animal cells by the presence of
    a. cell membranes.
    b. chloroplasts.
    c. vacuoles.
    d. lysosomes.

    Answer: B

19. Macrophages are a type of white blood cell specialized in engulfing and destroying foreign agents such as bacteria and debris. You would expect them to be high in
    a. lysosomes.
    b. peroxisomes.
    c. ribosomes.
    d. cristae.

    Answer: A

20. The characteristic of a cell that rules it out as being eukaryotic is that it
    a. contains chromosomes.
    b. has a nuclear membrane.
    c. uses oxygen.
    d. has ribosomes.

    Answer: B

21. The molecule unique to photosynthesizing cells is
    a. chlorophyll.
    b. cellulose.
    c. glycoprotein.
    d. collagen.

    Answer: A

22. Centrioles appear to play a role in the cellular process of
    a. cell division.
    b. metabolism.
    c. secretion.
    d. responding to stimuli.

    Answer: A

23. A flagella allows a cell to
    a. metabolize.
    b. reproduce.
    c. move in response to stimuli.
    d. secrete.

    Answer: C

24. Epithelial cells are involved in all of the following except
    a. secretion.
    b. protection.
    c. absorption.
    d. support.

    Answer: D

25. Tendons that connect muscles to bones prevent tearing because they stretch. They are considered to be
    a. connective tissue.
    b. nervous tissue.
    c. epithelial tissue.
    d. muscle tissue.

    Answer: A

26. The component of blood that transports oxygen to tissues is
    a. platelets.
    b. red blood cells.
    c. white blood cells.
    d. plasma.

    Answer: B

27. The component of blood that protects against infection is
    a. platelets.
    b. red blood cells.
    c. white blood cells.
    d. plasma.

    Answer: C

28. The cells of cartilage that secrete collagen are
    a. osteocytes.
    b. osteoclasts.
    c. chondrocytes.
    d. osteoblasts.

    Answer: C

29. Unlike most other tissues, cartilage lacks
    a. the ability to go through cell division.
    b. a noncellular matrix.
    c. cellular endoplasmic reticulum.
    d. blood vessels and nerves.

    Answer: D

30. Bone differs from cartilage in that it contains
    a. lacunae.
    b. collagen.
    c. mineral salts.
    d. a noncellular matrix.

    Answer: C

31. Cells that secrete bone matrix are called
    a. osteoclasts.
    b. osteoprogenitor cells.
    c. osteoblasts.
    d. chondrocytes.

    Answer: C

32. The structure of the neuron that releases a neurotransmitter is the
    a. axon.
    b. dendrite.
    c. cell body.
    d. neuroglia.

    Answer: A

33. An impulse traveling down a neuron is in the form of _____ energy.
    a. electrochemical
    b. light
    c. heat
    d. pressure

    Answer: A

34. Muscle cells have a high number of _____ to provide large amounts of energy for contraction.
    a. ribosomes
    b. lysosomes
    c. mitochondria
    d. endoplasmic reticulum

    Answer: C

35. The type of muscle tissue used to secrete sweat is
    a. smooth.
    b. cardiac.
    c. skeletal.
    d. myoepithelial.

    Answer: D

36. The evidence that best supports the endosymbiont theory is that
    a. prokaryotic cells have been observed being engulfed by eukaryotic cells.
    b. mitochondria and chloroplasts resemble bacteria.
    c. chloroplasts taken from cells can survive on their own.
    d. chloroplasts and mitochondria are not found in prokaryotic organisms.

    Answer: B

37. What advantage would endosymbiosis offer?
    a. Two cells can live cheaper than one.
    b. A nonphotosynthesizing cell could suddenly become autotrophic.
    c. Anaerobic cells could tolerate an oxygen-free atmosphere.
    d. Mobility.

    Answer: B

38. The organelle responsible for adding sugar to proteins and fats in preparing them for secretion is the
    a. smooth endoplasmic reticulum.
    b. rough endoplasmic reticulum.
    c. nucleolus.
    d. Golgi apparatus.

    Answer: D

39. RNA building blocks are stored in the
    a. nucleus.
    b. Golgi apparatus.
    c. nucleolus.
    d. rough endoplasmic reticulum.

    Answer: C

Chapter 5

1. A cell is identifiable by the unique array of _____ located on its cell surface.
    a. lipids
    b. sugar-proteins
    c. cilia
    d. nucleic acids

    Answer: B

2. Cancer cells are
    a. differentiated.
    b. foreign organisms.
    c. like embryonic cells.
    d. "sticky."

    Answer: C

3. The genetic markers on cells are known as
    a. antigens.
    b. antibodies.
    c. cytoskeletons.
    d. leukocytes.

    Answer: A

4. Cell membranes are different than cell walls because they
    a. regulate what passes into the cell.
    b. make up the outside of the cell.
    c. provide rigidity for the cell.
    d. protect the cell.

    Answer: A

5. The organic molecule that comprises cell membranes and is impermeable to water is
    a. glycoprotein.
    b. nucleic acid.
    c. phospholipid.
    d. protein.

    Answer: C

40. Cells that structurally support neurons are called
    a. myoepithelium.
    b. neuroglia.
    c. osteocytes.
    d. fibroblasts.

    Answer: B

6. Water-soluble molecules pass through the membrane by going through channels composed of
    a. glycoprotein.
    b. nucleic acid.
    c. phospholipid.
    d. protein.

    Answer: D

7. Saline solution consists of 0.85 grams of table salt dissolved in 100 milliliters of water. What do you call the salt?
    a. a solution
    b. the solvent
    c. the solute
    d. a mixture

    Answer: C

8. The free movement of potassium ions from high to low concentration through a cell membrane is referred to as
    a. facilitated diffusion.
    b. osmosis.
    c. active transport.
    d. passive diffusion.

    Answer: D

9. A red blood cell dropped in water swells to the point of breaking the membrane. This is caused by
    a. facilitated diffusion.
    b. osmosis.
    c. active transport.
    d. passive diffusion.

    Answer: B

10. In relationship to the concentration of the cell, water is considered to be
    a. hypertonic.
    b. hypotonic.
    c. isotonic.
    d. passive.

    Answer: B

11. Saline solution used in intravenous solutions in a hospital is the same concentration as the cells of the body. It is said to be
    a. hypertonic.
    b. hypotonic.
    c. isotonic.
    d. passive.

    Answer: C

12. A single-cell organism in a hypotonic solution can use its _____ to pump water out.
    a. cilia
    b. flagella
    c. contractile vacuole
    d. cell membrane

    Answer: C

13. Turgor pressure in plants maintains rigidity due to the presence of
    a. cell walls.
    b. cell membranes.
    c. contractile vacuoles.
    d. tubules.

    Answer: A

14. A diuretic will cause
    a. loss of water from a cell.
    b. uptake of water by a cell.
    c. an increase in the amount of water in urine.
    d. an increase in the water reabsorbed by the kidneys.

    Answer: C

15. A molecule can be aided in its transport across a membrane by a protein carrier. If this process requires no utilization of energy it is called
    a. facilitated diffusion.
    b. osmosis.
    c. active transport.
    d. passive diffusion.

    Answer: A

16. Calcium stored in bones is moved out of the circulatory system, where it is in relatively low concentration, to the highly concentrated bone matrix by
    a. facilitated diffusion.
    b. osmosis.
    c. active transport.
    d. passive diffusion.

    Answer: C

17. Active transport requires energy in the form of
    a. glucose.
    b. ATP.
    c. sodium.
    d. potassium.

    Answer: B

18. The means of moving a large molecule from high concentration outside the cell to low concentration inside without the aid of a carrier is called
    a. facilitated diffusion.
    b. exocytosis.
    c. passive diffusion.
    d. endocytosis.

    Answer: D

19. Secretion of materials such as lipids and neurotransmitters is by way of
    a. facilitated diffusion.
    b. exocytosis.
    c. passive diffusion.
    d. endocytosis.

    Answer: B

20. The cytoskeleton provides for all of the following functions except
    a. movement of the cell.
    b. movement of chromosomes within the cell.
    c. the shape of the cell.
    d. transport of materials into the cell.

    Answer: D

21. The common element found in flagella and cilia is
    a. tubulin.
    b. actin.
    c. myosin.
    d. phospholipid.

    Answer: A

22. Movement of flagella and cilia is powered by
    a. phospholipids.
    b. glucose.
    c. DNA.
    d. ATP.

    Answer: D

23. The protein molecules responsible for muscle contraction are
    a. actin and tubulin.
    b. myosin and tubulin.
    c. tubulin and ATP.
    d. actin and myosin.

    Answer: D

24. In sea urchins, successful fertilization depends upon the interaction of cellular membranes. _____ and _____ are the components essential for endocytosis of the male genetic material.
    a. Antigens, antibodies
    b. Microfilaments, proteins
    c. Glycoproteins, tubulin
    d. Microtubules, microfilaments

    Answer: B

25. After one sperm nucleus enters the sea urchin egg, the egg
    a. alters its membrane to prevent additional sperm from entering.
    b. dissolves the sperm nucleus on its surface.
    c. ejects additional sperm by exocytosis.
    d. secretes an enzyme that repels sperm.

    Answer: A

26. Membrane proteins are
    a. found only on the exterior surface.
    b. found only on the interior surface.
    c. capable of movement within the lipid bilayer.
    d. provide channels for lipid-soluble molecules.

    Answer: C

27. Surface molecules on cells cannot distinguish
    a. cells of one species from another.
    b. an individual within a species from others.
    c. a tissue from other tissues in an individual.
    d. like tissues within identical twins.

    Answer: D

28. Cross matching for tissue types is dependent upon identification of
    a. antibodies.
    b. antigens.
    c. phospholipids.
    d. carrier proteins.

    Answer: B

29. The human leukocyte antigen profile predicts
    a. cellular permeability.
    b. donor-recipient compatibility.
    c. specific diseases.
    d. active transport mechanisms.

    Answer: C

30. In the lipid portion of the cell membrane, the
    a. hydrophilic molecules are oriented towards each other.
    b. hydrophobic molecules are oriented towards each other.
    c. phosphate groups are found within the matrix.
    d. fatty acid chains are located on the surface.

    Answer: B

Chapter 6

1. Organisms need a continuous supply of energy because they
    a. cannot store it.
    b. cannot convert it to other forms.
    c. can only utilize it when it is directly available.
    d. lose it to the environment.

    Answer: D

2. All life forms on earth are interrelated by the similar ways they _____ energy for cellular functions.
    a. create
    b. obtain
    c. store
    d. convert

    Answer: B

3. Energy is defined as
    a. work x distance.
    b. force x distance.
    c. conversion of matter into different forms.
    d. ability to do work.

    Answer: C

4. An example of potential energy is the energy
    a. of muscle contraction.
    b. of transporting molecules.
    c. of chemical synthesis.
    d. stored in chemical bonds.

    Answer: D

5. An example of kinetic energy is
    a. a coiled spring.
    b. water behind a dam.
    c. a car moving down the highway.
    d. gasoline in the tank of a car.

    Answer: C

6. Which of the following reactions releases energy?
    a. lipids --> glycerol + fatty acids
    b. glycerol + fatty acids --> lipids
    c. glucose + fructose --> sucrose
    d. glucose + glucose --> starch

    Answer: A

7. Which molecule is the common currency of cellular energy transfer?
    a. ADP
    b. glucose
    c. ATP
    d. AMP

    Answer: C

8. Energy released from the breakdown of ATP may be used for
    a. facilitated diffusion.
    b. osmosis.
    c. diffusion.
    d. active transport.

    Answer: D

9. Energy stored in ATP is in the form of
    a. peptide bonds.
    b. ionic bonds.
    c. phosphate bonds.
    d. hydrogen bonds.

    Answer: C

10. Metabolism can be described as the
    a. use of ATP to produce starch from glucose.
    b. digestion of food in the small intestine.
    c. conversion of energy in the form of chemical bonds.
    d. loss of body heat.

    Answer: C

11. Chains of biological reactions are catalyzed by
    a. ATP.
    b. enzymes.
    c. ADP.
    d. hormones.

    Answer: B

12. An example of an anabolic reaction is
    a. glycerol + fatty acid --> lipid.
    b. protein + water --> histidine + proline + cystine.
    c. sucrose + water --> glucose + fructose.
    d. DNA + water --> purines + pyrimidines.

    Answer: A

13. Energy used to make ATP comes from the _____ in food molecules.
    a. atoms
    b. protons
    c. neutrons
    d. electrons

    Answer: D

14. In a negative feedback loop, a buildup of the _____ inhibits the reaction.
    a. reactant
    b. product
    c. enzymes
    d. hormone

    Answer: B

15. A molecule that is produced by one tissue and involved in metabolic reactions in another tissue is called a
    a. hormone.
    b. carrier.
    c. enzyme.
    d. coenzyme.

    Answer: A

16. Under the influence of insulin, glucose will
    a. be converted to glycogen in the liver.
    b. be released from the liver.
    c. increase in concentration in the blood.
    d. be released from the muscles.

    Answer: A

17. Biochemical pathways evolved in protocells because
    a. it was energetically inexpensive.
    b. the enzymes were already present.
    c. specific nutrients became limited.
    d. temperatures changed.

    Answer: C

18. BMR includes the energy required for
    a. movement of blood.
    b. skeletal muscle activity.
    c. digestive function.
    d. sleeping.

    Answer: A

19. An individual's BMR will increase with
    a. increased food intake.
    b. age.
    c. exercise.
    d. increased body fat.

    Answer: C

20. BMR is regulated by
    a. the pancreas.
    b. the thyroid gland.
    c. mental activity.
    d. heart rate.

    Answer: B

21. A _____ has the highest metabolic rate.
    a. rat
    b. horse
    c. dog
    d. elephant

    Answer: A

22. Most energy enters the biological world in the form of
    a. chemical bonds.
    b. heat.
    c. light.
    d. ATP.

    Answer: C

23. Organisms that extract energy from chemicals in the environment are
    a. heterotrophic.
    b. autotrophic.
    c. chemoautotrophic.
    d. eutrophic.

    Answer: C

24. Photosynthesis is used to store energy in the form of
    a. carbon dioxide.
    b. water.
    c. glucose.
    d. ATP.

    Answer: C

25. Visible light is necessary for photosynthesis because it
    a. makes chemical bonds.
    b. releases oxygen from carbon dioxide.
    c. combines CO_2 with hydrogen to form glucose.
    d. it excites electrons.

    Answer: D

26. Chlorophyll is contained within the
    a. stroma.
    b. grana.
    c. stroma lamellae.
    d. thylakoids.

    Answer: C

27. Electrons lost from chlorophyll during excitation are replaced with electrons from
    a. water.
    b. carbon dioxide.
    c. oxygen.
    d. glucose.

    Answer: A

28. Energy is transferred from the light reaction to the dark reaction by
    a. NADPH.
    b. glucose.
    c. ADP.
    d. water.

    Answer: A

29. The Calvin cycle results in
    a. the reduction of ADP.
    b. photolysis.
    c. the fixation of carbon dioxide.
    d. phosphorylation of ATP.

    Answer: C

30. Carbon dioxide enters plants through
    a. roots.
    b. stomata.
    c. stroma.
    d. grana.

    Answer: B

31. Energy enters glycolysis in the form of
    a. glucose.
    b. ATP.
    c. NADH.
    d. NADPH.

    Answer: A

32. Energy released from glycolysis is in the form of
    a. hydrogen ions.
    b. ATP.
    c. NAD.
    d. glucose.

    Answer: B

33. An organism that can only survive in the absence of oxygen is a(n)
    a. facultative anaerobe.
    b. aerobe.
    c. strict anaerobe.
    d. facultative aerobe.

    Answer: C

34. In the absence of oxygen, pyruvic acid is converted to
    a. glycogen.
    b. water.
    c. ethanol.
    d. ATP.

    Answer: C

35. The molecule that serves as the linkage between glycolysis and aerobic catabolic pathways is
    a. acetyl CoA.
    b. pyruvic acid.
    c. glucose.
    d. NADH.

    Answer: A

36. The energy carriers from Krebs cycle to the respiratory chain are
    a. acetyl CoA and ATP.
    b. NADH and $FADH_2$.
    c. NAD and ADP.
    d. oxygen and hydrogen.

    Answer: B

37. All of the carbon that enters aerobic respiration is released as carbon dioxide by the end of
    a. glycolysis.
    b. a respiratory chain reaction.
    c. Krebs cycle.
    d. fermentation.

    Answer: C

38. The net gain of ATP from breakdown of glucose to carbon dioxide and water is
    a. 16.
    b. 18.
    c. 32.
    d. 36.

    Answer: D

39. The molecule that serves as the final electron acceptor is
    a. water.
    b. hydrogen.
    c. carbon dioxide.
    d. oxygen.

    Answer: D

40. The metabolic pathway common to all cells is
    a. photosynthesis.
    b. fermentation.
    c. Krebs cycle.
    d. glycolysis.

    Answer: D

Chapter 7

1. Meiosis differs from mitosis and binary fission by producing daughter cells that
    a. are genetically identical to the parent cell.
    b. contain half the genetic material of the parent cell.
    c. contain the same amount of genetic material as the parent cell.
    d. are genetically identical to each other.

    Answer: B

2. Mitosis serves as the mode of reproduction in
    a. pole beans.
    b. bacteria.
    c. mammals.
    d. sea urchins.

    Answer: B

3. During interphase, the cell engages in
    a. mitosis.
    b. karyokinesis.
    c. cytokinesis.
    d. DNA replication.

    Answer: D

4. Proteins, lipids, and carbohydrates are synthesized during
    a. G_1 phase.
    b. S_1 phase.
    c. G_2 phase.
    d. M phase.

    Answer: A

5. During interphase the DNA
    a. is condensed into chromosomes.
    b. is unraveled.
    c. is condensed into sister chromatids.
    d. has observable centromeres.

    Answer: B

6. Microtubules that will be assembled into the spindle apparatus are synthesized during the
    a. S phase.
    b. G_1 phase.
    c. G_2 phase.
    d. M phase.

    Answer: A

7. Cats have 38 chromosomes. A cell arrested during prophase of mitosis would have _____ chromosomes and _____ sister chromatids.
    a. 38, 38
    b. 76, 76
    c. 19, 38
    d. 38, 76

    Answer: D

8. The nucleolus disappears during
    a. interphase.
    b. prophase.
    c. metaphase.
    d. anaphase.

    Answer: B

9. Spindle fibers attach to chromosomes during
    a. interphase.
    b. prophase.
    c. metaphase.
    d. anaphase.

    Answer: C

10. Corn has 20 chromosomes. At the end of anaphase, there would be _____ chromosomes.
    a. 40
    b. 20
    c. 10
    d. 80

    Answer: A

11. Cytokinesis involves the division of
    a. DNA.
    b. chromosomes.
    c. sister chromatids.
    d. cell membranes.

    Answer: D

12. Failure of a cell to go through cytokinesis results in multiple
    a. daughter cells.
    b. copies of chromosomes but one nucleus.
    c. nuclei.
    d. nucleoli.

    Answer: C

13. Bacterial cell division is not considered to be mitosis because
    a. cytokinesis is incomplete.
    b. there is only one strand of DNA.
    c. they lack nuclei.
    d. they do not contain organelles.

    Answer: C

14. Possible evidence for an undiscovered molecule that triggers mitosis is
    a. different rates of mitosis among different tissues.
    b. the Hayflick limit.
    c. cell fusion experiments.
    d. estrogen stimulation of cell division in the uterus.

    Answer: C

15. The Hayflick limit in mammalian cells averages _____ cell divisions.
    a. 65
    b. 35
    c. 40
    d. 50

    Answer: D

16. Cell types that do not divide very often include
    a. intestinal cells.
    b. skin cells.
    c. osteocytes in bone.
    d. nerve cells.

    Answer: D

17. Chemicals that are released by damaged tissue and cause healing are called
    a. hormones.
    b. growth factors.
    c. neurotransmitters.
    d. enzymes.

    Answer: B

18. Fibroblast growth factor results in production of
    a. epithelial cells.
    b. bone matrix.
    c. collagen.
    d. platelets.

    Answer: C

19. The purpose of cell division is to maintain
    a. a high surface area to volume ratio.
    b. proper enzyme concentration.
    c. a constant number of organelles.
    d. the integrity of the cell membrane.

    Answer: A

20. Cell division is limited in body tissues by
    a. waste product accumulation.
    b. crowding.
    c. a short supply of nutrients.
    d. the Hayflick limit.

    Answer: B

21. Plants cells divide
    a. in all tissues.
    b. only at the tips of roots and shoots.
    c. in the center of the stem.
    d. in the leaf margins.

    Answer: B

22. A half-full flask of bacteria with a doubling time of 20 minutes will be filled in
    a. 20 minutes.
    b. 40 minutes.
    c. 1 hour.
    d. 5 hours.

    Answer: A

23. Infectious bacteria that have short doubling times will
    a. show symptoms immediately.
    b. have a long incubation period.
    c. have a low mitotic rate.
    d. be treatable with antibiotics.

    Answer: A

24. In the _____ phase of the growth curve bacterial population will be at the maximum.
    a. lag
    b. log
    c. stationary
    d. decline

    Answer: C

25. Mitotic rates are highest during the _____ phase of the growth curve.
    a. lag
    b. log
    c. stationary
    d. decline

    Answer: B

26. Stem cells give rise to daughter cells in which
    a. both are identical stem cells.
    b. one is specialized and the other remains a stem cell.
    c. one is nonfunctional and the other remains a stem cell.
    d. both are specialized for a specific function.

    Answer: B

27. The cells of embryos are considered
    a. renewal cell populations.
    b. expanding cell populations.
    c. exponential cell populations.
    d. static cell populations.

    Answer: B

28. Cancer cells are characterized by
    a. a mitotic rate higher than embryonic cells.
    b. contact inhibition.
    c. differentiation.
    d. their transplantability.

    Answer: D

29. Benign tumors differ from malignant tumors in that they
    a. are metastatic.
    b. do not migrate.
    c. lack contact inhibition.
    d. are invasive.

    Answer: B

30. Oncogenes may cause cancer by
    a. ill-timed activation of cell division.
    b. differentiation of the cell.
    c. producing hormones.
    d. introducing transposons.

    Answer: A

31. The mode of action of cancer-causing agents such as chemicals, viruses, and radiation is to
    a. act as growth factors.
    b. act as hormones.
    c. cause DNA damage.
    d. cause membrane destruction.

    Answer: C

32. Transgenetic organisms are
    a.   the result of fusion of cells from two species.
    b.   individuals with oncogenes in every cell.
    c.   the result of fusion of two nuclei.
    d.   hybrids of two or more species.

    Answer: B

33. Burkitt's lymphoma is a cancer of
    a.   muscles.
    b.   lungs.
    c.   skin.
    d.   white blood cells.

    Answer: D

34. Infection with the Epstein-Barr virus appears to
    a.   cause nutrient deficiencies.
    b.   trigger chromosome breakage.
    c.   induce oncogene formation.
    d.   trigger antibody formation.

    Answer: B

Chapter 8

1. Sexual reproduction provides genetic variability that most benefits the species when
    a.   the environment is rapidly changing.
    b.   competition is low within the population.
    c.   the population is homogeneous.
    d.   the population is declining.

    Answer: A

2. The sequence of male reproductive structures from site of sperm production to ejaculation from the penis is
    a.   testes, urethra, vas deferens, epididymis.
    b.   urethra, vas deferens, epididymis, testes.
    c.   testes, epididymis, vas deferens, urethra.
    d.   vas deferens, epididymis, urethra, testes

    Answer: C

3. The male reproductive organ that secretes fructose sugar and prostaglandins in the seminal fluid is the
    a.   seminal vesicle.
    b.   bulbourethral gland.
    c.   Cowper's gland.
    d.   prostate gland.

    Answer: A

4. The typical number of sperm cells released in a single ejaculation of the human male is
    a.   1 million.
    b.   10 million.
    c.   100 million.
    d.   400 million.

    Answer: D

5. Sperm is stored in the
    a.   scrotum.
    b.   testes.
    c.   epididymis.
    d.   seminal vesicles.

    Answer: C

6. The testes are suspended within the scrotal sac outside of the body cavity because
    a.   they function as a sexual attractant.
    b.   they function in sexual identification.
    c.   of temperature limits on sperm development.
    d.   of male competition.

    Answer: C

7. The sequence of female reproductive structures from the site of ovum production to the exterior of the body is
    a.   fallopian tube, uterus, vagina, ovary.
    b.   uterus, vagina, ovary, fallopian tube.
    c.   ovary, uterus, fallopian tube, vagina.
    d.   ovary, fallopian tube, uterus, vagina.

    Answer: D

8. The site of implantation of the embryo is the
    a.   follicle cells.
    b.   vagina.
    c.   fallopian tubes.
    d.   uterus.

    Answer: D

9. If fertilization does not occur, the _____ is expelled as menstrual flow.
    a.   follicle cells
    b.   endometrium
    c.   vaginal lining
    d.   cervical epithelium

    Answer: B

10. The male reproductive organ that is homologous to the clitoris in females is the
    a.   scrotum.
    b.   testes.
    c.   penis.
    d.   seminal vesicle.

    Answer: C

11. Fertilization occurs in the
    a. fallopian tube.
    b. ovary.
    c. uterus.
    d. vagina.

    Answer: A

12. Sperm and egg are alike in that they both have
    a. a flagellum.
    b. 23 chromosomes.
    c. nutrients.
    d. cell walls.

    Answer: B

13. The somatic number of chromosomes in dogs is 78; their gametes contain _____ chromosomes.
    a. 78
    b. 39
    c. 156
    d. 46

    Answer: B

14. If the haploid chromosome number in mosquitoes is three, then the
    a. somatic cells contain nine chromosomes.
    b. somatic cells contain six chromosomes.
    c. gametes contain six chromosomes.
    d. gametes contain nine chromosomes.

    Answer: B

15. If the somatic number of chromosomes in a goldfish is 94, how many chromosomes will each of the daughter cells have following reduction division or meiosis I?
    a. 94
    b. 47
    c. 188
    d. 74

    Answer: B

16. Synapsis of homologs occurs during
    a. prophase I.
    b. metaphase I.
    c. prophase II.
    d. metaphase II.

    Answer: A

17. Crossing over, which results in new gene combinations, occurs between
    a. sister chromatids.
    b. alleles.
    c. homologs.
    d. histone proteins.

    Answer: B

18. Genes for brown eyes and blue eyes are called
    a. homologs.
    b. alleles.
    c. sister chromatids.
    d. homozygous.

    Answer: B

19. Crossing over leads to
    a. increased genetic variability.
    b. DNA stability.
    c. homozygous gametes.
    d. homogeneity within the population.

    Answer: A

20. The random separation of homologous chromosomes during meiosis I is the basis for the hereditary law of
    a. independent assortment.
    b. segregation.
    c. dominance/recessiveness.
    d. codominance.

    Answer: A

21. If an organism has three chromosomal pairs, the number of possible configurations during metaphase I is
    a. three.
    b. four.
    c. six.
    d. eight.

    Answer: D

22. During _____ the homologs separate.
    a. anaphase I
    b. telophase I
    c. prophase II
    d. metaphase II

    Answer: A

23. During the second interphase the chromosomes
    a. are replicated.
    b. become invisible.
    c. divide.
    d. are halved in number.

    Answer: B

24. In anaphase II _____ separate.
    a. homologs
    b. alleles
    c. sister chromatids
    d. unlike genes

    Answer: C

25. The daughter cells resulting from meiosis II are
    a. identical.
    b. diploid.
    c. four in number.
    d. somatic cells.

    Answer: C

26. The cells resulting from meiosis I in spermatogenesis are called
    a. spermatogonia.
    b. sperm.
    c. spermatids.
    d. primary spermatocytes.

    Answer: D

27. The acrosome of the sperm contains
    a. the genetic material.
    b. mitochondria.
    c. proteins that form the flagellum.
    d. enzymes that penetrate the ovum.

    Answer: D

28. Oogenesis results in
    a. one secondary oocyte and three polar bodies.
    b. four secondary oocytes.
    c. four ova.
    d. four polar bodies.

    Answer: A

Chapter 9

1. The secondary oocyte released from the ovary is protected by a layer of cells called the
    a. zona pellucida.
    b. corona radiata.
    c. amnion.
    d. lanugo.

    Answer: B

2. Capacitation of the sperm directly results in
    a. sperm moving towards the egg.
    b. the secondary oocyte undergoing meiosis II.
    c. the digestion of the zona pellucida.
    d. the digestion of the corona radiata.

    Answer: A

3. The female assists the movement of sperm by
    a. movement of flagella.
    b. contractions of the fallopian tube.
    c. cilia moving mucus secretions.
    d. abdominal contractions.

    Answer: C

4. Conception usually occurs in
    a. the uterus.
    b. the vagina.
    c. the ovary.
    d. the fallopian tube.

    Answer: D

5. Sperm touching the surface of the oocyte results in
    a. polyspermy.
    b. electrochemical changes that block polyspermy.
    c. electrochemical changes that induce mitotic division.
    d. immediate nuclear fusion.

    Answer: B

6. Fraternal twins are the result of
    a. two sperm and one oocyte.
    b. two oocytes and one sperm.
    c. two sperm and two oocytes.
    d. splitting of one fertilized oocyte.

    Answer: C

29. Meiosis in oogenesis is completed
    a. in the embryo.
    b. in a single follicle each menstrual cycle.
    c. after penetration of the sperm.
    d. at puberty.

    Answer: C

30. Meiosis differs from mitosis in that it
    a. results in separation of genetic material.
    b. results in daughter cells that are identical.
    c. results in haploid cells.
    d. occurs in all body cells.

    Answer: C

7. At the eight-cell stage of development, the cell mass is called a
    a. gastrula.
    b. zygote.
    c. blastomere.
    d. morula.

    Answer: C

8. The stage of development when implantation occurs is called the
    a. gastrula.
    b. morula.
    c. blastomere.
    d. blastocyst.

    Answer: D

9. In the blastocyst, the _____ develops into the embryo.
    a. chorion
    b. inner cell mass
    c. endometrium
    d. trophoblast

    Answer: B

10. Human chorionic gonadotropin (HCG) is responsible for the
    a. development of the embryo.
    b. implantation of the embryo.
    c. maintenance of the endometrium.
    d. growth of the inner cell mass.

    Answer: C

11. The placenta
    a. is a protective sac surrounding the embryo.
    b. is a tissue composed of both maternal and embryonic cells.
    c. provides a protective barrier from chemicals in the mother's blood.
    d. is reabsorbed after the formation of the chorionic membranes.

    Answer: B

12. Modern pregnancy tests detect _____ in the mother's urine.
    a. fetal cells
    b. alpha-fetoprotein
    c. HCG
    d. fetal antibodies

    Answer: C

13. The allantois
    a. contributes blood cells and gives rise to the fetal umbilical arteries and vein.
    b. is a bag of water that cushions the embryo, maintains a constant temperature and pressure, and protects the embryo.
    c. is the fusion of tissue from the endometrium of the mother with the chorionic villus of the fetus.
    d. is a white downy hair that forms during the fifth month of development.

    Answer: A

14. Amniocentesis is the
    a. extraction of fetal cells from the amnion.
    b. isolation of fetal antibodies from the uterus.
    c. isolation of fetal cells from the mother's urine.
    d. extraction of fetal cells from the chorionic villus.

    Answer: A

15. The umbilical cord
    a. prevents the embryo from exiting the womb.
    b. transports nutrients and oxygen from the mother to the embryo.
    c. contains the amnion.
    d. connects the yolk sac to the placental membranes.

    Answer: B

16. Embryonic induction refers to
    a. the specialization of one group of cells causing adjacent cells to specialize.
    b. formation of organs.
    c. development of the placental membranes.
    d. uterine implantation of the embryo.

    Answer: A

17. The primitive streak is important to the development of vertebrates because it
    a. gives rise to extraembryonic membranes.
    b. gives rise to the circulatory system.
    c. forms an anatomical reference point for other structures to develop.
    d. forms the brain and spinal cord.

    Answer: C

18. An embryo that has arms and legs represented only by small buds is approximately _____ weeks old.
    a. 3
    b. 4
    c. 5
    d. 6

    Answer: B

19. When all of the organ systems have formed, the unborn child is referred to as a(n)
    a. zygote.
    b. embryo.
    c. fetus.
    d. morula.

    Answer: C

20. External sexual identity can be determined by the _____ week of development.
    a. 8th
    b. 9th
    c. 10th
    d. 12th

    Answer: D

21. Body functions such as sucking, making fists, urination, and defecation become observable by the end of
    a. the first trimester.
    b. fetal development.
    c. the second trimester
    d. embryonic development.

    Answer: A

22. The vernix caseosa
    a. forms a mucus plug in the cervix.
    b. protects the developing skin of the fetus.
    c. is white downy hair.
    d. refers to the wrinkled skin of the fetus.

    Answer: B

23. The major growth of the fetus in length and weight occurs during the
    a. embryonic stage.
    b. first trimester.
    c. second trimester.
    d. third trimester.

    Answer: D

24. Normally labor is preceded by
    a. rupture of the amniotic sac.
    b. rotation and dropping of the fetus in the womb.
    c. the "bloody show."
    d. labor pains.

    Answer: B

25. Onset of the second stage of labor is signaled by
    a. rupture of the amniotic sac.
    b. uterine contractions.
    c. dilation of the cervix to 10 centimeters.
    d. expulsion of the placenta.

    Answer: C

26. _____ percent of secondary oocytes exposed to sperm result in a live birth.
    a. Fifty
    b. Seventy
    c. Twenty
    d. Thirty

    Answer: D

27. Older mothers are more likely to give birth to children with genetic defects because
   a. they have fewer viable primary oocytes.
   b. they have fewer viable secondary oocytes.
   c. their reproductive tract has undergone significant aging.
   d. the primary oocytes have had a longer time to accumulate genetic damage.

   Answer: D

28. The process of aging begins
   a. before birth.
   b. after the age of 10.
   c. after the age of 25.
   d. after the age of 35.

   Answer: A

29. The declining activity of the thymus through life is thought to play a key role in aging because of its role in
   a. regulating heart rate.
   b. regulating metabolism.
   c. immunity.
   d. hormone regulation.

   Answer: C

30. After age 30, body function declines about _____ every year.
   a. 1%
   b. 5%
   c. 10%
   d. 15%

   Answer: A

31. Testosterone, the male sex hormone, is highest in males around age
   a. 15.
   b. 18.
   c. 25.
   d. 30.

   Answer: B

32. Menopause, the cessation of menstruation, usually occurs for women in their
   a. 40s.
   b. 50s.
   c. 60s.
   d. 70s.

   Answer: B

33. Memory loss usually does not occur until age
   a. 50.
   b. 60.
   c. 70.
   d. 80.

   Answer: B

34. Increased risk of age-onset diabetes occurs at age 50 due to a decline in _____ activity.
   a. brain
   b. kidney
   c. physical
   d. pancreas

   Answer: D

35. A mechanism of passive aging that may account for alterations throughout the body is
   a. increased lipofuscin in cells.
   b. decreased elasticity.
   c. the declining ability to repair DNA damage.
   d. activation of an "aging gene."

   Answer: C

36. Free radicals are suspected of playing a role in aging systems because of their ability to
   a. decrease rates of chemical reactions.
   b. increase enzyme activity.
   c. increase metabolic reactions.
   d. make other molecules unstable.

   Answer: D

37. Evidence for the "programmed cell death" theory comes from
   a. the identification of "aging" genes.
   b. species that undergo metamorphism.
   c. the buildup of lipofuscin.
   d. decreased DNA repair.

   Answer: B

38. The mechanism underlying premature aging seen in progeria patients is apparently their
   a. inability to live past age 12.
   b. premature wrinkling and baldness.
   c. cells' inability to divide more than 30 times.
   d. death from heart attack.

   Answer: C

39. Alzheimer's disease is characterized by
   a. premature aging in childhood.
   b. severe memory loss as early as age 50.
   c. increased susceptibility to cancer.
   d. death by the age of 20.

   Answer: B

40. Overall it can be said of aging in Americans that
   a. alteration of habits will have no effect on longevity.
   b. it is solely an active process regulated by genes.
   c. social problems should increase in the near future.
   d. it is a passive phenomenon caused by breakdown of collagen, elastin, muscles, and membranes.

   Answer: C

Chapter 10

1. To be fertile a male must produce more than _____ sperm in a single ejaculation.
   a. 100 million
   b. 200 million
   c. 300 million
   d. 400 million

   Answer: D

2. Male infertility is often due to
   a. attack of the sperm by the female's white blood cells.
   b. high testosterone levels.
   c. varicose veins in the scrotum.
   d. bladder infection.

   Answer: C

3. Sperm quality can be restored in some cases by
   a. a warm bath.
   b. reducing stress.
   c. hormone treatment.
   d. increased sexual activity.

   Answer: C

4. For a sexually active woman under the age of 30 not using birth control, pregnancy usually occurs after
   a. 1 year.
   b. 1 month.
   c. 3 months.
   d. 9 months.

   Answer: C

5. Hormonal imbalance resulting in irregular ovulation can be due to
   a. endometriosis.
   b. an overactive thyroid gland.
   c. underproduction of prolactin.
   d. a tumor on the ovary.

   Answer: D

6. The drawback to using fertility drugs is they
   a. may result in multiple births.
   b. have negative side effects.
   c. are carcinogenic.
   d. are teratogenic.

   Answer: A

7. Blocked fallopian tubes are a problem because they result in
   a. fibroids.
   b. obstruction of passage of the egg.
   c. endometriosis.
   d. tubal pregnancy.

   Answer: B

8. Endometriosis is a condition where endometrial cells build up and then slough off the
   a. inside of the uterus.
   b. outside of the uterus.
   c. inside of the ovary.
   d. outside of the fallopian tube.

   Answer: B

9. Vaginal secretions may cause infertility if they
   a. contain antibodies against the male's sperm.
   b. are neutral in pH.
   c. contain excess hormones.
   d. do not have enough mucus.

   Answer: A

10. Fertility rates are highest for females in their
    a. teens.
    b. 20s.
    c. 30s.
    d. 40s.

    Answer: B

11. Spontaneous abortions often occur when
    a. the mother does not receive enough rest.
    b. the embryo or fetus is deformed.
    c. high HCG levels prevent implantation.
    d. the mother is physically active.

    Answer: B

12. An ectopic pregnancy occurs when the zygote implants in the
    a. uterus.
    b. cervix.
    c. vagina.
    d. fallopian tube.

    Answer: D

13. A baby is premature if the
    a. development period is less than 38 weeks.
    b. birth weight is under 4 pounds.
    c. reflexes are immature.
    d. baby is born before the 34th week.

    Answer: D

14. Respiratory distress syndrome is due to
    a. incomplete development of the lungs.
    b. absence of surfactin.
    c. immature respiratory centers in the brain.
    d. immature respiratory muscles.

    Answer: B

15. Low birth weight is a problem for full-term newborns because of
    a. immature sucking reflexes.
    b. immature breathing reflexes.
    c. heat loss.
    d. immature digestive function.

    Answer: C

16. Birth defects are caused by
    a. abnormal development of the placenta.
    b. malnourishment of the mother.
    c. viruses.
    d. low birth weight.

    Answer: C

17. The time during development when a specific body structure can be altered is the
    a. fetal period.
    b. embryonic period.
    c. critical period.
    d. preembryonic period.

    Answer: C

18. Two-thirds of all birth defects stem from a disruption during
    a. the first 8 weeks.
    b. the second trimester.
    c. the third trimester.
    d. conception.

    Answer: A

19. A genetically caused birth defect
    a. may be a product of teratogens.
    b. may be a result of an error in mitosis in development.
    c. may be passed on to the next generation.
    d. does not require action of the abnormal gene during development.

    Answer: C

20. Thalidomide has an effect on _____ development during the critical period.
    a. neural tube
    b. limb bud
    c. heart
    d. lung

    Answer: B

21. Children of mothers who contact German measles in the first trimester have a risk of
    a. reading disabilities.
    b. speech and hearing problems.
    c. juvenile onset diabetes.
    d. heart defects.

    Answer: D

22. Mothers of children born with small heads, misshapen eyes, and a flat face and nose probably _____ during pregnancy.
    a. took cocaine
    b. smoked marijuana
    c. drank alcohol
    d. smoked cigarettes

    Answer: C

23. Excessive vitamin A, sometimes taken in the form of Accutane, an antiacne medication, during pregnancy may result in
    a. hemorrhaging.
    b. spontaneous abortion.
    c. low birth weight.
    d. chromosomal defects.

    Answer: B

24. The controversy involved in neonatology is whether or not it is
    a. medically feasible to work on fetuses prior to birth.
    b. ethical to prolong life through medical intervention when quality of life is questionable.
    c. ethical to sacrifice a lab animal's life to prolong a human infant's life.
    d. ethical to overcome infertility through artificial means.

    Answer: B

25. Artificial insemination involves the introduction of _____ into the female reproductive tract.
    a. sperm
    b. oocytes or sperm
    c. oocytes
    d. embryos

    Answer: A

26. A medical treatment that is not yet available for a fetus is
    a. a blood transfusion.
    b. drainage of a blocked bladder.
    c. removal of excess fluid from the brain.
    d. a nutritional boost.

    Answer: A

27. The benefit of surrogate motherhood to agriculture is that it
    a. allows two animals to produce offspring that are geographically isolated.
    b. increases the number of genes passed on to offspring by a selected male.
    c. increases the number of genes passed on to offspring by a selected female.
    d. increases genetic variability in the population.

    Answer: C

28. In vitro fertilization differs from gamete intrafallopian transfer in that
    a. fertilization occurs in the natural environment.
    b. sperm and egg are donated and condensed in the laboratory.
    c. embryos are transplanted into the womb.
    d. it involves laproscopy.

    Answer: C

29. Which of these reproductive alternatives excludes the possibility that the child the woman is carrying is her own?
    a. artificial insemination
    b. in vitro fertilization
    c. gamete intrafallopian transfer
    d. embryo adoption

    Answer: D

30. The birth-control pill prevents conception by
    a. increasing the woman's immune response against sperm.
    b. alteration of the uterine lining.
    c. inducing menstruation.
    d. altering vaginal secretions so the local environment reduces the viability of sperm.

    Answer: B

31. Depo-Provera is similar to the birth-control pill in that it
    a. suppresses ovulation.
    b. induces menstruation.
    c. causes the mucus in the cervix to thicken, preventing sperm from getting through.
    d. contains the hormone estrogen.

    Answer: A

32. Gossypol, the male birth-control pill, works by
    a. producing sperm that are immotile.
    b. stopping the production of enzymes in the acrosome.
    c. stopping the production of enzymes that produce sperm.
    d. causing an immune response to the sperm.

    Answer: C

33. A vaccine against pregnancy will cause the woman's immune system to produce antibodies against
    a. her partner's sperm.
    b. her own oocytes.
    c. zygotes.
    d. embryos.

    Answer: A

34. Intrauterine devices work by
    a. blocking the passageway of sperm into the uterus.
    b. blocking the passageway of oocytes in the fallopian tubes.
    c. killing the sperm.
    d. not allowing implantation of the embryo.

    Answer: D

35. Condoms and diaphragms are alike in that they
    a. are used by the male partner.
    b. block the pathway of sperm.
    c. kill sperm.
    d. increase the risk of sexually transmitted disease.

    Answer: B

36. The rhythm method of birth control involves abstinence from sex during days _____ following the onset of menstruation.
    a. 7-12
    b. 12-16
    c. 16-21
    d. 21-28

    Answer: B

37. Tubal ligation, the means of surgical sterilization in females, involves alteration of the
    a. ovaries.
    b. uterus.
    c. cervix.
    d. fallopian tubes.

    Answer: D

38. First-trimester abortions involve
    a. salt-solution injections.
    b. prostaglandin suppositories.
    c. intrauterine scrapings.
    d. surgery.

    Answer: C

39. Sexually transmitted diseases
    a. are infectious only when symptoms are present.
    b. may be cured without medical intervention.
    c. have symptoms that may be similar to allergies.
    d. number in the hundreds.

    Answer: C

40. Pelvic inflammatory disease can lead to
    a. death.
    b. infertility.
    c. hemorrhoids.
    d. blisters on the genitalia.

    Answer: B

Chapter 11

1. Genes are made of
    a. protein.
    b. carbohydrates.
    c. ribonucleic acid (RNA).
    d. deoxyribonucleic acid (DNA).

    Answer: D

2. At the cellular level, genes are directly responsible for the synthesis of
    a. proteins.
    b. carbohydrates.
    c. fats.
    d. nucleic acids.

    Answer: A

3. When Mendel began his work, he knew about
   a. the laws of inheritance.
   b. the cell theory.
   c. meiosis.
   d. chromosomes.

   Answer: B

4. The law of segregation predicts that _____ will separate during the formation of gametes.
   a. two copies of a gene
   b. nonrelated genes
   c. sister chromatids
   d. nonhomologous chromosomes

   Answer: A

5. The process of _____ makes the law of segregation possible.
   a. mitosis
   b. meiosis
   c. crossing over
   d. cytokinesis

   Answer: B

6. Mendel's hypotheses regarding transmission of genetic material were not well accepted in his day probably because not enough was known about
   a. the scientific method.
   b. chromosomes.
   c. DNA.
   d. the cell theory.

   Answer: B

7. A gamete contains _____ copy (copies) of a particular gene.
   a. one
   b. two
   c. three
   d. four

   Answer: A

8. The tall and short genes described for Mendel's peas are considered to be
   a. homozygous.
   b. heterozygous.
   c. alleles.
   d. dominant.

   Answer: C

9. Both of Mendel's pure-breeding plants must have been
   a. homozygous.
   b. heterozygous.
   c. alleles.
   d. dominant.

   Answer: A

10. When one gene is said to be dominant over its allele, it
    a. shuts down the action of the other gene.
    b. masks the product of the other gene.
    c. alters the product of the other gene.
    d. will be passed on more frequently to the next generation.

    Answer: B

11. In Mendel's pea plants, a tall plant that has short offspring has the genotype
    a. *TT*.
    b. *Tt*.
    c. *tt*.
    d. *T*.

    Answer: B

12. A gene is considered mutant if
    a. its trait is less adapted to the environment.
    b. it is new in the population.
    c. it is harmful to the organism.
    d. it produces a nonfunctional protein.

    Answer: B

13. If you are considered the F2 generation, your grandparents would be the
    a. second filial generation.
    b. first filial generation.
    c. parental generation.
    d. second parental generation.

    Answer: C

14. What percent of the gametes of a plant with genotype *Tt* for height will contain the *t* gene?
    a. 25
    b. 50
    c. 75
    d. 100

    Answer: B

15. The expected phenotypic ratio of a cross between two heterozygotes for height in garden peas is
    a. 3 tall:1 short.
    b. 1 *TT*:2*Tt*:1 *tt*.
    c. All tall.
    d. 2 tall:2 short.

    Answer: A

16. To test the genetic identity of an unknown plant for height, a test cross is made with a plant with a genotype of
    a. *TT*.
    b. *Tt*.
    c. *tt*.
    d. *T*.

    Answer: C

17. In humans, the trait of straight little finger is controlled by the *S* gene and the trait of curved little finger is controlled by the *s* gene. What is the expected ratio of phenotypes from a cross between a person with curved finger and heterozygous straight finger?
    a. all straight fingers
    b. all curved fingers
    c. 3 straight fingers:1 curved finger
    d. 1 straight finger:1 curved finger

    Answer: D

18. Homozygous dark-colored mice bred with homozygous white mice result in all dark offspring. The genotype of the offspring is
    a. *DD*.
    b. *WW*.
    c. *Dd*.
    d. *dd*.

Answer: C

19. A parent plant with the genotype *RRYY* will produce gametes that have
    a. *RR* or *YY*.
    b. *RY*.
    c. *R* or *Y*.
    d. *RRYY*.

Answer: B

20. A parent plant with the genotype RrYy will produce gametes that have
    a. *RY*.
    b. *ry*.
    c. *RrYy*.
    d. *RY, Ry, rY, ry*.

Answer: D

21. The phenotypic ratio of the F1 generation produced by a cross between two heterozygous parents for two traits that are dominant/recessive results in
    a. 3:1.
    b. 1:2:1.
    c. 9:3:3:1.
    d. 1:1:1:1.

Answer: C

22. Independent assortment predicts that
    a. alleles separate during meiosis.
    b. nonrelated genes separate during meiosis.
    c. alleles stay together during meiosis.
    d. nonrelated genes stay together during meiosis.

Answer: B

23. Independent assortment predicts that a gamete from a parent plant with the genotype *RrYy* (seed shape and color) will
    a. have only one copy of the gene for seed shape.
    b. have two copies for each gene type.
    c. receive the *Y* gene if it receives the *R* gene.
    d. receive either the *Y* or *y* gene regardless of the seed-shape gene.

Answer: D

24. The ratio of phenotypes for seed shape and color of the F1 generation from a cross *RrYy* and *rryy*, where round and yellow is dominant over wrinkled and green, will be
    a. all round and yellow.
    b. 1 round green:2 round yellow:1 wrinkled green.
    c. 3 round yellow:1 wrinkled green.
    d. 1 round green:1 round yellow:1 wrinkled yellow:1 wrinkled green.

Answer: D

25. In rabbits, black fur is dependent on a dominant gene *(B)*, and brown is a recessive allele *(b)*. Long fur is determined by a dominant gene *(L)* and short by the recessive allele *(l)*. From a cross between a homozygous black, long-fur rabbit and a homozygous brown, short-fur rabbit, the ratio of phenotypes for the F1 generation will be
    a. 1 black, long:1 brown, short.
    b. all black, long.
    c. all brown, short.
    d. 9 black, long:3 black, short:3 brown, long:1 brown, short.

Answer: B

26. The portion of gametes produced by plant *BbCcYy* that would carry all recessive genes is
    a. 1/2
    b. 1/4
    c. 1/8
    d. 1/16

Answer: C

27. Using the law of probability predict the portion of offspring of genotype *BbCCYy* from a cross between parents *BbCcYy* and *bbCCyy*.
    a. 1/8
    b. 1/4
    c. 1/16
    d. 3/8

Answer: A

28. A lethal gene
    a. will show up in low frequency among the offspring.
    b. will result in death of the organism.
    c. is always recessive.
    d. is always heterozygous.

Answer: B

29. A male will have _____ copy (copies) of an X-linked gene.
    a. zero
    b. one
    c. two
    d. four

Answer: A

30. The gene for baldness, a sex-influenced trait in humans, is carried on
    a. the X chromosome.
    b. the Y chromosome.
    c. an autosomal chromosome.
    d. both X chromosomes.

Answer: C

31. In incomplete dominance
    a. the heterozygote is indistinguishable from the homozygous dominant condition.
    b. the heterozygote is different than both the homozygous conditions.
    c. both the alleles are observed in the heterozygous condition.
    d. the heterozygote is indistinguishable from the homozygous recessive condition.

    Answer: B

32. In carnations, petal color is determined by two genes, R = red pigment and R' = white pigment. The homozygous condition results in red or white petals but the heterozygous condition is pink. The ratio of phenotypes of the offspring from two pink plants would be
    a. all pink.
    b. 1 red:2 pink:1 white.
    c. all red.
    d. all white.

    Answer: B

33. Horses can have a color described as roan, which is produced by both white and red hairs. Since both the red and white genes are expressed this is called
    a. dominant/recessive.
    b. epistatic.
    c. codominance.
    d. incomplete dominance.

    Answer: C

34. In the condition sickle cell disease, a person who is heterozygous can show no symptoms to mild anemia. This variance in expression is called
    a. epistasis.
    b. variable expressive.
    c. pleiotropy.
    d. overdominance.

    Answer: B

35. Sex-limited genes are
    a. carried on either the X or Y chromosome.
    b. carried on autosomes but only expressed under the influence of hormones.
    c. only expressed in females in the homozygous condition.
    d. only expressed in males if the trait is dominant.

    Answer: B

36. In the royal families of medieval days, it was a common practice to keep the royal blood pure by arranged marriages between cousins. The phenomenon of _____ predicts that marriages between nonrelatives will produce healthier offspring.
    a. codominance
    b. dominance/recessiveness
    c. overdominance
    d. incomplete dominance

    Answer: C

37. A dominant gene that has 64% penetrance means that 64 out of 100 individuals
    a. with the gene express the trait.
    b. in the population express the trait if it is dominant.
    c. in the population express the trait if it is homozygous.
    d. are affected by the gene if it is in the heterozygous condition.

    Answer: A

38. A male child with the genes for PKU (phenylketonuria) is raised to adulthood under a stringent diet that prevents the gene from being activated. Consequently he lives an otherwise normal life and eventually becomes a father. Under these conditions, his offspring would have
    a. less chance of receiving the affected gene.
    b. less chance of expressing the gene.
    c. no chance of receiving the gene.
    d. the same chances of receiving and expressing the gene as they would have if the father had not received a special diet.

    Answer: D

39. Albinism is a human recessive disorder that results in unpigmented skin and eye problems. The unrelated problems caused by the single gene for albinism are called
    a. genetic heterogeneity.
    b. photocopies.
    c. expressivity.
    d. pleiotropy.

    Answer: D

40. Mice that are homozygous for the recessive albino allele may have genes for a variety of coat patterns on other chromosomes but only the albino condition is expressed. The ability of albino genes to shut off the expression of the color genes is called
    a. expressivity.
    b. penetrance.
    c. dominance.
    d. epistasis.

    Answer: A

Chapter 12

1. Biologists predicted that chromosomes carry more than one gene because
   a. there were more known inherited traits than the number of chromosomes observed in humans.
   b. they could see individual genes on the chromosomes.
   c. they observed chromosomes breaking into tiny pieces during meiosis, which they believed to be genes.
   d. a small organism like the fruit fly had nearly 1/10 of the number of chromosomes that humans did but had nowhere near 1/10 the number of traits.

   Answer: A

2. Linked genes
   a. dispute the law of segregation.
   b. cannot be epistatic.
   c. decrease the frequency of independent assortment during meiosis.
   d. support the law of independent assortment.

   Answer: C

3. If Mendel's traits for color and seed shape had been closely linked, the genotype ratio of the offspring from a cross of *YyRr* and *yyrr* would have been
   a. 1 *YyRr*:1 *yyrr*.
   b. 1 *YyRr*:1 *Yyrr*:1 *yyRr*:1 *yyrr*.
   c. all *YyRr*.
   d. all *yyrr*.

   Answer: A

4. The following cross for three traits found in plants is made: *AaBbCc* x *aabbcc*. The ratio of genotypes of the offspring is 1 *AaBbCc*:1 *aabbcc*:1 *AaBbcc*:1 *aabbCc*. Which genes are linked?
   a. *A* and *B*; *a* and *b*
   b. *B* and *C*; *b* and *c*
   c. *A* and *C*; *a* and *c*
   d. *A*, *B*, and *C*; *a*, *b*, and *c*

   Answer: A

5. In a cross of parents *YYRR* and *yyrr*, where *Y* and *R* are linked genes, offspring with the genotype *YyRr* represent
   a. a linkage group.
   b. the parental class.
   c. the recombinant class.
   d. gametes.

   Answer: C

6. The genes found in the parental class of the F2 generation from a cross of *AaBb* and *aabb*
   a. form a linkage group.
   b. are recombinants.
   c. are a product of crossing over.
   d. follow the law of independent assortment.

   Answer: A

7. Crossing over that results in increased genetic frequency involves
   a. exchanges between arms of sister chromatids in mitosis.
   b. exchanges between arms of homologous chromosomes in meiosis I.
   c. exchanges between arms of sister chromatids in meiosis II.
   d. the jumping of genes from one chromosome to another.

   Answer: B

8. The distance between linked genes can be determined by
   a. physically measuring them under magnification.
   b. sequencing the DNA.
   c. comparing the frequency of crossing over.
   d. counting the number in the parental class.

   Answer: C

9. Studies of three linked genes, *A*, *B*, and *C*, in a plant shows 15 crossovers between *A* and *B*, 3 crossovers between *B* and *C*, and 12 crossovers between *A* and *C*. Which of the following chromosome maps best represents these linked genes?
   a. *A B_____C*
   b. *B_C_____A*
   c. *C_A_____B*
   d. *A_C_____B*

   Answer: B

10. Genetic markers are
    a. radioactive labels that attach to a specific gene.
    b. closely linked genes to the gene of interest.
    c. antibodies that attach to a specific gene.
    d. recombinants.

    Answer: B

11. The Y chromosome carries
    a. no genetic material.
    b. only autosomal genes.
    c. a few autosomal genes and sex-determining genes.
    d. only sex-determining genes.

    Answer: C

12. In X-linked diseases
    a. females can only receive the gene from their fathers.
    b. females can only receive the gene from their mothers.
    c. males can only receive the gene from their fathers.
    d. males can only receive the gene from their mothers.

    Answer: D

13. The son of a woman who is a carrier for hemophilia and a normal man will have _____ chance of having the disease.
    a. 0%
    b. 25%
    c. 50%
    d. 100%

    Answer: C

14. The son of a man who has hemophilia and a normal woman will have _____ chance of having the disease.
    a. 0%
    b. 25%
    c. 50%
    d. 100%

    Answer: A

15. A male who has a single copy of the hemophilia gene is unable to clot blood whereas his sister, who also has a single copy of the hemophilia gene, is able to clot blood. This is because
    a. the sister has female hormones that do not trigger the gene.
    b. the pH is different in the female's blood.
    c. the gene has less expressivity in the female.
    d. the female also has a normal gene that produces clotting factor.

    Answer: D

16. Y-linked genes are passed only from
    a. father to son.
    b. mother to son.
    c. father to daughter.
    d. mother to daughter.

    Answer: A

17. A female who is a carrier for an X-linked disease will express the trait if
    a. the disease allele is recessive.
    b. the normal allele is on the inactivated X chromosome in the affected cell line.
    c. the disease allele is on the inactivated X chromosome in the affected cell line.
    d. she receives the disease allele from her mother.

    Answer: B

18. A Barr body is an inactive
    a. Y chromosome.
    b. extra chromosome.
    c. autosomal chromosome.
    d. X chromosome.

    Answer: D

19. A woman who is heterozygous for an X-linked trait will express the trait in
    a. all her tissues if the trait is dominant.
    b. all her tissues if the trait is recessive.
    c. some of her tissues, but not all.
    d. only in the reproductive tissues.

    Answer: C

20. Femaleness in humans is determined by
    a. the X chromosome.
    b. the Y chromosome.
    c. absence of the X chromosome.
    d. absence of the Y chromosome.

    Answer: D

21. A person who has two X chromosomes and one Y chromosome (Klinefelter syndrome) is
    a. a normal male.
    b. a male with underdeveloped sexual characteristics.
    c. a normal female.
    d. a female with reduced sexual characteristics.

    Answer: B

22. In birds the homogametic sex is the male, which means
    a. the male has two of the same sex chromosomes.
    b. the male has two different sex chromosomes.
    c. the female has two of the same sex chromosomes.
    d. the male has only one sex chromosome.

    Answer: A

23. Within the animal kingdom sex is determined by
    a. the presence of an X or Y chromosome.
    b. the presence of specific hormones.
    c. environmental, behavioral, and genetic factors.
    d. genes on autosomal chromosomes.

    Answer: C

24. A trait that is Y linked will
    a. skip a generation but be seen only in the male population.
    b. be seen in every generation but only in males.
    c. be seen only in females if in the homozygous condition.
    d. be seen in both sexes randomly.

    Answer: B

25. X inactivation occurs during
    a. puberty.
    b. embryonic development.
    c. fetal development.
    d. meiosis.

    Answer: B

26. A female who shows manifesting heterozygosity for an X-linked disease will have
    a. all of the symptoms of the disease.
    b. mild symptoms of the disease.
    c. no symptoms of the disease.
    d. inactivation of the disease gene.

    Answer: B

27. An individual with Klinefelter syndrome who is carrying a single copy of the X-linked gene for hemophilia will
    a. show manifesting heterozygosity.
    b. show no symptoms of the disease.
    c. have all the symptoms of the disease.
    d. inactivate the disease gene.

    Answer: A

28. A woman suspected of being a carrier for Lesch-Nyhan syndrome has hair cells examined from different areas of her head. It is determined that all of her hair cells do not produce the enzyme HGPRT. Since she does not have Lesch-Nyhan symptoms, it is determined that she is indeed a carrier. It must be that she is
   a. a carrier but X inactivation occurred very early in development.
   b. a carrier but X inactivation occurred late in normal development.
   c. homozygous.
   d. homozygous Lesch-Nyhan.

   Answer: A

29. There are viable human sex chromosome anomalies-- XXX, XXY, XO, and XYY. Why are there no YY individuals who survive?
   a. Too much testosterone makes for lethal conditions without an X chromosome to counterbalance.
   b. Sex-determining genes carried on the X chromosome are needed for survival.
   c. Two copies of autosomal genes carried on the Y chromosome gene are lethal.
   d. The X chromosome carries many autosomal genes required for viability.

   Answer: D

30. More males than females are conceived because
   a. Y-bearing sperm are less viable.
   b. X-bearing sperm are less viable.
   c. Y-bearing sperm can swim faster.
   d. Y-bearing sperm have more enzymes for penetration of the egg.

   Answer: C

31. The high spontaneous abortion rate of male zygotes is probably due to
   a. high mutation rates.
   b. an abnormal number of chromosomes.
   c. failure to produce human chorionic gonadotrophic hormone.
   d. X-linked lethal alleles.

   Answer: B

32. The tertiary sex ratio
   a. remains even over a lifetime.
   b. eventually results in men exceeding women.
   c. eventually results in women exceeding men.
   d. eventually results in five women for every man.

   Answer: D

33. The sex preselection method that is most reliable and acceptable is
   a. altering the pH of the vagina.
   b. timing of intercourse.
   c. abortion.
   d. artificial insemination.

   Answer: D

34. Sociologists predict that in regard to sex preselection the ratio of boys to girls will eventually be
   a. 1:1.
   b. 1:2.
   c. 2:1.
   d. 4:1.

   Answer: A

Chapter 13

1. Recombinant DNA technology has made possible the production of
   a. human insulin.
   b. cattle insulin.
   c. *E. coli* proteins.
   d. humans.

   Answer: A

2. Genes are made of
   a. amino acids.
   b. protein.
   c. polypeptides.
   d. nucleic acids.

   Answer: D

3. The building block of protein is the
   a. polymer.
   b. amino acid.
   c. glycogen.
   d. nucleotide.

   Answer: B

4. Proteins may be used as
   a. hormones, enzymes, genes, or pigment.
   b. pigment, hormones, enzymes, or insulation.
   c. hormones, a quick energy source, genes, or carriers.
   d. pigment, enzymes, hormones, or structure.

   Answer: D

5. The main source of amino acids is
   a. minerals.
   b. proteins.
   c. carbohydrates.
   d. lipids.

   Answer: B

6. The intermediate molecule that carries the direction for protein synthesis to the ribosome is the
   a. amino acid.
   b. ribosomal RNA.
   c. messenger RNA.
   d. transfer RNA.

   Answer: C

7. The genetic code lies in the _____ of DNA.
   a. phosphate
   b. nitrogen bases
   c. deoxyribose sugar
   d. ribose sugar

   Answer: B

8. Adenine binds with
   a. guanine.
   b. adenine.
   c. cytosine.
   d. thymine.

   Answer: D

9. DNA fits inside the nuclear membrane because it is
   a. stacked.
   b. composed of pieces.
   c. coiled in a double helix.
   d. circular.

   Answer: C

10. The human genome is _____ base pairs long.
    a. 10,000
    b. 1 million
    c. 2 million
    d. 3 billion

    Answer: D

11. The precursors of nucleotides in digestion come from
    a. nucleic acids.
    b. carbon.
    c. proteins.
    d. amino acids.

    Answer: A

12. A guanine nucleotide that is free-floating will attach to
    _____ of the DNA.
    a. guanine
    b. cytosine
    c. adenine
    d. thymine

    Answer: B

13. A nucleotide contains
    a. one amino acid, one phosphate, and one sugar
       molecule.
    b. one nitrogen base, one sugar, and one amino acid
       molecule.
    c. one nitrogen base, one sugar, and one phosphate
       molecule.
    d. one protein, one sugar, and one phosphate
       molecule.

    Answer: C

14. Semiconservative replication means that each daughter
    molecule will have
    a. two original strands of DNA.
    b. two newly synthesized strands of DNA.
    c. one newly synthesized strand and one original
       strand of DNA.
    d. four newly synthesized strands of DNA.

    Answer: C

15. Free-floating nucleotides are joined to the single
    strand of DNA
    a. swivelase.
    b. ligase.
    c. polymerase.
    d. gyrase.

    Answer: C

16. Gene amplification results in
    a. many copies of a gene.
    b. many copies of a gene product.
    c. exaggeration of the phenotype.
    d. many copies of a chromosome.

    Answer: A

17. The first step of gene replication is
    a. ligation of sugar and phosphate bonds.
    b. adding of nucleotides.
    c. opening of the replication forks.
    d. unwinding of the DNA helix.

    Answer: D

18. RNA differs from DNA in that it
    a. is double-stranded.
    b. contains deoxyribose sugar.
    c. contains thymine.
    d. contains uracil.

    Answer: D

19. The messenger RNA that will form from the DNA
    sequence ATGGCATAC is
    a. TACCGTATG.
    b. UACCGUAUG.
    c. AUGGCAUAC.
    d. GCAAUGCGU.

    Answer: B

20. Transcription differs from DNA replication in that
    a. only one strand of the DNA is replicated.
    b. the entire DNA molecule is read.
    c. thymine is used in place of uracil.
    d. it occurs in the nucleus.

    Answer: A

21. Messenger RNA
    a. serves as a workbench for amino acid synthesis.
    b. carries amino acids to the site of protein
       synthesis.
    c. serves as the blueprint for the synthesis of the
       protein.
    d. carries genes into the daughter cell.

    Answer: C

22. The nucleic acid that carries the anticodon is
    a. DNA.
    b. mRNA.
    c. rRNA.
    d. tRNA.

    Answer: D

23. The three-letter code of DNA that gives rise to the mRNA that binds to the anticodon AUG is
    a. ATG.
    b. UAC.
    c. AUG.
    d. TAC.

    Answer: A

24. For a mRNA strand that is 18 bases long, how many amino acids would be coded for if start and stop signals and leader are excluded?
    a. 6
    b. 9
    c. 18
    d. 36

    Answer: A

25. Which of these statements regarding mRNA is true?
    a. The sequence of nucleotides is the same for all members of a species.
    b. The sequence of nucleotides is the same for all genes in an individual.
    c. The percentages of nucleotides is the same for all members of a species.
    d. The percentages of nucleotides is the same for all genes in an individual.

    Answer: C

26. The position of the base that may be replaced in many cases without a subsequent change in the amino acids brought in is the
    a. first.
    b. second.
    c. third.
    d. second and third.

    Answer: C

27. The initiator transfer RNA is different from other transfer RNA because it
    a. contains methionine.
    b. is formylated.
    c. contains nitrogen.
    d. is a nucleic acid.

    Answer: B

28. In the process of elongation, the amino acid carried by the transfer RNA is released when
    a. the anticodon binds to the codon.
    b. it binds with the adjacent amino acid.
    c. mRNA binds with the ribosomal unit.
    d. the stop signal is read.

    Answer: B

29. The genetic code was "cracked" by putting _____ in a test tube with cellular components and determining which _____ was produced.
    a. mRNA, polypeptide
    b. transfer RNA, protein
    c. a gene, protein
    d. DNA, amino acid

    Answer: A

30. The final shape of a polypeptide chain is
    a. linear.
    b. circular.
    c. dependent upon interactions between its amino acids.
    d. dependent upon interactions between its genes.

    Answer: C

31. In bacteria, transcription and translation occur simultaneously, whereas in eukaryotes the two processes are separated by time because
    a. enzymes are not present for both.
    b. they are separated spatially.
    c. the product of transcription is the reactant of translation.
    d. so many more copies of the genes are needed in eukaryotic cells.

    Answer: B

32. The portion of the mRNA that is not translated is called the
    a. intron.
    b. exon.
    c. pre-mRNA.
    d. small nuclear RNA.

    Answer: A

33. Jumping genes are pieces of DNA that move from one
    a. organism to another.
    b. cell to another.
    c. site on a chromosome to another.
    d. species to another.

    Answer: C

34. Hybridomas are
    a. immortal immune cells that produce antibodies.
    b. immortal cancer cells.
    c. a product of recombinant DNA.
    d. large-scale cell cultures.

    Answer: A

35. Recombinant DNA technology for the production of human insulin combines DNA of the human species with the DNA from
    a. viruses.
    b. bacteria.
    c. yeast.
    d. protozoa.

    Answer: B

36. The Asilomar meeting of scientists was set up to
    a. fight antibiotechnology groups.
    b. sell biotechnology ideas to industry.
    c. form regulations for use of products of biotechnology.
    d. form regulations for conducting biotechnology research.

    Answer: D

37. The tool used in recombinant technology for cutting DNA is the
    a. vector.
    b. plasmid.
    c. restriction enzyme.
    d. polymerase.

    Answer: C

38. In recombinant technology, human genes are translated into human proteins in
    a. bacterial cells.
    b. eukaryotic cells.
    c. human tissue culture.
    d. yeast cultures.

    Answer: A

39. When a multicelled organism receives a foreign gene it is called a
    a. transgenic.
    b. transmutant.
    c. transposon.
    d. transducer.

    Answer: A

40. Human gene libraries contain
    a. DNA sequences for each of the human genes known.
    b. pieces of human DNA, but they are not necessarily genes.
    c. human restriction enzymes.
    d. human proteins.

    Answer: B

Chapter 14

1. The chromosome number of a human who is triploid is
    a. 23.
    b. 46.
    c. 69.
    d. 92.

    Answer: C

2. A sequence of DNA that reads
   ATGGCAC
   TACCGTG would read, after an inversion,
    a. TACCGTG
       ATGGCAC.
    b. CACGGTA
       GTGCCAT.
    c. CACGTA
       CACGTA.
    d. ATGGCAC
       GTGCCAT.

    Answer: B

3. Johnny has Down syndrome. His condition is caused by the presence of an extra piece of chromosome 21, which is attached to chromosome 14. This genetic anomaly is called
    a. trisomy.
    b. monosomy.
    c. translocation.
    d. aneuploidy.

    Answer: C

4. In the genetic condition cri-du-chat, part of the arm of chromosome 5 is missing. Geneticists refer to this as
    a. aneuploidy.
    b. monosomy.
    c. euploidy.
    d. deletion.

    Answer: D

5. If a Down syndrome person (trisomy 21) were to reproduce with a normal homozygous person, what would be the chances of having a child with Down syndrome?
    a. 0%
    b. 25%
    c. 50%
    d. 100%

    Answer: C

6. A gamete that results from nondisjunction will have
    a. 22 or 24 chromosomes.
    b. 23 chromosomes.
    c. 46 chromosomes.
    d. 23 sister chromatids.

    Answer: A

7. Most cases of Down syndrome are caused by
    a. monosomy 21.
    b. trisomy 21.
    c. trisomy 13.
    d. trisomy 18.

    Answer: B

8. Down syndrome is identifiable by
    a. characteristic facial features, a protruding tongue, and creased palms.
    b. a pinched face and a cry like a cat.
    c. a webbed neck.
    d. extreme height.

    Answer: A

9. The frequency of Down syndrome increases with maternal age probably because the
    a. womb is no longer able to maintain adequate nourishment.
    b. mother's hormone levels are not adequate.
    c. egg has lost nutrients from being arrested in meiosis for so many years.
    d. egg has undergone chemical damage while being arrested in meiosis.

    Answer: D

10. Down syndrome has been associated with
    a. cancer.
    b. progeria.
    c. Alzheimer's.
    d. senile dementia.

    Answer: C

11. The most devastating genetic problems result from
    a. monosomy of autosomes.
    b. trisomy of autosomes.
    c. monosomy of sex chromosomes.
    d. trisomy of sex chromosomes.

    Answer: A

12. The symptoms of Alzheimer's disease include
    a. heart and kidney defects.
    b. a suppressed immune system.
    c. memory loss.
    d. creased palms.

    Answer: C

13. Women with Turner syndrome
    a. have underdeveloped reproductive systems.
    b. are severely mentally retarded.
    c. have symptoms of Alzheimer's disease.
    d. have heart and kidney defects.

    Answer: A

14. Triplo-X symptoms include
    a. lower-than-average intelligence.
    b. sterility.
    c. menstrual irregularities.
    d. masculine traits.

    Answer: C

15. Genetic studies of XYY syndrome eventually supported earlier claims that these individuals were
    a. aggressive and antisocial.
    b. of lower intelligence.
    c. had effeminate traits.
    d. above-average height.

    Answer: D

16. If a single nonreciprocal translocation occurs during meiosis, the resulting gametes will
    a. be 50% normal, 25% with extra genetic material, and 25% with a deletion of genetic material.
    b. all be normal.
    c. be 50% normal and 50% with the right amount of genetic material but with translocations.
    d. be 50% with extra genetic material and 50% with a deletion of genetic material.

    Answer: A

17. A person who is a translocation carrier
    a. may produce gametes with deletions or duplications of genes.
    b. shows symptoms, depending on which chromosomes are involved.
    c. produces normal gametes.
    d. is unable to reproduce because all resulting embryos will abort.

    Answer: A

18. Most inherited diseases are
    a. sex-linked recessive.
    b. sex-linked dominant.
    c. autosomal recessive.
    d. autosomal dominant.

    Answer: C

19. Carriers for a genetic disease are
    a. homozygous if the gene is dominant.
    b. heterozygous if the gene is recessive.
    c. homozygous if the gene is recessive.
    d. heterozygous if the gene is dominant.

    Answer: B

20. Autosomal dominant diseases are passed less frequently than other kinds of disease genes because they
    a. skip a generation.
    b. are usually lethal in the heterozygous condition.
    c. are lethal only in the homozygous condition.
    d. rarely rise by mutation.

    Answer: B

21. A carrier of a recessive X-linked trait is a
    a. female who is homozygous.
    b. female who is heterozygous.
    c. male who receives a dominant gene.
    d. male who receives a recessive gene.

    Answer: B

22. A disease that is not ethnic related is
    a. Tay-Sach's disease.
    b. sickle cell disease.
    c. Down syndrome.
    d. cystic fibrosis.

    Answer: C

23. Huntington disease is a dominant autosomal disease. Unfortunately, onset of symptoms usually occurs after a person has reproduced. What are the chances that a person with the disease will produce a child with the disease if mated with a normal homozygous person?
    a. 0%
    b. 25%
    c. 50%
    d. 100%

    Answer: C

24. Populations with a high frequency of genetic disorders result most often
    a. in a geographical area that has selective pressure for certain disorders.
    b. from exposure to high amounts of DNA-damaging agents.
    c. from defective individuals reproducing.
    d. from being reproductively isolated.

    Answer: D

25. Most inherited diseases are caused by
    a. trisomy.
    b. monosomy.
    c. single-gene disorders.
    d. multiple-gene disorders.

    Answer: C

26. Inborn errors of metabolism are usually a result of
    a. pH imbalance.
    b. defective enzymes.
    c. organelle malfunction.
    d. membrane malfunction.

    Answer: B

27. Heterozygotes with most inborn errors of metabolism can be detected because they have
    a. minor symptoms of the disease.
    b. defective enzymes.
    c. an imbalance of blood pH.
    d. reduced nutrient uptake in the gut.

    Answer: B

28. Point mutations refer to
    a. duplication or deletion of chromosomes.
    b. duplication or deletion of genes.
    c. inversions of DNA sequences.
    d. alterations in nucleotides.

    Answer: D

29. Sickle cell disease is caused by a(n)
    a. missing enzyme.
    b. extra chromosome.
    c. extra globin chain.
    d. defective globin chain.

    Answer: D

30. Symptoms of sickle cell disease result from
    a. hemoglobin sticking to itself, resulting in a change in cell shape.
    b. the inability of the hemoglobin to bind with oxygen.
    c. the lack of absorption of iron from the diet.
    d. persistence of fetal hemoglobin.

    Answer: A

31. Treatment for people with beta-thalassemia involves activation of _____ globulin genes.
    a. beta
    b. alpha
    c. fetal
    d. embryonic

    Answer: C

32. Orphan diseases refer to genetic diseases that
    a. are high among orphans.
    b. are extremely rare.
    c. are untreatable and death occurs early in adulthood.
    d. result in death after childbirth.

    Answer: A

33. Alpha thalassemia is a result of
    a. one or more deletions of alpha gene globin.
    b. duplications of alpha globulin genes.
    c. the production of fetal globin after birth.
    d. production of embryonic globin after birth.

    Answer: A

34. Tourette's syndrome took a long time to discover because it
    a. was extremely rare.
    b. appeared to be a behavioral problem, not a medical one.
    c. was found in a geographically isolated though large population.
    d. resulted in spontaneous abortions.

    Answer: B

35. A trait that runs in families but is not controlled by Mendelian genetic mechanisms is
    a. sickle cell disease.
    b. dwarfism.
    c. breast cancer.
    d. albinism.

    Answer: C

36. It is believed that neural tube defects are caused by
    a. teratogenic agents.
    b. a single gene.
    c. mutagenic agents.
    d. both genetic and environmental factors.

    Answer: D

37. Intelligence appears to be controlled by
    a. a single gene.
    b. multiple genes.
    c. environment.
    d. both multiple genes and environment.

    Answer: D

38. If a trait shows 100% concordance in MZ and DZ twins it means that it is
    a. a dominant genetic trait.
    b. environmentally controlled.
    c. both environmentally and genetically controlled.
    d. a recessive genetic trait.

    Answer: B

39. Limitations of twin studies come from the fact that
    a. twins are treated differently than nontwin siblings.
    b. identical twins are often kept in closer contact than fraternal twins.
    c. investigators are biased because they know which kind of twins they are dealing with.
    d. environmental factors introduce too many problems when comparing MZ to DZ twins.

    Answer: B

Chapter 15

1. Genetic information from pedigrees is easiest to decipher if the families
   a. are large.
   b. have serial marriages.
   c. have blended families.
   d. have used artificial insemination.

   Answer: A

2. The main use of pedigrees is for
   a. diagnosis of a disease.
   b. determining if a fetus has a genetic disease.
   c. choosing mates.
   d. deciding whether to reproduce.

   Answer: A

3. The chromosomes in a karyotype are arranged
   a. from mother or father.
   b. from largest to smallest.
   c. by shape.
   d. by amount of dye uptake.

   Answer: B

4. Chromosomes cultured for producing karyotypes are stopped during mitosis because they are
   a. most condensed then.
   b. in the form of chromatin.
   c. found in the greatest number compared with nondividing cells.
   d. not housed in a nuclear membrane.

   Answer: C

5. Karyotypes are used to identify
   a. single-gene defects.
   b. multiple-gene defects.
   c. aneuploidy.
   d. point mutations.

   Answer: C

6. Each band of a chromosome corresponds to about _____ genes.
   a. 2
   b. 20
   c. 200
   d. 2,000

   Answer: B

7. Paired chromosomes in karyotypes
   a. are identical in genetic makeup.
   b. carry the same alleles but are not identical.
   c. carry nonalleles.
   d. are from the same parent.

   Answer: B

8. Fragile X syndrome is an inherited condition of mental retardation affecting
   a. only males who receive one copy of the fragile X chromosome.
   b. either sex if they receive one copy of the fragile X chromosome.
   c. females who receive three copies of the X chromosome.
   d. males who receive two copies of the X chromosome.

   Answer: A

9. The advantage to parents of knowing whether their child's Down syndrome was a result of fragile X or trisomy 21 would be that they could better predict
   a. the level of retardation that the child could attain.
   b. the inheritability of the condition.
   c. which parent contributed to the defect.
   d. the life span of the child.

   Answer: B

10. Chemical or radiation exposure is detectable by observation of
    a. frazzled chromosomes in karyotypes.
    b. fragile X chromosomes.
    c. chromosomes that will not stain.
    d. broken chromosomes.

    Answer: D

11. The closer two species are related the more similar are the
    a. number of chromosomes.
    b. banding pattern of chromosomes.
    c. location of centromeres.
    d. size and shape of chromosomes.

    Answer: B

40. Twin studies are most reliable when
    a. fraternal twins are raised together.
    b. identical twins are raised together.
    c. fraternal twins are raised apart.
    d. identical twins are raised apart.

    Answer: D

12. An ultrasound exam for prenatal diagnosis may reveal to prospective parents
    a. chromosomal defects.
    b. single-gene defects.
    c. enzyme defects.
    d. physical deformities.

    Answer: D

13. Amniocentesis involves the removal of cells from the
    a. fluid surrounding the embryo.
    b. embryo itself.
    c. placenta.
    d. uterus.

    Answer: A

14. The drawback to chorionic villus sampling compared with amniocentesis is that
    a. chromosomal studies cannot be done.
    b. biochemical assays of defective enzymes cannot be done.
    c. samples are drawn through the abdominal wall of the mother.
    d. it cannot be done until late in the pregnancy.

    Answer: B

15. DNA probes are
    a. molecular scissors that can cut DNA.
    b. complementary strands of DNA to a specific gene or portion of a gene.
    c. a specific gene or portion of a gene of recombinant bacteria.
    d. a virus that can carry a synthesized gene into a cell.

    Answer: B

16. DNA probes are identifiable in the Southern blotting test by
    a. a staining process.
    b. the presence of radioactivity.
    c. doing a karyotype.
    d. microscopy.

    Answer: B

17. RFLPs reveal the presence of abnormal alleles by differences in _____ with normal alleles.
    a. karyotypes
    b. staining pattern
    c. fragment length of DNA
    d. levels of radioactivity

    Answer: C

18. The presence of an abnormal gene in an individual that has already been identified in a population is best diagnosed by the technique of
    a. karyotyping.
    b. RFLPs.
    c. gene silencing.
    d. DNA probes.

    Answer: D

19. A genetic marker refers to a
    a. known piece of DNA next to an unknown gene.
    b. diseased gene that is associated with a particular banding pattern in karyotypes.
    c. series of recombinant DNA that will attach to the diseased gene.
    d. restriction enzyme that will cut only in the diseased allele.

    Answer: A

20. The disadvantage of using the genetic marker approach is
    a. the diseased gene sequence must be known.
    b. whole families must be diagnosed.
    c. restriction enzymes must be radioactively labeled.
    d. large amounts of tissue samples must be taken from the individual.

    Answer: B

21. The farther away the genetic marker is from the disease gene, the greater risk that the
    a. probe will not bind with the complementary strand.
    b. probe will not attach to the disease gene.
    c. marker and disease genes will be separated by crossover.
    d. probe will attach to the normal allele.

    Answer: C

22. The first genetic marker identified was for
    a. cystic fibrosis.
    b. Alzheimer's disease.
    c. neurofibromatosis.
    d. Huntington disease.

    Answer: D

23. The presence of high levels of alpha-fetoprotein in a pregnant mother's blood may indicate that the fetus has
    a. a neural tube defect.
    b. Tay-Sachs disease.
    c. phenylketonuria.
    d. Huntington disease.

    Answer: A

24. The genetic disease for which all American children are screened at birth is
    a. Huntington disease.
    b. PKU.
    c. Sickle cell disease.
    d. Tay-Sachs disease.

    Answer: B

25. Mandatory screening for sickle cell disease is recommended for populations at high risk because
    a. decisions on future pregnancies can be made by carrier parents.
    b. treatment can be begun at birth.
    c. affected children can be more closely watched for infections.
    d. it is so common.

    Answer: C

26. Genetic screening at the workplace is done
    a. to determine high-risk employees as insurance carriers.
    b. to give employees with a genetic disease suitable work.
    c. as a benefit for potential parents.
    d. to determine contact with dangerous chemicals.

    Answer: D

27. A genetic counselor's job is to give
    a. genetic information to the family.
    b. advice as to when an abortion is appropriate.
    c. advice as to when certain medical treatments are appropriate.
    d. advice as to when prepregnancy couples should adopt children rather than have their own.

    Answer: A

28. As it becomes possible to identify more carriers of genetic diseases and as inherited disorders become more detectable, it will
    a. become increasingly important to protect the individual's rights.
    b. lead to more freedom for the individual to make ethical decisions.
    c. reduce the need for the courts to intervene in ethical questions.
    d. reduce the number of ethical questions raised by society.

    Answer: A

29. The technique proposed to reduce the risk of an affected parent of a genetic disease passing on the trait to his or her offspring is
    a. gene silencing.
    b. nonheritable gene therapy.
    c. heritable gene therapy.
    d. treatment of the phenotype.

    Answer: C

30. The technique that alters the genetic material of somatic cells of a victim by inserting normal genes into affected cells is referred to as
    a. gene silencing.
    b. nonheritable gene therapy.
    c. heritable gene therapy.
    d. treatment of the phenotype.

    Answer: B

Chapter 16

1. The portion of the neuron that does most of the metabolic work is the
   a. axon.
   b. cell body.
   c. dendrite.
   d. synapse.

   Answer: B

2. The process that conducts an action potential away from the cell body is the
   a. dendrite.
   b. axon.
   c. synaptic terminal.
   d. synaptic bulb.

   Answer: B

3. The structure of a neuron that normally receives information from other neurons or receptors is the
   a. axon.
   b. cell body.
   c. dendrite.
   d. synapse.

   Answer: C

4. Neurons that carry messages from the skin to the spinal cord have long
   a. axons.
   b. nerve fibers.
   c. dendrites.
   d. synapses.

   Answer: C

5. Neurons that connect one neuron to another are called
   a. afferent neurons.
   b. efferent neurons.
   c. interneurons.
   d. effector neurons.

   Answer: C

6. The action potential is
   a. an electrochemical change caused by ions moving across the cell membrane.
   b. very similar to the electrical charge moving down a wire.
   c. a passive process found in motor neurons that activates effectors.
   d. a chemical change caused by enzymes on the surface of the cell membrane.

   Answer: A

7. During the resting state the neuron is
   a. depolarized.
   b. polarized.
   c. slightly positive on the inside with respect to the outside.
   d. not charged in this inactive state.

   Answer: B

8. The electrical charge of a neuron results from
   a. the equal distribution of sodium and potassium ions inside and outside the cell.
   b. the unequal distribution of sodium and potassium ions in the cell body and neuronal processes.
   c. differences in concentration of sodium and potassium ions inside and outside the cell.
   d. accumulation of excess negative ions inside the cell.

   Answer: C

9. The property of a cell membrane that admits some substances and not others is called
   a. selective permeability.
   b. voltage gate.
   c. sodium-potassium pump.
   d. electrical gradient.

   Answer: A

10. Ions distribute themselves across a membrane by
    a. electrical gradients.
    b. concentration gradients.
    c. electrical and concentration gradients.
    d. active transport.

    Answer: C

11. Blocking the activity of the $Na+/K+$ pump causes
    a. an immediate inability to propagate an action potential.
    b. hyperpolarization of the resting membrane potential.
    c. gradual depolarization of the resting membrane potential.
    d. $Na+$ to accumulate outside the cell.

    Answer: C

12. Neurons use the most energy when
    a. maintaining the resting potential.
    b. conducting an action potential.
    c. releasing neurotransmitters.
    d. transmitting an action potential by saltatory conduction.

    Answer: A

13. An action potential begins when
    a. $Na+$ starts to enter the neuron.
    b. $K+$ starts to leave the neuron.
    c. the sodium gate opens.
    d. the resting membrane potential falls to zero.

    Answer: A

14. At the peak of the action potential, the inside of the cell is
    a. neutral.
    b. positively charged.
    c. negatively charged.
    d. at zero potential.

    Answer: B

15. At the peak of the action potential the membrane becomes
   a. more permeable to Na+.
   b. less permeable to K+.
   c. less permeable to Na+.
   d. more permeable to both Na+ and K+.

   Answer: C

16. During the action potential the outward flow of K+ causes the
   a. cell to depolarize.
   b. activation of the Na+/K+ pump.
   c. cell to return to the resting potential.
   d. sodium gates to close.

   Answer: C

17. A stimulus to a neuron causes
   a. a change of pH in the cell.
   b. the release of neurotransmitters.
   c. activation of the Na+/K+ pump.
   d. a change in permeability to Na+.

   Answer: D

18. Only one action potential is initiated when the
   a. Na+ gates are open.
   b. Na+ and K+ gates are open.
   c. cell is at the resting potential.
   d. cell is hyperpolarized.

   Answer: B

19. The action potential travels along the length of the neuron by
   a. electrons moving within the cytoplasm.
   b. ions rapidly entering and leaving the cell.
   c. ions traveling on the surface of the membrane.
   d. ions rapidly moving within the cytoplasm.

   Answer: B

20. The intensity of a stimulus can be perceived at the conscious level because strong stimuli cause
   a. larger action potentials.
   b. the action potentials to travel faster.
   c. a higher frequency of action potentials.
   d. the cells to release more neurotransmitter at the synapses.

   Answer: C

21. You can tell the difference between touch and sound because
   a. there are different neurotransmitters for each type of stimuli.
   b. there are different receptors for each type of stimuli.
   c. the sensory neurons are connected to different parts of the brain.
   d. the frequency of action potentials is different for each type of stimuli.

   Answer: C

22. The speed of conduction of an action potential in a nerve fiber is
   a. the same for all neurons.
   b. slow when the diameter of the axon is large.
   c. slower when the fiber has a myelin sheath.
   d. rapid when the diameter of the axon is large.

   Answer: D

23. Myelin sheaths surrounding vertebrate nerve fibers
   a. serve to insulate the fibers from surrounding neurons.
   b. increase the conduction velocity of the action potential.
   c. assist in the process of integration.
   d. are composed of a dense layer of protein.

   Answer: B

24. In myelinated neurons the action potential
   a. travels in a continuous wave over the surface of the membrane.
   b. "jumps" from node to node.
   c. travels at a slower conduction velocity.
   d. is propagated through the Schwann cells.

   Answer: B

25. The gray matter of the nervous system
   a. lacks myelin and is involved in the process of integration.
   b. has a large number of Schwann cells surrounding the neurons.
   c. is lacking in cell bodies.
   d. is specialized for rapidly propagating action potentials.

   Answer: A

26. Neurotransmitters are contained in
   a. the postsynaptic membrane.
   b. the synaptic cleft.
   c. synaptic vesicles.
   d. synaptic knobs.

   Answer: C

27. Neurotransmitters are released from a neuron when
   a. the action potential arrives at the dendrite.
   b. calcium ions enter the cell and cause exocytosis of synaptic vesicles.
   c. Na+ triggers a change in permeability of the postsynaptic membrane.
   d. the action potential "jumps" across the synaptic cleft.

   Answer: B

28. Neurotransmitters
   a. bind to receptors and change the permeability of the postsynaptic membrane.
   b. diffuse across the synaptic cleft and bind to enzymes on the presynaptic membrane.
   c. remain in the synaptic cleft to ensure that the postsynaptic membrane is stimulated.
   d. speed up the propagation of action potentials from one neuron to another.

   Answer: A

29. It is necessary to remove the neurotransmitter from the synaptic cleft because they
    a. will physically block the space and prevent the transmission of the action potential.
    b. will continuously activate the postsynaptic cell.
    c. require a large amount of energy to produce and must therefore be recycled.
    d. will block the transmission of action potentials unless they are repackaged in the synaptic vesicles.

    Answer: B

30. Neurotransmitters that decrease the probability of an action potential from developing
    a. are found in excitatory synapses.
    b. are not destroyed in the synaptic cleft.
    c. increase permeability to negative ions.
    d. cause the postsynaptic membrane to become depolarized.

    Answer: C

31. Whether a neuron transmits an action potential depends on the
    a. location of the synapses on the dendrites and cell body.
    b. sum of excitatory and inhibitory synaptic activity.
    c. concentration of the released neurotransmitters.
    d. number of synapses present.

    Answer: B

32. The existence of both inhibitory and excitatory synapses on a single neuron
    a. insures that action potentials will be initiated in the postsynaptic cell.
    b. allows for fine tuning of the generation of action potentials in this cell.
    c. demonstrates that the presynaptic neuron has two different neurotransmitters.
    d. insures that this neuron cannot generate action potentials.

    Answer: B

33. In humans the filtering of unimportant stimuli is dependent upon
    a. the activity of inhibitory synapses.
    b. the presence of enzymes to deactivate excitatory neurotransmitters.
    c. the activation of the reticular system in the brain.
    d. chemical blocking of excitatory synapses.

    Answer: A

34. A psychoactive drug that blocks the action of a neurotransmitter is called a(n)
    a. second messenger.
    b. agonist.
    c. antagonist.
    d. enzyme.

    Answer: C

35. Second messengers such as cAMP can alter the effect of neurotransmitters on specific neurons by
    a. changing the magnitude of the action potential.
    b. changing the neurotransmitter released.
    c. altering the ion channels, which affect the duration of the action potential.
    d. allowing the neurotransmitter to accumulate in the synaptic cleft.

    Answer: C

36. Endogenous depression is caused by low levels of
    a. GABA.
    b. cAMP.
    c. noradrenaline.
    d. L-dopa.

    Answer: C

37. A drug that blocks the breakdown of noradrenaline in the brain would be expected to cause
    a. depression.
    b. alertness.
    c. schizophrenia.
    d. sleep.

    Answer: B

38. The existence of endorphins was predicted when
    a. a radioactively labeled morphine was found to bind to specific receptors of nerve cells.
    b. morphine was found to be released from specific nerve cells.
    c. experimental mammals were shown to become addicted to morphine.
    d. morphine was demonstrated to have mood-elevating and pain-reducing characteristics.

    Answer: A

39. Acupuncture is thought to be effective in reducing pain by
    a. blocking the initiation of action potentials in pain-receptive neurons.
    b. increasing the release of neurotransmitters in the pain-receptive neurons.
    c. stimulating the release of endorphins from the pituitary gland.
    d. acting directly on the spinal neurons involved in the transmission of the pain message.

    Answer: C

40. Multiple sclerosis, a disease with the symptoms of muscle weakness and loss of coordination, is caused by
    a. destruction of the muscle tissue by a virus.
    b. destruction of the connective tissue between muscles and bones.
    c. changes in the myelin sheaths of neurons, blocking the conduction of action potentials.
    d. changes in the release of neurotransmitters that activate muscles.

    Answer: C

41. Excess dopamine in the brain causes
    a. Alzheimer's disease.
    b. Parkinson's disease.
    c. schizophrenia.
    d. depression.

    Answer: C

Chapter 17

1. A rapid, involuntary response to a stimulus is a property of the
   a. central nervous system.
   b. peripheral nervous system.
   c. ganglia.
   d. sensory neurons.

   Answer: A

2. Sensory information is delivered to the spinal cord via the peripheral nerve entering the
   a. posterior portion.
   b. anterior portion.
   c. medial portion.
   d. autonomic ganglia.

   Answer: A

3. The vertebral column of the lower back contains
   a. the spinal cord.
   b. autonomic ganglia.
   c. nerves and spinal fluid.
   d. the spinal cord and spinal fluid.

   Answer: C

4. The spinal column grows
   a. slower than the spinal cord.
   b. at the same rate as the spinal cord.
   c. faster than the spinal cord.
   d. out and around the peripheral nerves.

   Answer: C

5. Spinal anesthetics
   a. block the transmission of sensory impulses to the spinal cord.
   b. can be injected anywhere into the spinal column because of the presence of cerebral spinal fluid.
   c. can only be injected into the thoracic portion of the spinal column.
   d. prevent pain by blocking the impulses of the motor neurons in the spinal nerves.

   Answer: A

6. The spinal cord conducts information to and from the brain via
   a. neurons in the gray matter.
   b. myelinated tracts in the periphery of the cord.
   c. glial cells in the periphery of the cord.
   d. interneurons in the gray matter.

   Answer: B

7. The spinal cord
   a. conducts information to and from the brain from the periphery.
   b. is capable of integration and responding by spinal reflexes.
   c. rapidly transmits all sensory information to the brain for integration.
   d. generates spinal reflexes, which are initiated and integrated by specific centers in the brain.

   Answer: B

8. In a reflex arc, the cell body of the motor neuron is located in the
   a. anterior gray matter of the spinal cord.
   b. autonomic ganglia of the spinal cord.
   c. descending myelinated tracts from the brain.
   d. gray matter in the motor cortex of the brain.

   Answer: A

9. A spinal reflex
   a. requires a functional brain.
   b. will signal the brain of the event.
   c. does not conduct information to the brain.
   d. involves only sensory and motor neurons of the spinal cord.

   Answer: B

10. The brain consumes _____ of the blood glucose for a supply of energy.
    a. 5%
    b. 10%
    c. 15%
    d. 20%

    Answer: C

11. Permanent brain damage may occur if the oxygen supply is cut off for more than _____ minutes.
    a. 2
    b. 5
    c. 10
    d. 15

    Answer: B

12. Blood pressure, heart rate, and breathing are controlled in the
    a. cerebrum.
    b. cerebellum.
    c. medulla.
    d. pons.

    Answer: C

13. The left side of the brain controls the right side of the body; the neurons cross over in the
    a. cerebrum.
    b. cerebellum.
    c. medulla.
    d. pons.

    Answer: D

14. The gray matter of the midbrain plays a role in
    a. connecting the cerebrum to other structures.
    b. vision.
    c. controlling some aspects of respiration.
    d. sneezing and swallowing.

    Answer: B

15. The portion of the brain involved in the refinement and coordination of muscular movements is the
    a. midbrain.
    b. cerebrum.
    c. cerebellum.
    d. pons.

    Answer: C

16. Many conscious activities involve subconscious components that are governed by the
    a. midbrain.
    b. cerebrum.
    c. cerebellum.
    d. pons.

    Answer: C

17. The relay station for sensory input to the cerebrum is the
    a. hypothalamus.
    b. reticular activating system.
    c. thalamus.
    d. midbrain.

    Answer: C

18. Homeostasis is maintained in large part by activities of the
    a. hypothalamus.
    b. reticular activating system.
    c. thalamus.
    d. midbrain.

    Answer: A

19. The important link between the endocrine system and the nervous system is the
    a. hypothalamus.
    b. reticular activating system.
    c. thalamus.
    d. midbrain.

    Answer: A

20. Screening or filtering sensory information before it reaches the cerebrum is the function of the
    a. hypothalamus.
    b. reticular activating system.
    c. thalamus.
    d. midbrain.

    Answer: B

21. Intelligence, learning, perception, and emotion are controlled by the
    a. cerebellum.
    b. brain stem.
    c. cerebrum.
    d. hippocampus.

    Answer: C

22. The _____ forms a connecting link between the two cerebral hemispheres.
    a. amygdala
    b. thalamus
    c. pons
    d. corpus callosum

    Answer: D

23. Experiments that have intentionally damaged regions of the cerebral cortex in experimental animals and studies on loss of function in brain-damaged humans show that
    a. the cortex has no specific organization.
    b. the cortex has two identical hemispheres.
    c. certain behaviors are found within specific regions.
    d. sensory and motor functions are not discretely organized.

    Answer: C

24. The left hemisphere specializes in _____ abilities.
    a. spatial
    b. mathematical
    c. intuitive
    d. artistic

    Answer: B

25. If a set of keys is placed in the left hand of a split-brain person that individual will be able to
    a. immediately describe the object.
    b. point to a picture of the object.
    c. describe the object but not name it.
    d. feel the object but not recognize it.

    Answer: B

26. Short-term memory is thought to be based on
    a. chemical changes of the neuronal membranes.
    b. self-stimulating electrical circuits.
    c. establishment of new synaptic connections.
    d. changes in membrane-bound proteins.

    Answer: B

27. New facts can become long-lasting memories by long-term synaptic potentiation, which occurs in the
    a. amygdala.
    b. cerebellum.
    c. hippocampus.
    d. basal ganglia.

    Answer: C

28. The basis of skill memories seems to involve
    a. the formation of new electrical circuits in the cerebellum.
    b. protein synthesis in the cerebellum and basal ganglia.
    c. calcium ions as regulators of neuron shape.
    d. the formation of new synaptic connections in the amygdala.

    Answer: B

29. Removal of the amygdala and hippocampus in a young man resulted in loss of the ability to
    a. remember events that occurred a long time ago.
    b. perform complex puzzles.
    c. convert short-term memories into long-term memory.
    d. remember what took place before his surgery.

    Answer: C

30. The nerves of the autonomic nervous system send impulses to the
    a. skeletal muscles.
    b. sensory organs.
    c. sensory receptors.
    d. cardiac and smooth muscles.

    Answer: D

31. Somatic nerves arise from the
    a. brain only as cranial nerves.
    b. brain and spinal cord.
    c. skeletal muscles.
    d. visceral organs.

    Answer: B

Chapter 18

1. Perceptions are formed by
   a. sensory receptors gathering stimuli.
   b. sensory organs interpreting stimuli.
   c. integrating sensory information by the brain.
   d. the increase of sensory nerve impulses to the spinal cord.

   Answer: C

2. Sensory receptors
   a. are selective for a particular energy form.
   b. produce different forms of action potentials for different stimuli.
   c. all utilize the same sensory pathways.
   d. form the quality of sensations transmitted to the brain.

   Answer: A

3. The intensity of sensory stimuli is coded by the
   a. size of the action potential.
   b. location of the sensory organ.
   c. frequency of action potentials.
   d. transmission of the receptor potential to the spinal cord and brain.

   Answer: C

32. The _____ nervous system prepares the body to face emergencies.
    a. somatic
    b. parasympathetic
    c. sympathetic
    d. peripheral

    Answer: C

33. If neurons are damaged in the central nervous system
    a. they can regenerate.
    b. adjacent undamaged neurons will take over their functions.
    c. Schwann cells at the damaged site will grow, providing a pathway for the damaged neuron.
    d. new axons will sprout and reform the correct connections.

    Answer: B

34. In the aging brain
    a. neurotransmitter levels increase.
    b. postsynaptic receptors become more sensitive to neurotransmitters.
    c. some synaptic connections deteriorate.
    d. reasoning and mental alertness decrease rapidly.

    Answer: C

35. Nervous tissue is first organized into a brain in the
    a. phylum Cnidaria.
    b. flatworms.
    c. annelids.
    d. arthropods.

    Answer: B

4. Many sensory receptors
   a. detect only changes in environmental stimuli.
   b. increase the size of receptor potentials to constant stimuli.
   c. increase the rate of action potentials with constant stimuli.
   d. increase the rate of receptor potentials to constant stimuli.

   Answer: A

5. Sensations are
   a. limited by the five senses.
   b. produced by blending together the different senses by the brain.
   c. specific to a single sense organ.
   d. limited by the small number of sense organs.

   Answer: B

6. The flavor of food is produced by the
   a. action of chemoreceptors on the tongue.
   b. activation of chemoreceptors of the tongue and nose.
   c. combination of many senses including, chemoreception, temperature, sight, and touch.
   d. smell and appearance of the food.

   Answer: C

7. For a chemical to be detected by smell it must
   a. be fat soluble.
   b. be a gas or in very small particles.
   c. activate taste receptors as well as olfactory receptors.
   d. be a protein.

   Answer: B

8. Complex odor sensations depend upon
   a. thousands of different receptor molecules in the nose.
   b. combining the activity of a few basic receptor molecules by the brain.
   c. thousands of different neurons projecting to the limbic system.
   d. different receptor potentials of the olfactory chemoreceptors.

   Answer: B

9. Burning your tongue with hot coffee may result in
   a. permanent loss of taste buds.
   b. temporary loss of taste buds.
   c. increase in sensitivity of the taste buds.
   d. loss of sensory neurons activated by the taste buds.

   Answer: B

10. The four basic taste sensations are
    a. sweet, sour, bitter, and salty.
    b. sweet, tart, sharp, and salty.
    c. sweet, tart, bland, and salty.
    d. sweet, smooth, bitter, and salty.

    Answer: A

11. Photoreceptors
    a. form images.
    b. respond to light.
    c. transmit action potentials to the brain.
    d. transmit receptor potentials to the brain.

    Answer: B

12. Compound eyes form
    a. course mosaic images.
    b. simple black-and-white images.
    c. multiple views of the world integrated by the brain.
    d. simple images of the world integrated at the retina.

    Answer: C

13. Light enters the human eye through the
    a. sclera.
    b. retina.
    c. cornea.
    d. choroid layer.

    Answer: C

14. The _____ absorbs light and prevents reflections within the eye.
    a. sclera
    b. cornea
    c. choroid coat
    d. iris

    Answer: C

15. The _____ is responsible for eye color.
    a. ciliary muscle
    b. ciliary body
    c. iris
    d. pupil

    Answer: D

16. In dim light it is easier to see an object when
    a. looking directly at it.
    b. looking to one side of the object.
    c. the cones are stimulated.
    d. the pupil is constricted.

    Answer: B

17. The cornea and lens of the eye are nourished by
    a. the vitreous humor.
    b. capillaries of the ciliary body.
    c. the aqueous humor.
    d. capillaries of the pupil.

    Answer: C

18. When looking at faraway objects the
    a. ciliary muscle is relaxed.
    b. pupil is constricted.
    c. lens is more curved.
    d. pupil is dilated.

    Answer: A

19. In farsighted persons the
    a. eyeball is elongated.
    b. image is focused behind the retina.
    c. lens curves too much and the image is formed ahead of the retina.
    d. image is projected right-side up.

    Answer: B

20. Light entering a rod cell of the retina splits the _____ pigment molecule.
    a. scotopsin
    b. retinal
    c. rhodopsin
    d. vitamin A

    Answer: C

21. Color vision is possible because
    a. different pigments absorb light of different wavelengths.
    b. a single pigment absorbs light of different wavelengths.
    c. rhodopsin is sensitive to different wavelengths of light.
    d. scotopsin is sensitive to different wavelengths of light.

    Answer: A

22. Depth perception in humans is possible because
    a. we can focus each eye independently.
    b. the visual fields of each eye overlap.
    c. each eye has a distinct and different visual field.
    d. each eye can accommodate independently.

    Answer: B

23. Sense of hearing depends on
    a. proprioceptors.
    b. mechanoreceptors.
    c. stretch receptors.
    d. pressure receptors.

    Answer: B

24. The bones of the middle ear
    a. amplify vibrations.
    b. cause the round window to vibrate.
    c. transmit pressure waves.
    d. cause the tympanic membrane to vibrate.

    Answer: A

25. When the narrow region of the basilar membrane at the base of the cochlea vibrates the brain interprets it as _____ sound.
    a. low-frequency
    b. high-frequency
    c. loud
    d. low-intensity

    Answer: B

26. Exposure to loud noises can result in permanent hearing loss by
    a. damaging the tympanic membrane.
    b. destroying hair cells.
    c. damaging the bones of the middle ear.
    d. causing fluid to accumulate in the middle ear.

    Answer: B

27. Rotation of the head is detected by the
    a. ampulla.
    b. vestibule.
    c. utricle.
    d. saccula.

    Answer: A

28. Position of the head with respect to gravity is detected by the
    a. semicircular canals.
    b. ampulla.
    c. vestibule.
    d. otoliths.

    Answer: C

29. Firm pressure is detected by
    a. Pacinian corpuscles.
    b. free nerve endings.
    c. Meissner's corpuscles.
    d. proprioceptors.

    Answer: A

30. Temperature receptors
    a. adapt rapidly.
    b. respond to temperatures in excess of 45° C.
    c. respond to temperatures below 10° C.
    d. are modified Pacinian corpuscles.

    Answer: A

31. Animals with sensory structures containing an iron oxide material are sensitive to
    a. gravity.
    b. magnetic fields.
    c. temperature changes.
    d. body position with reference to the sun.

    Answer: B

Chapter 19

1. Hormones regulate tissues and systems by altering
   a. cellular activity.
   b. neurotransmitter release.
   c. muscle contraction.
   d. cell secretions.

   Answer: A

2. Hormones are produced by _____ glands.
   a. exocrine
   b. endocrine
   c. apocrine
   d. holocrine

   Answer: B

3. Hormones target specific cells only because
   a. there are specific receptors on the target cell.
   b. the hormone is actively transported by these cells.
   c. the hormone is taken up by facilitated transport.
   d. the hormone is able to diffuse into these cells only.

   Answer: A

4. Peptide hormones normally form a complex with the target cell, which
   a. releases ATP.
   b. stimulates cell growth and division.
   c. releases an intracellular messenger.
   d. converts ATP to ADP.

   Answer: C

5. Peptide hormones
   a. are lipid soluble.
   b. pass easily into cells.
   c. activate genes.
   d. bind to receptors on cell membranes.

   Answer: D

6. Reabsorption of calcium in the kidney tubules is stimulated by the _____ hormone.
   a. thyroid
   b. parathyroid
   c. steroid
   d. peptide

   Answer: B

7. Steroid hormones
   a. are water soluble.
   b. can enter target cells.
   c. bind to receptor molecules on the membrane.
   d. activate enzymes.

   Answer: B

8. Which of the following is an example of a negative feedback loop?
   a. Eating causes an increase in blood-glucose levels.
   b. Exercising causes an increase in muscle mass.
   c. Dieting causes a decrease in body weight.
   d. Sweating maintains body temperature in a hot environment.

   Answer: D

9. The brain controls the level of many hormones through the _____ and pituitary gland.
   a. thalamus
   b. hypothalamus
   c. limbic system
   d. amygdala

   Answer: B

10. A _____ hormone increases the secretion of another hormone from a different gland.
    a. tropic
    b. peptide
    c. steroid
    d. stimulating

    Answer: A

11. How does the anterior lobe of the pituitary gland differ from the posterior lobe?
    a. The anterior lobe is smaller.
    b. The anterior lobe is directly connected to the hypothalamus.
    c. The anterior lobe synthesizes its own hormones.
    d. The anterior lobe stores hormones from the hypothalamus.

    Answer: C

12. Neurosecretory cells in the hypothalamus
    a. produce hormones that are stored in the anterior pituitary.
    b. are functional neurons that produce hormones.
    c. produce tropic hormones that stimulate the posterior pituitary.
    d. are controlled by tropic hormones from the posterior pituitary.

    Answer: B

13. Growth hormone
    a. is released from the anterior pituitary when the levels of glycogen fall.
    b. affects protein and fat metabolism but not carbohydrate metabolism.
    c. stimulates cells to take up amino acids, mobilize fats, and release glycogen from the liver.
    d. stimulates cells to divide by altering the levels of nucleic acids.

    Answer: C

14. The gonadotropic hormone _____ stimulates the development of the testes and maturation of sperm cells.
    a. LH
    b. prolactin
    c. ICSH
    d. PSH

    Answer: C

15. The tropic hormone _____ stimulates the thyroid to release its hormones.
    a. FSH
    b. GH
    c. ACTH
    d. TSH

    Answer: D

16. Antidiuretic hormone causes
    a. an increase in blood pressure.
    b. an increase in urine formation.
    c. osmoreceptors of the hypothalamus to release vasopressin.
    d. reabsorption of sodium from the urine.

    Answer: A

17. Drinking large quantities of water
    a. increases the release of ADH.
    b. decreases the release of ADH.
    c. decreases the formation of urine.
    d. increases the release of vasopressin.

    Answer: B

18. Iodine is required for the synthesis of
    a. calcitonin.
    b. thyroxin.
    c. growth hormone.
    d. ACTH.

    Answer: B

19. Increased levels of thyroid hormones will
    a. slow metabolic rate.
    b. result in weight gain and low blood pressure.
    c. increase cellular respiration.
    d. slow heart rate and reduce body temperature.

    Answer: C

20. Decreased levels of estrogen in women following menopause may cause osteoporosis by
    a. decreasing the sensitivity of bone-forming cells to parathyroid hormone.
    b. increasing the sensitivity of bone-forming cells to parathyroid hormone.
    c. increasing calcitonin levels.
    d. decreasing calcitonin levels.

    Answer: B

21. The sympathetic nervous system causes the release of _____ from the adrenal glands.
    a. glucocorticoids
    b. mineralocorticoids
    c. epinephrine
    d. aldosterone

    Answer: C

22. The hormone _____ maintains the level of sodium ions in the blood.
    a. aldosterone
    b. ADH
    c. cortisol
    d. ACTH

    Answer: A

23. The hormone _____ stimulates the breakdown of proteins into amino acids and releases glucose from the liver.
    a. insulin
    b. cortisol
    c. ADH
    d. aldosterone

    Answer: B

24. Cushing's syndrome is characterized by
    a. redistribution of body fat.
    b. weight loss, mental fatigue, and weakness.
    c. imbalance of ions in the blood.
    d. darkening of the skin.

    Answer: A

25. The rate of nutrient absorption in the digestive system is controlled by
    a. glucogon.
    b. glucocorticoids.
    c. insulin.
    d. somatostatin.

    Answer: D

26. A person with diabetes mellitus injects too much insulin into the body; the physiological response would appear as
    a. excess sugar in the blood.
    b. hypoglycemia.
    c. mobilization of fats.
    d. breakdown of glycogen to glucose in the liver.

    Answer: B

27. Stress can cause symptoms that may be mistaken for
    a. diabetes insipidus.
    b. hypoglycemia.
    c. hyperglycemia.
    d. thyroid deficiency.

    Answer: B

28. The hormone _____ is released from the ovaries and controls secretions associated with pregnancy.
    a. estrogen
    b. progesterone
    c. FSH
    d. LHRH

    Answer: B

29. If too little LHRH is secreted by the hypothalamus, the ovaries will
    a. produce too many follicles.
    b. produce excess estrogen.
    c. produce excess progesterone.
    d. not produce a mature follicle.

    Answer: D

30. When a fertilized ovum implants in the uterine lining it
    a. produces HCG.
    b. elevates the mother's progesterone levels.
    c. lowers the mother's progesterone levels.
    d. stimulates the production of LHRH.

    Answer: A

31. In males the late stage of sperm development and secretion of testosterone depends upon the hormone
    a. LTH.
    b. ISCH.
    c. LHRH.
    d. FSH.

    Answer: B

32. The pineal gland seems to be involved in controlling
    a. responses to changing seasons.
    b. changes in pigmentation.
    c. changes in adrenal function.
    d. blood-glucose levels.

    Answer: A

33. Hormones are produced and released from
    a. neural tissue.
    b. distinct glands.
    c. many cells of the body.
    d. exocrine glands.

    Answer: C

34. Atrial natriuretic factor, a hormone that in part regulates blood pressure, is released from the
    a. pineal gland.
    b. anterior pituitary gland.
    c. heart.
    d. posterior pituitary gland.

    Answer: C

35. The release of prostaglandins results when
    a. individuals suffer from stress.
    b. cell membranes are disturbed or injured.
    c. individuals are exposed to cold.
    d. the hypothalamus releases neurosecretory hormones.

    Answer: B

36. Prostaglandins
    a. have limited effects on local tissues.
    b. only alter smooth muscle function.
    c. have a wide variety of physiological effects.
    d. are produced by specific endocrine glands.

    Answer: C

37. _____ are chemicals that influence the physiology or behavior of individuals of the same species.
    a. Hormones
    b. Prostaglandins
    c. Peptides
    d. Pheromones

    Answer: D

38. Individuals who use synthetic versions of the male hormone testosterone will
    a. increase their performances for the long term.
    b. decrease their short-term performances.
    c. possibly suffer long-term damage to kidney, liver, and heart.
    d. increase their fertility.

    Answer: C

39. The human female is unique in her sexually receptive behavior and is referred to as a(n) _____ ovulator.
    a. induced
    b. cyclic
    c. seasonal
    d. daily

    Answer: B

Chapter 20

1. Hydrostatic skeletons do not provide
    a. support.
    b. body shape.
    c. sites for muscle attachment.
    d. protection.

    Answer: C

2. Exoskeletons are limiting to the organism because they
    a. must be shed for growth to continue.
    b. provide only external support.
    c. provide limited muscle attachment sites.
    d. store calcium salts outside the body.

    Answer: A

3. Animals with endoskeletons
    a. are very mobile.
    b. are unable to grow without molting.
    c. have a high percentage of body mass in the skeletal system.
    d. have limited or no protection.

    Answer: A

4. Cartilage
    a. is found only in joints in the adult organism.
    b. is not very flexible.
    c. forms the rib cage of vertebrate animals.
    d. forms the skeleton in the embryo.

    Answer: D

5. _____ forms a strong network that allows cartilage to bear great weight.
    a. Elastin
    b. Hyaluronic acid
    c. Collagen
    d. Proteoglycan

    Answer: C

6. Cartilage consists primarily of
    a. elastin.
    b. collagen.
    c. cells.
    d. water.

    Answer: D

7. The great strength of bone is due to the presence of
   a. collagen.
   b. minerals.
   c. osteocytes.
   d. elastin.

   Answer: A

8. The osteocytes of bone are located in
   a. Haversian canals.
   b. canaliculi.
   c. lacunae.
   d. marrow.

   Answer: C

9. Spongy bone is filled with
   a. red marrow.
   b. yellow marrow.
   c. osteocytes.
   d. fat and a few blood cells.

   Answer: A

10. Blood cells and platelets are manufactured in
    a. yellow marrow.
    b. compact bone.
    c. red marrow.
    d. the hollow shaft of long bones.

    Answer: C

11. In the process of ossification,
    a. minerals are lost from bone.
    b. cartilage is replaced by bone tissue.
    c. collagen and elastin are reabsorbed from bone.
    d. cartilage grows out over the surface of bones.

    Answer: B

12. Cartilage
    a. cells transform into osteocytes.
    b. expands into flat plates in the formation of bone.
    c. lacks a blood supply.
    d. migrates out of the matrix to the surface of bone.

    Answer: C

13. Bones are able to elongate during childhood because of
    a. growth at the epiphyseal plates.
    b. constantly growing spongy bone.
    c. growth of compact bone in the center of the shaft.
    d. cartilage replacement of spongy bone at the ends.

    Answer: A

14. The first reaction to a bone fracture is the
    a. deposition of dense connective tissue.
    b. invasion by cartilage cells.
    c. formation of a blood clot.
    d. growth of spongy bone.

    Answer: C

15. Bones are
    a. passive sites of mineral storage.
    b. constantly broken down and undergoing renewal.
    c. only metabolically active in the marrow.
    d. storage sites for sodium and potassium salts.

    Answer: B

16. Bone releases calcium under the stimulation of
    a. calcitonin.
    b. parathyroid hormone.
    c. vitamin D.
    d. estrogen.

    Answer: B

17. Spaces within the bones of the skull are termed
    a. sutures.
    b. fontanels.
    c. sinuses.
    d. foramina.

    Answer: C

18. The bones in the skull of a newborn child
    a. are fused together.
    b. contain dense connective tissue at the sutures.
    c. contain cartilage and are flexible.
    d. contain more blood vessels than those of an adult.

    Answer: B

19. Ribs attach to the _____ vertebrae.
    a. cervical
    b. sacral
    c. thoracic
    d. lumbar

    Answer: C

20. The abnormal curvature of the spine that results in an S or C shape is termed
    a. lordosis.
    b. scoliosis.
    c. spina bifida.
    d. slipped disk.

    Answer: B

21. In humans _____ pairs of ribs attach to the sternum.
    a. 2
    b. 7
    c. 10
    d. 12

    Answer: C

22. The _____ bones make up the pectoral girdle.
    a. humerus and scapula
    b. scapula and clavicle
    c. clavicle and humerus
    d. humerus, radius, and ulna

    Answer: B

23. The female pelvis is _____ than the corresponding male.
   a.   narrower in the front
   b.   smaller in diameter
   c.   broader
   d.   more elastic

   Answer: C

24. The phalanges of the foot are _____ than those of the hand.
   a.   longer
   b.   shorter
   c.   fewer in number
   d.   greater in number

   Answer: B

25. Synovial joints
   a.   have interlocking projections preventing movement.
   b.   permit only slight movement.
   c.   contain a fluid-filled capsule.
   d.   are found between the cranial bones.

   Answer: C

26. Bursa are
   a.   found in nonmovable joints.
   b.   formed from osteocytes.
   c.   fluid-filled sacs that reduce friction.
   d.   precursors to the formation of ligaments.

   Answer: C

27. Rheumatoid arthritis is
   a.   caused by cartilage wearing away from the bone ends.
   b.   an inflammation of the synovial membranes.
   c.   an inflammation of the bursae.
   d.   breakdown of collagen produced by cartilage cells.

   Answer: B

28. Osteoporosis in women may be due to
   a.   declining levels of estrogen.
   b.   increasing levels of calcitonin.
   c.   larger bone mass than men.
   d.   shorter life spans than men.

   Answer: A

29. Exercise may lead to
   a.   increased bone mass.
   b.   decreased bone mass.
   c.   osteoporosis.
   d.   arthritis.

   Answer: A

30. The bowed legs seen in the disease rickets are due to a lack of
   a.   calcitonin.
   b.   parathyroid hormone.
   c.   vitamin D.
   d.   calcium in the diet.

   Answer: C

Chapter 21

1. Smooth muscle cells
   a.   are under voluntary control.
   b.   are long and tapered.
   c.   have intercalated disks.
   d.   are found in the heart.

   Answer: B

2. Skeletal muscle cells
   a.   have intercalated disks.
   b.   have one nucleus.
   c.   move food through the digestive tract.
   d.   appear striated under the microscope.

   Answer: D

3. Skeletal muscles have a connective tissue sheath surrounding many bundles of long cells called
   a.   myofibrils.
   b.   sarcomeres.
   c.   muscle fibers.
   d.   myofilaments.

   Answer: C

4. Thick myofilaments
   a.   are made up of actin.
   b.   are twisted in a double strand.
   c.   contain troponin and tropomyosin.
   d.   are made up of myosin.

   Answer: D

5. Each myosin filament is surrounded by _____ actin filaments.
   a.   two
   b.   three
   c.   four
   d.   six

   Answer: D

6. The striations of skeletal muscle are due to the orderly arrangement of
   a.   thick and thin filaments.
   b.   the T-tubules.
   c.   the sarcoplasmic reticulum.
   d.   muscle fibers.

   Answer: A

7. Actin filaments slide past myosin filaments during muscle contraction because
   a.   actin binds to ATP.
   b.   of the rocking motion of the myosin head on the actin filaments.
   c.   troponin binds to tropomyosin.
   d.   myosin binds to troponin.

   Answer: B

8. The connection of a myosin head to actin is termed
   a. enzyme reaction.
   b. troponin-tropomyosin linkage.
   c. cross-bridge.
   d. ATP linkage.

   Answer: C

9. The message to contract from the motor nerve cell to the muscle is
   a. ATP.
   b. calcium.
   c. dystrophin.
   d. acetylcholine.

   Answer: D

10. Increased levels of _____ cause tropomyosin to move on actin filaments, exposing the site of attachment for myosin heads.
    a. acetylcholine
    b. calcium
    c. dystrophin
    d. ATP

    Answer: B

11. Myosin is activated for cross-bridge formation by binding to
    a. ATP.
    b. dystrophin.
    c. acetylcholine.
    d. calcium.

    Answer: A

12. Relaxation of a muscle occurs when
    a. ATP is no longer available.
    b. calcium returns to the sarcoplasmic reticulum.
    c. acetylcholine is transported into the motor neuron.
    d. dystrophin is broken down by cellular enzymes.

    Answer: B

13. In muscle cells ATP is needed to
    a. release calcium from the sarcoplasmic reticulum.
    b. break the cross-bridges of actin and myosin.
    c. form the cross-bridges of actin and myosin.
    d. activate the T-tubule system.

    Answer: B

14. When muscle activity begins, ATP can be regenerated in muscle cells by the presence of
    a. glucose.
    b. glycogen.
    c. creatine phosphate.
    d. fatty acids.

    Answer: C

15. Muscle fatigue can be attributed to accumulation of
    a. ADP.
    b. creatine phosphate.
    c. lactic acid.
    d. pyruvate.

    Answer: C

16. The oxygen debt after extensive exercise is due to the accumulation of
    a. ADP.
    b. creatine phosphate.
    c. lactic acid.
    d. pyruvate.

    Answer: C

17. The strength of muscle contraction depends upon the number of _____ activated.
    a. sarcomeres
    b. myofibrils
    c. myofilaments
    d. muscle fibers

    Answer: D

18. In a motor unit that exerts fine control over muscle movement the motor neuron contacts
    a. many muscle fibers.
    b. few muscle fibers.
    c. few myofibrils.
    d. many myofibrils.

    Answer: B

19. A second stimulus to a muscle fiber immediately following the refractory period results in
    a. no response.
    b. a twitch.
    c. a second, more powerful contraction.
    d. tetanus.

    Answer: C

20. A smooth continuous contraction is called
    a. a twitch.
    b. summation.
    c. tetanus.
    d. muscle refractory period.

    Answer: C

21. Slow twitch-fatigue resistant muscle fibers differ from fast twitch-fatigue resistant fibers in that they
    a. have a larger supply of ATP.
    b. have more myosin.
    c. have more myoglobin.
    d. are able to split ATP faster.

    Answer: C

22. An athlete with a high percentage of fast twitch muscle fibers would excel at
    a. marathon running.
    b. weight lifting.
    c. long-distance cycling.
    d. endurance sports.

    Answer: B

23. The muscle attachment to the bone that does not move is called the
    a. fulcrum.
    b. lever.
    c. insertion.
    d. origin.

    Answer: D

24. The closer the point of muscle insertion to the movable joint the _____ to lift the weight.
    a. less effort needed
    b. more effort needed
    c. neither; effort will not change
    d. neither; effort will be less but the movement will be faster

    Answer: B

25. When you straighten your arm at the elbow the
    a. biceps contract.
    b. triceps contract.
    c. biceps push the forearm outward.
    d. triceps push the forearm outward.

    Answer: B

26. Muscle spindles are
    a. the major contractile portion of skeletal muscles.
    b. receptors that control muscle tone.
    c. receptors that inhibit muscle tone.
    d. essential in initiating relaxation of skeletal muscles.

    Answer: B

27. In exercise-induced hypertrophy in humans the muscle increases in the
    a. number of muscle spindles.
    b. number of muscle fibers.
    c. size of the muscle fibers.
    d. size and number of muscle fibers.

    Answer: C

Chapter 22

1. Many very small organisms can get by without a circulatory system because
    a. they are relatively inactive.
    b. their internal cellular environment is tolerant of wide changes in composition.
    c. their surface to volume ratio is high.
    d. they have hemoglobin inside their body cells for oxygen binding.

    Answer: C

2. In a closed circulatory system
    a. blood bathes the cells before returning to the heart.
    b. blood is contained in sinuses in the body tissues.
    c. all cells of the body are very close to the smallest blood vessels.
    d. hemoglobin is carried in the fluid portion of the blood.

    Answer: C

28. Compared to sedentary individuals the muscles of conditioned athletes do not contain more
    a. ATP.
    b. blood vessels and myoglobin.
    c. mitochondria.
    d. muscle spindles.

    Answer: D

29. The RICE plan for treatment of injuries to muscles, ligaments, or joints stands for
    a. recuperation, ice, compression, evaluation.
    b. rest, ice, compression, elevation.
    c. relaxation, incubation, cold, evaluation.
    d. rest, incubation, cold, evaluation.

    Answer: B

30. In muscular dystrophy the absence of the protein dystrophin results in the inability of
    a. calcium to be released from the sarcoplasmic reticulum.
    b. troponin to bind to calcium.
    c. myosin to bind to actin.
    d. ATP to bind to myosin.

    Answer: C

3. The fluid portion of blood does not leak out of the vessels because
    a. the pH of the blood is regulated at 7.2.
    b. of the high concentration of albumin.
    c. of the low concentration of sodium ions in the tissues.
    d. of the high pressure in the vessels established by heart contractions.

    Answer: B

4. The connective tissue matrix of blood is termed
    a. formed elements.
    b. plasma.
    c. platelets.
    d. red and white blood cells.

    Answer: B

5. Blood provides protection from injury by clotting, which requires _____ protein.
    a. globulin
    b. albumin
    c. fibrinogen
    d. lipo-

    Answer: C

6. Red blood cells live for only about 4 months because they
   a. are repeatedly pounded and squeezed through the capillaries.
   b. have no nucleus or mitochondria.
   c. are selectively destroyed by the spleen.
   d. are packed with hemoglobin.

   Answer: B

7. Carbon monoxide
   a. binds weakly with hemoglobin.
   b. destroys red blood cells.
   c. blocks oxygen binding to hemoglobin.
   d. blocks the release of carbon dioxide from hemoglobin.

   Answer: C

8. Low oxygen in the atmosphere results in
   a. fewer red blood cells.
   b. the release of erythropoietin from the kidneys.
   c. bone marrow producing more white blood cells.
   d. fewer red blood cells with more hemoglobin.

   Answer: B

9. Anemia
   a. results when there are too many white blood cells.
   b. limits the body's ability to deliver oxygen to cells.
   c. results when there are too many red blood cells.
   d. results from too much hemoglobin in the red blood cells.

   Answer: B

10. Individuals have different ABO blood types because of the presence of different
    a. forms of hemoglobin in the red blood cells.
    b. antigens on red blood cell membranes.
    c. albumins in the plasma.
    d. globins in the plasma.

    Answer: B

11. White blood cells originate in the
    a. spleen.
    b. liver.
    c. bone marrow.
    d. lymphatic tissue.

    Answer: C

12. White blood cells
    a. are confined within the circulatory system.
    b. produce antigens.
    c. cause inflammation.
    d. protect against anemia.

    Answer: C

13. The initiation of a blood clot from injury depends upon the action of
    a. fibrinogen.
    b. thromboplastin.
    c. platelets.
    d. prothrombin.

    Answer: C

14. The normal flow of blood is from arteries to
    a. veins to capillaries to venules to arterioles.
    b. arterioles to capillaries to venules to veins.
    c. arterioles to venules to capillaries to veins.
    d. venules to capillaries to arteries to veins.

    Answer: B

15. Blood pressure and blood distribution to tissues are controlled by the
    a. heart.
    b. arteries.
    c. arterioles.
    d. capillaries.

    Answer: C

16. The site of exchange of gases, nutrients, and waste products is the
    a. venules.
    b. arteries.
    c. capillaries.
    d. sinuses.

    Answer: C

17. The metabolic needs of the tissues are met by the opening and closing of the _____, which direct blood flow.
    a. venules
    b. capillaries
    c. arterioles
    d. sinuses

    Answer: C

18. There are one-way valves in
    a. arteries.
    b. arterioles.
    c. capillaries.
    d. veins.

    Answer: D

19. The first vessels that branch off the aorta supply blood to the
    a. lungs.
    b. heart.
    c. kidneys.
    d. brain.

    Answer: B

20. Blood from the digestive system
    a. returns directly to the heart through the inferior vena cava.
    b. first goes to the kidneys and then returns to the heart.
    c. goes to the liver before returning to the heart.
    d. goes to the spleen and then returns to the heart through the inferior vena cava.

    Answer: C

21. Blood pressure reaches a maximum during
    a. ventricular contraction.
    b. ventricular relaxation.
    c. atrial contraction.
    d. atrial relaxation.

    Answer: A

22. Blood pressure
    a. is constant throughout the body.
    b. is highest in the vena cava.
    c. decreases with distance from the heart.
    d. is lowest in the capillaries.

    Answer: C

23. Blood flow slows down in the capillaries because of the
    a. low pressure at this point.
    b. large cross-sectional area.
    c. constriction of the arterioles.
    d. dilation of the arterioles.

    Answer: B

24. Cardiac muscle cells are similar to skeletal muscles in that they
    a. have a single nucleus.
    b. form a branching network.
    c. are striated.
    d. are organized into motor units.

    Answer: C

25. A heart murmur could mean
    a. backflow of blood through the heart valves.
    b. weak ventricular contractions.
    c. weak atrial contractions.
    d. low blood pressure in the aorta.

    Answer: A

26. The heart is considered to be two pumps because
    a. there are two atria and two ventricles.
    b. the right side pumps blood to the body and the left side to the lungs.
    c. the right side pumps blood to the lungs and the left side to the body.
    d. alternating periods of contraction and relaxation occur.

    Answer: C

27. The flow of blood through the heart is
    a. right atrium to left atrium to lungs to right ventricle to left ventricle to body.
    b. left atrium to right atrium to lungs to right ventricle to left ventricle to body.
    c. right atrium to right ventricle to lungs to left atrium to left ventricle to body.
    d. left atrium to left ventricle to lungs to right atrium to right ventricle to body.

    Answer: C

28. The heart beat begins in the
    a. intercalated disks.
    b. atrioventricular node.
    c. sinoatrial node.
    d. Purkinje fibers.

    Answer: C

29. Heart rate increases when the activity of the _____ nervous system predominates.
    a. parasympathetic
    b. sympathetic
    c. autonomic
    d. peripheral

    Answer: B

30. Heart tissue is nourished by blood flow
    a. through the atrial and ventricular chambers.
    b. through capillaries within the muscle.
    c. through the sinuses of the heart.
    d. from the venous return via the inferior and superior vena cava.

    Answer: B

31. When blood pressure is low the vasomotor center in the medulla causes
    a. vasodilation.
    b. a decrease in heart rate.
    c. vasoconstriction.
    d. decreased blood flow returning to the heart.

    Answer: C

32. Long-term aerobic exercise will
    a. increase resting blood pressure.
    b. increase resting heart rate.
    c. lower high-density lipoproteins (HDLs).
    d. lower resting blood pressure.

    Answer: D

33. The lymphatic system returns fluid to the
    a. digestive system.
    b. kidneys.
    c. tissue spaces.
    d. blood.

    Answer: D

34. Lymph nodes
    a. are involved in the absorption of dietary fats.
    b. filter cellular debris and bacteria.
    c. are located in small arterioles.
    d. are located in small venules.

    Answer: B

35. Decreased blood flow to the heart muscle will most likely cause
    a. atherosclerosis.
    b. angina pectoris.
    c. congestive heart failure.
    d. hypertension.

    Answer: B

Chapter 23

1. In the presence of oxygen, cells can produce _____ molecules of ATP from glucose.
   a. 2
   b. 36
   c. 3
   d. 12

   Answer: B

2. The end result of aerobic metabolism is the conversion of glucose to
   a. lactic acid.
   b. carbon dioxide and water.
   c. carbon dioxide and ATP.
   d. carbon dioxide.

   Answer: B

3. The respiratory surface must be
   a. protected from the environment.
   b. large enough to meet the metabolic needs of the organism.
   c. contained within a body cavity.
   d. directly linked to a closed circulatory system.

   Answer: B

4. _____ water has the highest oxygen content.
   a. Warm
   b. Cold
   c. Salt
   d. Stagnant

   Answer: B

5. Terrestrial organisms compared to aquatic animals require
   a. a larger respiratory surface.
   b. the movement of the media over the respiratory membrane.
   c. a smaller respiratory surface.
   d. the same relative size respiratory surface.

   Answer: C

6. Large organisms that use the body surface as a respiratory surface
   a. are very sluggish or inactive.
   b. have an effective circulatory system to the skin.
   c. are cylindrical to aid diffusion.
   d. have no hemoglobin in their blood.

   Answer: B

7. Insects have
   a. lungs.
   b. gills.
   c. trachea.
   d. a body surface that serves as a respiratory membrane.

   Answer: C

8. Water generally flows over the gills of fish by moving in
   a. and out of the mouth.
   b. and out of the operculum.
   c. the mouth and out under the operculum.
   d. under the operculum and out the mouth.

   Answer: C

9. The flow of blood in the gills is
   a. very slow.
   b. in the opposite direction from the flow of water.
   c. regulated by the oxygen content of the water.
   d. regulated by the carbon dioxide content of the water.

   Answer: B

10. An advantage of lungs in terrestrial animals is that the
    a. respiratory surface can be kept moist.
    b. air is easily moved in and out of the system.
    c. air can be filtered.
    d. air can flow in one direction through the system.

    Answer: A

11. Before reaching the lungs air
    a. is warmed and purified in the nose.
    b. passes through the trachea and then the larynx.
    c. passes from the pharynx to the trachea.
    d. is cooled on the moist surfaces of the nose.

    Answer: A

12. The vocal cords are located in the
    a. pharynx.
    b. larynx.
    c. trachea.
    d. primary bronchi.

    Answer: B

13. Food is prevented from entering the respiratory system by action of the
    a. glottis.
    b. pharynx.
    c. epiglottis.
    d. tongue.

    Answer: C

14. The trachea does not collapse with changes in air pressure because of
    a. the strong muscular walls.
    b. the presence of cartilaginous rings.
    c. its large size.
    d. its small diameter.

    Answer: B

15. Inhaled particles that enter the bronchial tree are removed by
    a. absorption.
    b. phagocytosis.
    c. the action of cilia.
    d. the action of smooth muscles.

    Answer: C

16. Bronchioles differ from the bronchi in that they
    a. have cilia.
    b. have smooth muscles in the walls.
    c. do not have cartilage in the walls.
    d. are involved in gas exchange.

    Answer: C

17. Drugs that relieve asthma attacks work by
    a. stimulating smooth muscles to contract.
    b. stimulating deep breathing.
    c. relaxing bronchial smooth muscles.
    d. reducing inflammation of the nasal passages.

    Answer: C

18. Gas exchange occurs in the
    a. bronchioles.
    b. bronchi.
    c. trachea.
    d. alveoli.

    Answer: D

19. Alveoli do not collapse because
    a. smooth muscles hold them open.
    b. of cartilaginous rings.
    c. of the presence of surfactant.
    d. of the surface tension of water on the alveolar surface.

    Answer: C

20. Infantile respiratory distress syndrome in premature babies is caused by
    a. abnormal circulation to the lungs.
    b. lack of surfactant.
    c. inflammation of the pleural membranes.
    d. insufficient development of the respiratory muscles.

    Answer: B

21. Air is moved into the lungs when the
    a. diaphragm relaxes.
    b. volume of the thoracic cavity decreases.
    c. rig cage is pulled downward.
    d. pressure in the thoracic cavity is less than air pressure.

    Answer: D

22. Expiration is caused by
    a. contraction of the abdominal muscles.
    b. contraction of the muscles attached to the rib cage.
    c. increasing pressure in the thoracic cavity.
    d. increasing the volume of the thoracic cavity.

    Answer: C

23. The air exchange in the respiratory system when breathing quietly is called the
    a. vital capacity.
    b. tidal volume.
    c. residual volume.
    d. inspiratory volume.

    Answer: B

24. The percentage of oxygen carried in the blood bound to hemoglobin is
    a. 3%.
    b. 9%.
    c. 39%.
    d. 93%.

    Answer: D

25. Each hemoglobin molecule can carry _____ molecule(s) of oxygen.
    a. one
    b. two
    c. three
    d. four

    Answer: D

26. The most important factor determining whether hemoglobin will bind to oxygen is the
    a. blood acidity.
    b. partial pressure of carbon dioxide.
    c. partial pressure of oxygen.
    d. temperature of the blood.

    Answer: C

27. When there is a serious requirement for oxygen by the tissues, the rate of oxygen delivery can increase _____ = fold without an increase in blood flow.
    a. 2
    b. 3
    c. 6
    d. 12

    Answer: B

28. The largest percentage of carbon dioxide transported in the blood is
    a. in the form of dissolved gas.
    b. in the form of bicarbonate ion.
    c. bound to hemoglobin.
    d. dissolved in the red blood cells.

    Answer: B

29. Carbon monoxide
    a. binds more readily to hemoglobin than oxygen.
    b. dissolves in the plasma and is carried to the tissues.
    c. blocks inspiration centers in the brain.
    d. causes an increase in respiratory rate.

    Answer: A

30. The respiratory center for rhythmic breathing is located in the
    a. pons.
    b. hypothalamus.
    c. medulla.
    d. cerebral cortex.

    Answer: C

31. The most important factor in regulating breathing rate is the
    a. blood level of oxygen.
    b. blood temperature.
    c. blood level of carbon dioxide.
    d. activity of the apneustic center.

    Answer: C

32. Chemoreceptors in the _____ are the most sensitive to changes in carbon dioxide levels of the blood.
    a. aortic bodies
    b. carotid bodies
    c. medulla
    d. cerebral cortex

    Answer: C

33. Prolonged hyperventilation at rest will
    a. increase the blood oxygen level.
    b. decrease the blood carbon dioxide level.
    c. decrease the blood pH.
    d. increase the blood level of bicarbonate ion.

    Answer: B

34. Sudden infant death syndrome (SIDS) may be caused by failure of carotid and aortic bodies to respond to _____ in the blood.
    a. low carbon dioxide
    b. high carbon dioxide
    c. low oxygen
    d. high oxygen

    Answer: C

35. _____ decrease(s) mucus secretion and combats watery eyes and sneezing in a person with a cold.
    a. Decongestants
    b. Antihistamines
    c. Analgesics
    d. Aspirin

    Answer: B

Chapter 24

1. The substance found in highest concentration in the blood leaving the digestive tract is
    a. starch.
    b. cholesterol.
    c. protein.
    d. amino acids.

    Answer: D

2. The basic units of fats are
    a. amino acids and glycerol.
    b. glycogen and fatty acids.
    c. glycerol and fatty acids.
    d. glycogen and fatty acids.

    Answer: B

36. Bronchitis
    a. is an acute infection accompanying a cold.
    b. may be caused by viruses or bacteria.
    c. is characterized by inflammation of the alveoli.
    d. cannot be treated with antibiotics.

    Answer: B

37. In a person with emphysema the
    a. bronchial tree becomes inflamed.
    b. lungs become more elastic.
    c. alveoli become overinflated and burst.
    d. disease is acute and nonprogressive.

    Answer: C

38. The main symptom of emphysema is
    a. an expanded chest.
    b. a deep cough that produces gray or greenish yellow phlegm.
    c. shortness of breath.
    d. fever and fatigue.

    Answer: C

39. In the first stages of tuberculosis
    a. the body seals the bacteria in a fibrous connective tissue capsule.
    b. cavities form in the lungs.
    c. the immune system destroys the bacteria.
    d. the bacteria invade the brain, kidneys, and bone tissue.

    Answer: A

40. The lungs are
    a. insensitive to airborne irritants.
    b. subject to airborne irritants that may later cause disease.
    c. protected from airborne irritants by the presence of smooth muscles in the bronchioles.
    d. protected from airborne irritants by the action of the epiglottis.

    Answer: B

3. The substance required to split any macromolecule of food into micromolecules is
    a. pepsin.
    b. bile.
    c. water.
    d. acid.

    Answer: C

4. Filter feeders capture food by
    a. grabbing food with large teeth.
    b. surrounding their prey with cellular membranes.
    c. capturing food in webbed fingers.
    d. swallowing large amounts of food and water.

    Answer: D

5. All animals have specialized compartments for digestion because this
   a. increases the efficiency of the digestive system.
   b. keeps the digestive enzymes from digesting the organism.
   c. prevents waste from coming in contact with other parts of the organism.
   d. prevents toxins from coming in contact with other parts of the organism.

   Answer: B

6. Food vacuoles of low invertebrates are very similar to _____ in higher organisms.
   a. ribosomes
   b. perisomes
   c. lysosomes
   d. polysomes

   Answer: C

7. In animals that have digestive systems with a single opening, the cavity doubles as the
   a. circulatory system.
   b. excretory system.
   c. respiratory system.
   d. skeletal system.

   Answer: A

8. The organ that aids in digestion but is not part of the gastrointestinal tract is the
   a. esophagus.
   b. liver.
   c. large intestine.
   d. mouth.

   Answer: B

9. Digestive enzymes are produced and secreted by the mouth and the
   a. liver.
   b. spleen.
   c. pancreas.
   d. large intestine.

   Answer: C

10. Mucus is secreted into the digestive tract for the purpose of
    a. digesting food.
    b. breaking down fat.
    c. transporting nutrients.
    d. protection from digestion itself.

    Answer: D

11. Peristalsis aids digestion by
    a. moving food quickly through the tract.
    b. churning the food to mix it with enzymes.
    c. adding water to the tract.
    d. squeezing the food forward.

    Answer: B

12. Mechanical digestion increases the rate of chemical digestion because it
    a. increases the surface area of food particles.
    b. increases the secretion of enzymes by the mouth.
    c. increases the secretion of water by the mouth.
    d. reduces the size of the food so it passes down the digestive tract easily.

    Answer: A

13. Salivary amylase secreted by the mouth begins the digestion of
    a. protein.
    b. carbohydrates.
    c. lipids.
    d. nucleic acids.

    Answer: B

14. No digestion takes place in the
    a. mouth.
    b. esophagus.
    c. stomach.
    d. small intestine.

    Answer: B

15. The rugae of the stomach are involved in the
    a. movement of the bolus.
    b. secretion of digestive enzymes.
    c. stretching of the stomach.
    d. mechanical digestion of the bolus.

    Answer: C

16. Foods high in _____ remain in the stomach the longest time.
    a. fats
    b. carbohydrates
    c. proteins
    d. starch

    Answer: A

17. The secretion of hydrochloric acid into the stomach
    a. activates the enzyme pepsin.
    b. breaks down butterfat molecules.
    c. breaks down glycerol.
    d. hydrolyzes protein.

    Answer: A

18. The release of gastrin from the stomach lining results in
    a. secretion of gastric juices.
    b. breakdown of fats.
    c. increase of mechanical digestion.
    d. increase of micronutrient absorption.

    Answer: A

19. Ulcers result from too much _____ in the stomach.
    a. mucus
    b. pepsinogen
    c. hydrochloric acid
    d. gastrin

    Answer: C

20. The enzyme involved in the breakdown of fat to glycerol and fatty acids is
    a. trypsin.
    b. chymotrypsin.
    c. bile.
    d. lipase.

    Answer: D

21. Fat emulsification results from the action of
    a. lipase.
    b. trypsin.
    c. chymotrypsin.
    d. bile.

    Answer: D

22. Individuals with lactose intolerance lack
    a. bile.
    b. lactase.
    c. maltase.
    d. amylase.

    Answer: B

23. Secretin causes the release of
    a. sodium bicarbonate.
    b. hydrochloric acid.
    c. bile.
    d. intestinal enzymes.

    Answer: A

24. Cholecystokinin causes the release of
    a. sodium bicarbonate.
    b. hydrochloric acid.
    c. bile.
    d. intestinal enzymes.

    Answer: C

25. A decrease in the number of villi in the small intestine would result in a decrease in the
    a. absorption of nutrients.
    b. rate of digestion.
    c. production of enzymes.
    d. production of secretion.

    Answer: A

26. Amino acids move into the circulatory system via
    a. facilitated diffusion.
    b. passive diffusion.
    c. osmosis.
    d. active transport.

    Answer: D

27. Fatty acids move into the circulatory system via
    a. facilitated diffusion.
    b. passive diffusion.
    c. osmosis.
    d. active transport.

    Answer: B

28. In order for fat to enter the lymphatic system, it must first be
    a. emulsified.
    b. broken into fatty acids and glycerol.
    c. coated with carrier proteins.
    d. reassembled into proteins.

    Answer: C

29. The normal flora of bacteria in the colon may
    a. cause diarrhea.
    b. extract vitamins needed by the body.
    c. break down food into smaller particles.
    d. cause infection.

    Answer: B

30. The main function of the colon is to
    a. finish the digestion of food.
    b. reabsorb water used in digestion.
    c. produce digestive enzymes.
    d. absorb nutrients.

    Answer: B

31. Sodium bicarbonate is released by the pancreas to
    a. neutralize hydrochloric acid.
    b. digest fats.
    c. activate pepsinogen.
    d. regulate blood-sugar levels.

    Answer: A

32. Besides producing bile, the liver
    a. regulates salt/water balance.
    b. fights infection.
    c. stores extra sugar as glycogen.
    d. secretes digestive enzymes.

    Answer: C

33. Jaundice is caused by abnormal amounts of
    a. fat.
    b. secretin.
    c. cholesterol in the blood.
    d. bile.

    Answer: D

34. Hepatitis is the inflammation of the
    a. small intestine.
    b. pancreas.
    c. liver.
    d. spleen.

    Answer: C

35. Bile is stored in the
    a. liver.
    b. pancreas.
    c. spleen.
    d. gallbladder.

    Answer: D

Chapter 25

1. The Cro-Magnons had a healthier diet than many Americans because they had low intake of
   a. fiber.
   b. fat.
   c. carbohydrates.
   d. protein.

   Answer: B

2. Nutrition is the study of
   a. foods required for a healthier body.
   b. nutrients and their fate in the body.
   c. the caloric content of various kinds of nutrients.
   d. the combination of various food groups to maintain a balanced diet.

   Answer: B

3. Vitamins and minerals are called micronutrients because they are
   a. tiny molecules compared with the macronutrients.
   b. absorbed without digestion.
   c. required in small amounts.
   d. helpful but not necessary.

   Answer: C

4. Nonessential nutrients refer to some of the
   a. amino acids.
   b. carbohydrates.
   c. nucleic acids.
   d. fats.

   Answer: A

5. A kilocalorie refers to the amount of energy it takes to
   a. maintain body temperature at 98.6°F.
   b. raise 1 gram of water 1°C.
   c. convert carbohydrates into simple sugars.
   d. get water to boil in a bomb calorimeter.

   Answer: B

6. If you are an average female eating 2,100 kilocalories per day and maintain a constant body weight, how many kilocalories are burned?
   a. 1,200
   b. 1,800
   c. 2,100
   d. 2,800

   Answer: C

7. When determining how much of a particular micronutrient is required for the average diet, scientists consider
   a. whether artificial vitamins are available.
   b. whether or not it is stored in the body.
   c. where it is used in the body.
   d. how much is lost through excretion.

   Answer: B

8. Which of these foods is highest in kilocalories?
   a. meat (mostly protein)
   b. butter (mostly fat)
   c. candy (mostly sugar)
   d. potatoes (mostly starch)

   Answer: B

9. The United States Department of Agriculture and the Department of Health and Human Services recommend that Americans
   a. eat a few kinds of good foods.
   b. decrease intake of starch.
   c. increase intake of fiber.
   d. increase protein intake.

   Answer: C

10. The exchange system in dieting allows for an exchange of
    a. one protein food for a carbohydrate food and vice versa.
    b. one fat food for one protein and one carbohydrate.
    c. two carbohydrates for one fat food.
    d. one carbohydrate food for another carbohydrate food.

    Answer: D

11. Secondary nutrient deficiencies include those caused by
    a. shortages of vitamins in the diet.
    b. shortages of minerals in the diet.
    c. inborn metabolic conditions.
    d. insufficient water uptake.

    Answer: C

12. Individuals with micronutrient deficiencies
    a. gradually show symptoms.
    b. can die almost immediately.
    c. are rarely identified.
    d. may need a blood transfusion.

    Answer: B

13. Pica is a micronutrient disorder caused by insufficient _____ in the diet.
    a. iron
    b. sodium
    c. vitamin C
    d. copper

    Answer: D

14. In starvation conditions, protein is converted to
    a. starch.
    b. fat.
    c. sugar.
    d. glycogen.

    Answer: C

15. One of the first tissues to be dismantled during starvation is
    a. skeletal muscle.
    b. brain.
    c. heart.
    d. lungs.

    Answer: A

16. The condition of marasmus is caused by a lack of
    a. protein.
    b. fat.
    c. carbohydrates.
    d. all nutrients.

    Answer: D

17. Kwashiorkor is a form of starvation caused by a lack of
    a. protein.
    b. fat.
    c. carbohydrates.
    d. all nutrients.

    Answer: A

18. Anorexia nervosa is a starvation condition caused by
    a. insufficient vitamins.
    b. insufficient protein.
    c. low calorie intake.
    d. a diet restricted to carbohydrates.

    Answer: C

19. The bulimic individual differs from the anorexic in that
    a. she eats normal amounts of food.
    b. she self-induces vomiting.
    c. she is overweight rather than underweight.
    d. it is not a psychological disorder.

    Answer: A

20. Obesity is defined as being above _____ of "ideal" weight.
    a. 10%
    b. 15%
    c. 20%
    d. 25%

    Answer: C

21. Obesity is particularly hard on the _____ system.
    a. respiratory
    b. circulatory
    c. nervous
    d. skeletal

    Answer: B

22. Lean tissue is low in
    a. carbohydrates.
    b. water.
    c. fat.
    d. protein.

    Answer: C

23. To lose weight, one should
    a. decrease protein intake by half.
    b. eliminate fats completely.
    c. reduce all food intake proportionately.
    d. maintain normal amounts of vitamins and minerals.

    Answer: D

24. Diets should be avoided that
    a. are very high in proteins.
    b. reduce carbohydrate intake by one-third.
    c. reduce fat intake to one-half.
    d. are based on a variety of foods.

    Answer: A

25. To gain weight an individual should
    a. increase calorie intake but eat balanced meals.
    b. drink alcohol, which is high in calories.
    c. eat large amounts of carbohydrates.
    d. eat large amounts of protein.

    Answer: A

Chapter 26

1. It is important for an animal's body temperature to remain within specific limits because temperature
    a. influences the behavior of the organism.
    b. can change reproductive strategies.
    c. influences the specificity of enzymes binding to specific substrate molecules.
    d. influences the rate of chemical reactions that sustain life.

    Answer: D

2. Extreme temperatures can alter the shape of
    a. nucleic acids.
    b. proteins.
    c. carbohydrates.
    d. lipids.

    Answer: B

3. An example of an animal that loses or gains heat from its surroundings by moving into areas where the temperature is suitable is a
    a. bacterium.
    b. lizard.
    c. polar bear.
    d. human.

    Answer: B

4. An example of an animal considered to be an endotherm is a
   a. lizard.
   b. grasshopper.
   c. human.
   d. bacterium.

   Answer: C

5. Ectothermy is an advantageous solution to temperature regulation because the
   a. animal's temperature is independent of the environment.
   b. animal's temperature is dependent upon continuous behavioral responses.
   c. metabolic cost to maintain body temperature is low.
   d. animal's activity is dependent upon environmental conditions.

   Answer: C

6. An example of an organism that uses both endothermy and ectothermy to regulate body temperature is a
   a. bacterium.
   b. bee.
   c. human.
   d. lizard.

   Answer: B

7. An animal that daily enters the sleeplike state of _____ allows its body temperature to fall towards that of the surroundings.
   a. hypnosis
   b. torpor
   c. hibernation
   d. ectothermy

   Answer: B

8. Humans do not use _____ to regulate body temperature.
   a. behavior
   b. metabolism
   c. vasoconstriction
   d. torpor

   Answer: D

9. Piloerection is not an effective strategy in _____ to regulate body temperature.
   a. hummingbirds
   b. humans
   c. polar bears
   d. lizards

   Answer: B

10. A young person is more likely to survive a "drowning" incident if the water is
    a. warm.
    b. cold.
    c. fresh.
    d. salty.

    Answer: B

11. Above 35°C most heat is lost from the human body through the mechanism of
    a. vasodilation.
    b. panting.
    c. slowing metabolic rate.
    d. sweating.

    Answer: D

12. In humans the center for the control of body temperature is located in the
    a. medulla.
    b. thalamus.
    c. hypothalamus.
    d. pituitary gland.

    Answer: C

13. A fever of 99°F may be insignificant in the
    a. mid-morning.
    b. afternoon.
    c. evening.
    d. early morning.

    Answer: B

14. Of the forms of nitrogenous wastes generated by protein metabolism the one that is the least expensive in terms of energy expenditure is
    a. urea.
    b. uric acid.
    c. ammonia.
    d. guano.

    Answer: C

15. Adult amphibians and mammals excrete _____ as their primary nitrogenous waste.
    a. uric acid
    b. ammonia
    c. urea
    d. guano

    Answer: C

16. In humans, uric acid is derived from
    a. proteins.
    b. purine bases.
    c. urea.
    d. ammonia.

    Answer: B

17. If water loss exceeds _____ of the body weight, the person becomes deaf and delirious and can no longer feel pain.
    a. 1%
    b. 5%
    c. 10%
    d. 12%

    Answer: C

18. The major source of water loss is
    a. through the skin.
    b. through excretion.
    c. from the lungs.
    d. in the feces.

    Answer: B

19. Urine is prevented from reentering the kidneys by
    a. muscular contractions of the ureters.
    b. one-way valves in the ureters.
    c. sphincters in the ureters just below the kidneys.
    d. convolutions of the ureter as it travels to the bladder.

    Answer: B

20. The bladder is stimulated to contract when _____ milliliters of urine accumulates.
    a. 100
    b. 300
    c. 600
    d. 900

    Answer: B

21. The kidneys do not change the _____ of blood.
    a. composition
    b. pH
    c. volume
    d. plasma proteins

    Answer: D

22. The functional unit of the kidney is the
    a. peritubular capillaries.
    b. nephron.
    c. loop of Henle.
    d. collecting duct.

    Answer: B

23. Urine formed in the kidney collects in the _____ before entering the ureter.
    a. cortex
    b. medulla
    c. nephron
    d. pelvis

    Answer: D

24. Approximately _____ liters of blood enter the kidneys each day.
    a. 800
    b. 1,600
    c. 2,400
    d. 5,000

    Answer: B

25. The kidney does not utilize _____ in the processing of blood to form urine.
    a. filtration
    b. reabsorption
    c. secretion
    d. exocytosis

    Answer: D

26. Blood leaves the glomerulus via the
    a. afferent arteriole.
    b. efferent arteriole.
    c. renal venule.
    d. renal vein.

    Answer: B

27. Blood is _____ by the glomerulus into Bowman's capsule.
    a. secreted
    b. reabsorbed
    c. filtered
    d. actively transported

    Answer: C

28. The fluid that enters Bowman's capsule does not contain
    a. carbohydrates.
    b. fatty acids.
    c. proteins.
    d. urea.

    Answer: C

29. The volume of fluid entering Bowman's capsule each day is _____ liters.
    a. 90
    b. 180
    c. 360
    d. 1,600

    Answer: B

30. The four functional regions of the renal tubule in order from Bowman's capsule are
    a. proximal convoluted tubule, loop of Henle, distal convoluted tubule, collecting duct.
    b. proximal convoluted tubule, distal convoluted tubule, loop of Henle, collecting duct.
    c. loop of Henle, proximal convoluted tubule, distal convoluted tubule, collecting duct.
    d. distal convoluted tubule, proximal convoluted tubule, loop of Henle, collecting duct.

    Answer: A

31. Selective reabsorption occurs primarily in the
    a. distal convoluted tubule.
    b. proximal convoluted tubule.
    c. loop of Henle.
    d. collecting duct.

    Answer: B

32. Water is conserved and the urine is concentrated by the action of the
    a. distal convoluted tubule.
    b. proximal convoluted tubule.
    c. loop of Henle.
    d. collecting duct.

    Answer: C

33. The primary mechanism in the kidney for concentration of urine is
    a. filtration.
    b. active transport of sodium ions.
    c. secretion of water.
    d. reabsorption of chloride ions.

    Answer: B

34. The countercurrent multiplier system in the kidney moves _____ between the limbs of the loop of Henle.
   a. sodium ions and water
   b. sodium ions
   c. water
   d. chloride ions

   Answer: A

35. An animal that produces very concentrated urine has
   a. long loops of Henle.
   b. short loops of Henle.
   c. vestigial loops of Henle.
   d. loops of Henle that do not enter the medullary portion of the kidney.

   Answer: A

36. The pH of the blood is maintained primarily by the action of the
   a. distal convoluted tubule.
   b. proximal convoluted tubule.
   c. loop of Henle.
   d. collecting duct.

   Answer: A

37. If the collecting duct becomes more permeable to water the urine
   a. volume decreases.
   b. volume increases.
   c. concentration decreases.
   d. volume remains the same.

   Answer: A

38. Diabetes insipidus results from too
   a. little ADH.
   b. much ADH.
   c. little insulin.
   d. much insulin.

   Answer: A

39. Ethyl alcohol decreases
   a. renin production.
   b. aldosterone production.
   c. sodium ion reabsorption.
   d. ADH production.

   Answer: D

40. Aldosterone
   a. increases secretion of sodium ions in the distal convoluted tubule.
   b. enhances reabsorption of sodium ions in the distal convoluted tubule.
   c. increases filtration in the glomerulus.
   d. decreases blood pressure by increasing sodium retention.

   Answer: B

Chapter 27

1. Nonspecific defenses of the body do not include
   a. unpunctured skin.
   b. mucus.
   c. tears.
   d. antibodies.

   Answer: D

2. Inflammation at the site of injury
   a. creates an environment that is hostile to microbes.
   b. concentrates toxins.
   c. decreases swelling.
   d. decreases blood flow.

   Answer: A

3. Reducing a fever with aspirin will
   a. help fight off a cold virus.
   b. increase the body's ability to resist bacterial infection.
   c. hinder the immune system in fighting a cold.
   d. increase phagocytic activity.

   Answer: C

4. A body's defenses recognize itself from bacteria because
   a. of the difference in genetic material.
   b. bacteria have cell walls.
   c. of the unique arrangement of molecules on the cell surface.
   d. of the difference in permeability of the cell membranes.

   Answer: C

5. Molecules that elicit a response from the immune system are called
   a. antibodies.
   b. antigens.
   c. glycoproteins.
   d. lipoproteins.

   Answer: B

6. Lymphocytes are produced in the
   a. spleen.
   b. tonsils.
   c. lymph nodes.
   d. bone marrow.

   Answer: D

7. One of the first cell types to respond when the body is attacked by bacteria is the
   a. B cell.
   b. macrophage.
   c. T cell.
   d. mast cell.

   Answer: B

8. Macrophages stimulate other lymphocytes by exposing them to
   a. an antigen bound to a protein badge.
   b. antibodies.
   c. histamine.
   d. interferon.

   Answer: A

9. The _____ secretes proteins termed antibodies.
   a. B cell
   b. T cell
   c. NK cell
   d. clone cell

   Answer: A

10. B cells that have become sensitive to a foreign substance will develop into
    a. plasma cells and memory cells.
    b. T cells.
    c. NK cells.
    d. macrophages.

    Answer: A

11. A cold or flu lasts about 10 days because
    a. it takes that long for B cells to mature into plasma cells.
    b. it takes that long to activate memory cells.
    c. macrophages require that long to grow and divide.
    d. T cells take that long to produce appropriate antibodies.

    Answer: A

12. Antibodies are large, complex
    a. lipids.
    b. nucleic acids.
    c. proteins.
    d. carbohydrates.

    Answer: C

13. Antibodies bind to specific antigens because of the
    a. three-dimensional shape of the variable regions.
    b. three-dimensional shape of the constant regions.
    c. presence of light and heavy chains.
    d. disulfide linkages between the chains.

    Answer: A

14. The large number of antibodies produced in the body result from
    a. being coded by individual genes.
    b. moving genes on the chromosome during B cell development.
    c. adding new genes to the developing B cells.
    d. adding new chromosomes during development of the B cells.

    Answer: B

15. T cells acquire the ability to recognize particular nonself cells and molecules in the
    a. spleen.
    b. bone marrow.
    c. thymus.
    d. lymph nodes.

    Answer: C

16. The _____ form of T cells stimulates B cells to produce antibodies.
    a. helper
    b. killer
    c. suppressor
    d. NK

    Answer: A

17. Cytolysin is produced by _____ T cells.
    a. helper
    b. killer
    c. suppressor
    d. NK

    Answer: B

18. When cancer cells arise in the body they are destroyed by _____ cells.
    a. helper T
    b. killer T
    c. B
    d. NK

    Answer: D

19. The fetus does not exhibit active immunity because
    a. it is not exposed to foreign cells and biochemicals.
    b. it would attack cells and biochemicals from the mother.
    c. it would attack its own cells.
    d. the thymus does not develop until 24 months following birth.

    Answer: B

20. Infections affect the elderly more severely than young people because
    a. they have more memory cells than T cells.
    b. the immune system declines with age.
    c. their phagocytic cells respond more slowly.
    d. they have lost the capacity for passive immunity.

    Answer: B

21. AIDS *cannot* be transmitted by
    a. blood contact with an infected individual.
    b. a child born of an infected mother.
    c. skin contact.
    d. sexual intercourse when there is a break in the tissue or bleeding.

    Answer: C

22. The AIDS virus is replicated in the
    a. B cell.
    b. plasma cell.
    c. helper T cell.
    d. suppressor T cell.

    Answer: C

23. In severe combined immune deficiency
    a. T cells do not function.
    b. B cells do not function.
    c. neither T nor B cells function.
    d. the thymus gland does not produce T cells.

    Answer: C

24. A virus that incorporates proteins from the host cell surface may cause
    a. severe combined immune deficiency disease.
    b. autoantibodies to be produced.
    c. cancer.
    d. AIDS.

    Answer: B

25. In myasthenia gravis the
    a. myelin coat around neurons is attacked.
    b. the synovial membranes become inflamed.
    c. the neuromuscular junctions are destroyed.
    d. the pancreatic beta cells are destroyed.

    Answer: C

26. An overly sensitive response of the immune system to substances that are not a threat to health results in
    a. production of allergens.
    b. an allergic response.
    c. production of antihistamine.
    d. a decrease in the circulating allergy mediators.

    Answer: B

27. An individual can be desensitized to allergy-producing substances by taking
    a. antihistamines.
    b. allergy mediators.
    c. small doses of the allergen.
    d. an injection of mast cells.

    Answer: C

28. Anaphylactic shock is normally the result of an allergic reaction to
    a. insect stings.
    b. dust mites.
    c. pollen.
    d. hair.

    Answer: A

29. A vaccine causes the production of
    a. T cells.
    b. macrophages.
    c. antibodies.
    d. committed memory cells.

    Answer: D

30. A vaccine does not normally cause illness because the toxic organism is
    a. stripped of its surface molecules.
    b. killed or damaged.
    c. stripped of its genetic material.
    d. in a mutated form.

    Answer: B

31. Supervaccines can be produced by
    a. injecting many different toxic organisms at one time.
    b. using surface proteins from many toxic organisms in a single virus.
    c. using the DNA from many different toxic organisms in a single injection.
    d. fusing many different viruses together.

    Answer: B

32. Measles is still a serious disease in this country because
    a. many individuals are not vaccinated.
    b. the measles organism has mutated into a more powerful form.
    c. the measles organism has changed its surface proteins.
    d. measles affects adults who were vaccinated after 1980.

    Answer: A

33. The success of organ transplants can be predicted from
    a. family histories.
    b. blood types of the individuals.
    c. red blood cell counts of the individuals.
    d. white blood cell counts of the individuals.

    Answer: B

34. Cyclosporin is used in organ transplants because it suppresses
    a. T cells that recognize foreign proteins.
    b. B cells that produce antibodies.
    c. T cells that recognize cancer cells.
    d. both T and B cells.

    Answer: A

35. Monoclonal antibodies are produced by fusing
    a. a plasma cell with a cancerous white blood cell.
    b. a T cell with a plasma cell.
    c. T cells with macrophages.
    d. B cells with NK cells.

    Answer: A

36. Monoclonal antibodies can be used to detect foreign substances in the body by binding to
    a. T cells.
    b. B cells.
    c. the molecule that reacts with a color change.
    d. radioisotopes.

    Answer: C

37. Monoclonal antibodies can be used to treat cancer by binding to the tumor cells and attracting
    a. T cells.
    b. B cells.
    c. macrophages.
    d. memory cells.

    Answer: C

38. Biotherapy is used to treat
    a. AIDS.
    b. cancer.
    c. allergies.
    d. autoimmune diseases.

    Answer: B

39. Many effective cancer treatments in the future will use
    a. immune system cells and biochemicals.
    b. radiation.
    c. surgery.
    d. antiinflammatory agents.

    Answer: A

40. Individual activated B cells fused with cancer cells form
    a. killer cells.
    b. hybridomas.
    c. memory cells.
    d. interleukin-activated cells.

    Answer: B

Chapter 28

1. Which of the following is *not* generated from plants?
   a. medicines
   b. dyes
   c. CO_2
   d. fragrances

   Answer: C

2. People who were hunter-gatherers moved often to
   a. get enough variety in their diets.
   b. get enough food to keep from starving.
   c. keep away from predators.
   d. get to areas less populated with people.

   Answer: B

3. The first spark of civilization occurred around
   a. 100 B.C.
   b. 1000 B.C.
   c. 10,000 B.C.
   d. 100,000 B.C.

   Answer: C

4. The first spark of civilization occurred when
   a. food availability became high in regions cleared by receding ice sheets.
   b. humans learned to use fire to cook their food.
   c. humans developed a language.
   d. humans learned to cultivate food in warm regions.

   Answer: A

5. Food supply was greatly increased about 10,000 years ago due to the
   a. increased wild vegetation from warming climates.
   b. domestication of animals.
   c. increased technology in tool making for hunting.
   d. increased communications between tribes in regards to the whereabouts of edible plants.

   Answer: B

6. Early humans began altering the genetics of some animal and plant species by
   a. killing off all those that were not wanted.
   b. cross-pollination and cross-breeding to get new varieties.
   c. increasing the mutation rate by applying chemicals extracted from other plants.
   d. giving a reproductive advantage to those they wanted.

   Answer: D

7. Though agriculture eventually arose on different continents, evidence indicates it first occurred in
   a. the Middle East.
   b. Eastern Europe.
   c. the Western Mediterranean.
   d. Central Europe.

   Answer: A

8. Most of the food consumed by humans today is in the form of
   a. animal products.
   b. seeds or legumes (beans, lentils, peas, peanuts, and soybeans).
   c. grains from grasses such as wheat.
   d. roots or tubers of plants such as potatoes, beets, and carrots.

   Answer: C

9. The plant products highest in protein content are
   a. grains or grasses.
   b. seeds or legumes.
   c. roots and tubers.
   d. leafy plants.

   Answer: B

10. To get the spectrum of required amino acids in the diet, a vegetarian must eat
    a. leafy vegetables and seeds of legumes.
    b. seeds of legumes and cereals.
    c. oils and cereals.
    d. leafy vegetables and cereals.

    Answer: B

11. Biochemicals that we use as spices were used by the plant as
    a. protection against herbivores.
    b. coenzymes.
    c. protection against dehydration.
    d. attracting pollinates for reproduction.

    Answer: A

12. The three leading cereals that are staples in the human diet are
    a. oats, corn, and rice.
    b. wheat, barley, and corn.
    c. oats, barley, and rice.
    d. wheat, corn, and rice.

    Answer: D

13. The high starch supply provided by cereal grains comes from the seed's
    a. embryo.
    b. endosperm.
    c. aleurone.
    d. pericarp.

    Answer: B

14. Modern varieties of wheat plants vary genetically from each other in their
    a. number of chromosomes.
    b. number of sets of chromosomes.
    c. great variation in the kinds of genes.
    d. expression of traits.

    Answer: B

15. The first tetraploid in the wheat family likely arose from
    a. cross-breeding between two different species.
    b. in-breeding of two closely related individual plants.
    c. mutations caused by chemicals in the plant's environment.
    d. environmental stress such as extreme heat or cold.

    Answer: A

16. Modern-day corn kernels are much larger and tastier than their ancestors because of
    a. nondisjunction.
    b. hybrid vigor.
    c. inbreeding.
    d. mutation.

    Answer: B

17. The International Rice Research Institute (IRRI) was founded in 1961 in the Philippines to
    a. promote the consumption of rice as a food staple in the Western world.
    b. produce hybrids that could be grown under most conditions.
    c. maintain the world's rice varieties.
    d. produce hybrids that were high in yield.

    Answer: C

18. Plant banks provide all of the following benefits to humanity except for serving as a source of
    a. genetic material for fashioning new varieties.
    b. variants in the event of disaster to world crops.
    c. endangered or extinct varieties.
    d. food for starving populations.

    Answer: D

19. To eliminate a repeat of the potato famine in Ireland
    a. many species of potatoes are grown simultaneously.
    b. research in agricultural genetics strives to produce a mold-immune species.
    c. no imports of uninspected foreign potatoes are allowed.
    d. only varieties that show hybrid vigor are grown.

    Answer: A

20. Amaranth has been proposed as a major food source because it has a higher percentage of _____ than the other grains.
    a. fats
    b. carbohydrates
    c. nucleic acids
    d. proteins

    Answer: D

21. Natural products chemists
    a. combine naturally occurring chemicals to form new compounds.
    b. study new natural chemicals extracted from living matter to discover new uses for them.
    c. extract natural chemicals from living matter for use in products rather than artificially synthesizing them in the laboratory.
    d. produce new chemicals from petroleum products.

    Answer: B

22. The ability of malaria to spread is dependent on vegetation since
    a. mosquitos need a wet environment like a rice patty to breed.
    b. certain species of plants serve as habitats for the mosquito.
    c. certain species of plants can harbor the malaria virus.
    d. mosquitos need to feed on specific plants during the adult portion of their life cycle.

    Answer: A

23. Alkaloids extracted from plants are used by humans most often as
    a. herbicides.
    b. insecticides.
    c. painkillers.
    d. spices.

    Answer: C

24. The chemical extracted from plants in 1834 that was effective for malaria victims was
    a. opium.
    b. morphine.
    c. codeine.
    d. quinine.

    Answer: D

25. Deforestation of tropical rain forests in South America may result in
    a. an increase in cultured species that produce chemicals that can be used by humans.
    b. the extinction of plant species whose biochemicals have not been assessed for use by humans.
    c. the destruction of "natural pharmacies" of plant species since natural biochemicals cannot be synthesized in the laboratory.
    d. the destruction of sites where new species of plants can evolve that may produce new biochemicals.

    Answer: B

Chapter 29

1. To keep a fir tree in your yard from growing taller than 6 feet you should prune off the
   a. apical meristem.
   b. lateral meristem.
   c. intercalary meristem.
   d. primary tissue.

   Answer: A

2. Plants grow to their full size by the
   a. first third of their life span.
   b. second third of their life span.
   c. second half of their life span.
   d. end of their life span.

   Answer: C

3. Grasses unlike other plants will continue to grow at the top of the shoot because of the
   a. apical meristem.
   b. lateral meristem.
   c. intercalary meristem.
   d. primary tissue.

   Answer: C

4. Parenchyma cells are capable of all the following functions except
   a. storage.
   b. photosynthesis.
   c. cell division.
   d. elongation.

   Answer: D

5. Sclerenchyma cells can be differentiated from other tissues by their
   a. elongated shape.
   b. fibrous nature.
   c. chloroplasts.
   d. secondary cell walls.

   Answer: D

6. The cells in ground tissue that are green with high amounts of chlorophyll are the
   a. collenchyma.
   b. chlorenchyma.
   c. sclerenchyma.
   d. sclereids.

   Answer: B

7. Cuticle on the surface of the plant is for
   a. photosynthesis.
   b. support of underlying tissues.
   c. protection from water loss.
   d. gas exchange.

   Answer: C

8. The passageway into the leaf for gas exchange is provided by
   a. stomata.
   b. epidermal cells.
   c. cuticle.
   d. sclereids.

   Answer: A

9. During the day when a plant is photosynthesizing, _____ is diffusing into the leaf and _____ is diffusing out.
   a. nitrogen, oxygen
   b. oxygen, carbon dioxide
   c. carbon dioxide, nitrogen
   d. carbon dioxide, oxygen

   Answer: D

10. The thorns on a rosebush are composed of
    a. guard cells.
    b. stomata.
    c. trichomes.
    d. xylem.

    Answer: C

11. Xylem cells that are dead at maturity are capable of
    a. transporting sugar.
    b. photosynthesis.
    c. regulation of water loss.
    d. conduction of water.

    Answer: D

12. Tracheids have thick walls for protection against
    a. predators.
    b. water loss.
    c. collapsing under water.
    d. exploding under pressure.

    Answer: C

13. The solutes that move through sieve pores are
    a. chlorophyll.
    b. salts.
    c. sugars.
    d. water.

    Answer: C

14. The sap moving through phloem cells moves
    a. by diffusion.
    b. by osmosis.
    c. under negative pressure.
    d. under positive pressure.

    Answer: D

15. Axillary buds contain the
    a. apical meristem.
    b. lateral meristem.
    c. intercalary meristem.
    d. cortex.

    Answer: B

16. The cortex is mostly involved in
    a. food conduction.
    b. water conduction.
    c. storage.
    d. protection from invasion by predators.

    Answer: C

17. Monocot and dicot stems differ in their
    a. epidermis.
    b. arrangement of vascular bundles.
    c. kinds of tissue present.
    d. amount of storage.

    Answer: B

18. The growing region in the stem of a dicot is
    a. the dermis tissue.
    b. the cortex.
    c. between the phloem and xylem cells of the vascular ring.
    d. the pith.

    Answer: C

19. The modified stem that is capable of asexual reproduction is the
    a. stolon.
    b. thorn.
    c. tendril.
    d. tuber.

    Answer: A

20. The form and arrangement of leaves on a stem varies from species to species in order to
    a. maximize sun exposure.
    b. reduce water loss.
    c. reduce predation.
    d. increase gas exchange.

    Answer: A

21. The vascular tissue in plants is analogous to the _____ system in animals.
    a. respiratory
    b. circulatory
    c. integument
    d. excretory

    Answer: B

22. The tissue where most of photosynthesis occurs in the leaf is the
    a. spongy mesophyll.
    b. palisade mesophyll.
    c. vascular tissue.
    d. epidermis.

    Answer: B

23. Tendrils are found on
    a. stalky plants.
    b. climbing plants.
    c. cacti.
    d. shrubby plants.

    Answer: B

24. Cotyledons differ from most leaves in that they
    a. are nonphotosynthetic.
    b. store water.
    c. store energy.
    d. support the plant.

    Answer: C

25. Leaf abscission occurs
    a. in all trees.
    b. only in deciduous trees at the end of the growing season.
    c. only in evergreens throughout the year.
    d. only in evergreens at the end of the growing season.

    Answer: A

26. Most roots are unable to
    a. store nutrients.
    b. absorb water.
    c. provide support.
    d. photosynthesize.

    Answer: D

27. Roots absorb oxygen for
    a. photosynthesis.
    b. aerobic respiration.
    c. transpiration.
    d. anaerobic respiration.

    Answer: B

28. A carrot is a(n)
    a. fibrous root.
    b. adventitious root.
    c. taproot.
    d. tendril.

    Answer: C

29. The root tip is protected from the destruction of ploughing through the ground by
    a. a tough epidermis.
    b. a very slow rate of growth.
    c. the making and sloughing of cells.
    d. release of water to soften the soil.

    Answer: C

30. The main function of tiny root hairs is to increase
    a. water absorption.
    b. anchorage in the soil.
    c. oxygen absorption.
    d. mineral absorption.

    Answer: A

31. Root hairs arise from the
    a. zone of cellular division.
    b. zone of cellular elongation.
    c. meristematic tissue.
    d. zone of cellular maturation.

    Answer: D

32. The movement of the root through the ground is due to changes in the
    a. zone of cellular division.
    b. zone of cellular elongation.
    c. meristematic tissue.
    d. zone of cellular maturation.

    Answer: B

33. Branch roots are derived from the
    a. epidermis.
    b. cortex.
    c. pericycle.
    d. vascular tissue.

    Answer: A

34. The macromolecules most often found stored in roots are
    a. proteins.
    b. fats.
    c. nucleic acids.
    d. carbohydrates.

    Answer: D

35. Pneumatophores are different than other root structures in that they
    a. poke through the ground.
    b. absorb high amounts of iron.
    c. absorb oxygen.
    d. absorb calcium.

    Answer: A

36. Plants infected with members of the genus *Rhizobium* benefit from the association since the bacteria provide them with a source of
    a. oxygen.
    b. carbon.
    c. nitrogen.
    d. calcium.

    Answer: C

37. The ability to put on girth through secondary growth probably evolved in response to a plant's need to compete for
    a. water.
    b. minerals.
    c. light.
    d. carbon dioxide.

    Answer: C

38. Tree rings are made up of
    a. secondary xylem.
    b. secondary phloem.
    c. periderm.
    d. cork cambium.

    Answer: A

39. The densest part of the tree stem is the heartwood because
    a. it contains more support fibers.
    b. the tissue has been compressed by pressure.
    c. it is nearest to the vascular cambium.
    d. it has more tracheids.

    Answer: A

40. Most of the cork cells that protect surfaces of roots and stems are
    a. high in structural fiber.
    b. high in tracheids.
    c. high in waxy cells.
    d. loosely packed for gas exchange.

    Answer: C

Chapter 30

1. Bryophytes are small compared to tracheophytes because they
   a. have shorter roots.
   b. have less chlorophyll.
   c. lack a vascular system.
   d. have different leaf structure.

   Answer: C

2. Besides dehydration, when terrestrial plants evolved to come on land they had to deal with the problem of
   a. overheating.
   b. getting enough sunlight.
   c. getting gametes.
   d. getting minerals.

   Answer: C

3. Evidence that an organism has alternation of generations would come from the production of
   a. sperm and eggs.
   b. spores.
   c. spores and gametes.
   d. zygotes.

   Answer: C

4. Which of the following reflects the sequence of events in alternation of generations?
   a. spore-gamete-gametophyte-zygote-spore
   b. sporophyte-gamete-gametophyte-zygote
   c. sporophyte-gamete-gametophyte-spore-gamete
   d. gametophyte-zygote-gamete-sporophyte

   Answer: A

5. Megaspores give rise to
   a. female sporophytes.
   b. male gametophytes.
   c. male sporophytes.
   d. female gametophytes.

   Answer: D

6. Which of the following is diploid in plants?
   a. gametophyte
   b. sporophyte
   c. gamete
   d. spore

   Answer: B

7. The function of sepals is to
   a. protect the inner flora.
   b. attract pollinators.
   c. provide nutrients.
   d. support upper structure.

   Answer: B

8. Androeciums are associated with
   a. carpels.
   b. ovules.
   c. stamens.
   d. style.

   Answer: C

9. The stigma serves to
   a. protect the egg.
   b. attract pollinators.
   c. collect pollen.
   d. produce pollen.

   Answer: C

10. Which of the cells found in the anther is diploid?
    a. sperm
    b. tube nuclei
    c. microsporangia
    d. microspores

    Answer: C

11. Pollen enters the ovary through the
    a. stamen.
    b. style.
    c. stigma.
    d. anthers.

    Answer: C

12. The embryo sac is the
    a. megametophyte.
    b. microspore.
    c. microgametophyte.
    d. megasporangium.

    Answer: A

13. The megaspore mother cell goes through meiosis to produce _____ microspores.
    a. one
    b. two
    c. three
    d. four

    Answer: D

14. Self-pollination
    a. results in the production of clones.
    b. is asexual reproduction.
    c. results in more variability than outcrossing.
    d. occurs in uniform environments.

    Answer: D

15. Outcrossing is crucial in a changing environment because
    a. there is intense competition.
    b. there is a need for adaptation.
    c. possible mates are few in number.
    d. it increases the number of seeds produced.

    Answer: B

16. The animals most important in the role of pollination are
    a. mammals.
    b. birds.
    c. insects.
    d. spiders.

    Answer: C

17. Most pollinators are attracted to plants by their production of
    a. sap.
    b. nectar.
    c. pollen.
    d. seeds.

    Answer: B

18. Floral preference by insects is most often determined by
    a. protein content of pollen.
    b. availability of nectar.
    c. floral color.
    d. ease of obtaining food.

    Answer: C

19. The advantage for a plant of having a specific pollinator is that it
    a. better ensures that the pollinator will visit the plant.
    b. is cheaper than making many attractors for different pollinators.
    c. increases the likelihood of self-pollination.
    d. ensures that the pollen is delivered to the same species.

    Answer: D

20. Wind pollination works best for plants
    a. where pollinators are scarce.
    b. that grow close together.
    c. that are self-pollinators.
    d. that exist in small populations.

    Answer: B

21. Wind-pollinated angiosperms
    a. are very fragrant.
    b. are brightly colored.
    c. have ultraviolet light markings.
    d. are high pollen producers.

    Answer: D

22. The pollen tube provides a pathway from the _____ to the _____.
    a. style, anther.
    b. anther, style.
    c. stigma, ovary.
    d. ovary, egg.

    Answer: C

23. The endosperm forms from
    a. fusion of polar nuclei and sperm.
    b. developing embryonic tissue.
    c. the pollen tube.
    d. the zygote.

    Answer: A

24. Cotyledons are embryonic
    a. roots.
    b. stems.
    c. leaves.
    d. flowers.

    Answer: C

25. The cotyledon in monocots
    a. absorbs the endosperm and becomes thick.
    b. transfers food from the endosperm to the embryo.
    c. is photosynthetic and produces food for the embryo.
    d. cushions the embryo from blows to the seed.

    Answer: B

26. The radicle refers to the embryonic
    a. leaf.
    b. root.
    c. flower.
    d. stem.

    Answer: B

27. The fruit forms from the
    a. ovary.
    b. stigma.
    c. style.
    d. petal.

    Answer: A

28. Seeds remain dormant when
    a. there is not enough light.
    b. minerals are lacking in the soil.
    c. there is a high concentration of carbon dioxide in the air.
    d. water is unavailable in the soil.

    Answer: D

29. The growing of the pollen tube triggers the production of
    a. seeds.
    b. fruit.
    c. ethylene.
    d. pollen.

    Answer: C

30. Besides protecting seeds from desiccation, fruits
    a. provide nourishment for the embryo.
    b. attract pollinators.
    c. distribute the seeds.
    d. are required for germination.

    Answer: A

31. The first step of germination requires the uptake of
    a. carbon dioxide.
    b. oxygen.
    c. water.
    d. light.

    Answer: C

32. The first structure to rise through the soils in dicots is the
    a. epicotyl.
    b. coleoptile.
    c. hypocotyl.
    d. cotyledon.

    Answer: C

33. Conifers differ from angiosperms in that the seeds are
    a. haploid.
    b. winged.
    c. not covered with protective structures.
    d. produced in larger quantities.

    Answer: C

34. The ovule in a pine cone is located
    a. at the base.
    b. on the upper surface of each scale.
    c. on the bottom of each scale.
    d. at the very top.

    Answer: B

35. Fertilization is different in pines than it is in angiosperms in that
    a. no pollen tube is produced.
    b. the pollen is mature when it is released.
    c. two sperm cells are produced.
    d. the process takes far longer.

    Answer: D

36. Angiosperms differ from gymnosperms in that
    a. they have single fertilization.
    b. the haploid tissue of the female gametophyte nourishes the embryo.
    c. they bear fruit.
    d. ovules are unprotected.

    Answer: C

Chapter 31

1. Plants adapt to their environment by
    a. movement.
    b. growth.
    c. releasing chemicals into the soil.
    d. mutation.

    Answer: B

2. Plant hormones differ from animal hormones in that they
    a. are produced in one tissue and affect another.
    b. allow response to the environment.
    c. can elicit a response from many tissues.
    d. are not produced in specialized tissues.

    Answer: D

3. The way in which plant hormone action is expressed is dependent upon
    a. the presence of other hormones.
    b. temperature.
    c. pH.
    d. ions present in tissue.

    Answer: A

4. Auxin causes cell elongation by
    a. increasing the uptake of water.
    b. cell division.
    c. stretching cell walls.
    d. increasing turgor pressure.

    Answer: C

37. Asexual reproduction
    a. maintains a favorable combination of genes.
    b. increases genetic variation.
    c. works best in unfavorable environmental conditions.
    d. is more expensive for the plant.

    Answer: A

38. A clone is not propagated from
    a. leaves.
    b. roots.
    c. stems.
    d. flowers.

    Answer: D

39. Adventitious buds produce
    a. new roots.
    b. aerial shoots.
    c. new branches.
    d. stolons.

    Answer: B

40. Tubers produce nodes that are capable of producing
    a. leaves.
    b. stolons.
    c. shoots.
    d. runners.

    Answer: C

5. The herbicide 2,4-D, a synthetic compound similar to auxin, kills plants by
    a. stunting growth.
    b. reducing reproduction.
    c. blocking photosynthesis.
    d. growing too fast.

    Answer: D

6. Gibberellin differs from auxin in that it stimulates cell
    a. elongation in immature regions of the stem.
    b. elongation after a 1-hour delay.
    c. division in immature regions of the stem.
    d. differentiation.

    Answer: B

7. Cytokinins are found in highest concentrations in the
    a. stems.
    b. leaves.
    c. roots.
    d. flowers.

    Answer: C

8. Cytokinins stimulate
    a. cell elongation.
    b. increased uptake of nutrients.
    c. photosynthesis.
    d. cell division.

    Answer: D

9. Ethylene causes
   a. fruit to ripen.
   b. shoots to grow.
   c. flowers to bud.
   d. abscission.

   Answer: A

10. Abscisic acid inhibits
    a. plant growth.
    b. fruit ripening.
    c. senescence.
    d. abscission.

    Answer: C

11. Senescence begins with
    a. increased levels of ethylene.
    b. increased levels of abscisic acid.
    c. decreased levels of auxin.
    d. decreased levels of gibberellin.

    Answer: C

12. The movement of a root downward is caused by an accumulation of
    a. auxin in the underside of the roots.
    b. auxin on the upper side of the roots.
    c. abscisic acid on the underside of the root.
    d. abscisic acid on the upper side of the root.

    Answer: B

13. Phototropism refers to movement
    a. towards shade.
    b. away from shade.
    c. towards light.
    d. away from light.

    Answer: C

14. Phototropism is caused by
    a. sensitive cells in the stem responding to light.
    b. sensor cells in the stem responding to shade.
    c. auxin migrating to the shady side of the stem.
    d. auxin migrating to the sunny side of the stem.

    Answer: C

15. The photoreceptor molecule for phototropism is
    a. chlorophylla.
    b. flavin.
    c. chlorophyll.
    d. caratenoids.

    Answer: B

16. Geotropism is a plant's response to
    a. sunlight.
    b. lack of sunlight.
    c. gravity.
    d. an object.

    Answer: C

17. Thigmotropism results in
    a. leaves turning towards the light.
    b. roots moving towards gravity.
    c. tendrils wrapping around an object.
    d. tendrils moving toward light.

    Answer: C

18. Thigmotropism results from specialized cells that detect
    a. light.
    b. lack of light.
    c. contact with an object.
    d. movement.

    Answer: C

19. Receptors in plants that respond with seismonastic movement are in animals most like
    a. glands.
    b. cones and rods of the eye.
    c. chemoreceptors on the tongue.
    d. pressure sensors on the skin.

    Answer: D

20. Seismonastic movements of plants allow for
    a. increased exposure to sunlight.
    b. decreased exposure to predators.
    c. decreased exposure to weather.
    d. decreased dehydration.

    Answer: B

21. The Venus's-flytrap distinguishes an insect from a falling leaf by
    a. chemical composition.
    b. weight.
    c. movement.
    d. temperature.

    Answer: C

22. Seismonastic movement in leaves is caused by rapid
    a. cell division.
    b. expansion of epidermal cells.
    c. cell wall elongation.
    d. cell differentiation.

    Answer: B

23. Movement of leaves in nyctinasty is initiated by movement by
    a. ions.
    b. auxin.
    c. Ht ions.
    d. water.

    Answer: A

24. Thigmomorphogenesis can be caused by
    a. reduced light.
    b. reduced water.
    c. high wind and hail.
    d. low nutrients.

    Answer: C

25. Thigmomorphogenesis can be induced in the laboratory by spraying plants with
    a. auxins.
    b. ethylene.
    c. abscisic acid.
    d. gibberellins.

    Answer: B

26. Photoperiodism is a plant's ability to detect the
    a. amount of light in a day.
    b. intensity of light.
    c. infrared waves of light.
    d. far-red waves of light.

    Answer: A

27. Photoperiodism allows a plant to
    a. maximize reproductive effort.
    b. store food for winter.
    c. do vegetative growth.
    d. uptake nutrients.

    Answer: A

28. Day-neutral plants flower
    a. during short hours of light.
    b. in any amount of light.
    c. in reduced intensity of light.
    d. during long periods of darkness.

    Answer: B

29. Long-day plants actually flower when they have
    a. long periods of interrupted light.
    b. long periods of interrupted darkness.
    c. short periods of interrupted light.
    d. short periods of interrupted darkness.

    Answer: D

30. Plants that exhibit obligate photoperiodism probably have a(n)
    a. long flowering time in the summer.
    b. intermediate flowering time.
    c. very limited flowering time.
    d. range of flowering times in a season.

    Answer: C

31. Scientists have determined that plants need _____ in order to flower.
    a. light
    b. darkness
    c. temperature
    d. water

    Answer: B

32. The conversion of phytochrome from its inactive form (Pr) to its active form (Pfr) is by
    a. heat.
    b. freezing.
    c. light.
    d. darkness.

    Answer: C

33. A delay in flowering until reproductive maturity ensures that
    a. pollinators will be available.
    b. enough structural support is available for the fruit.
    c. enough food is available for seed production.
    d. enough moisture is available for fruit production.

    Answer: C

34. Phytochrome in seeds allows them to evaluate if enough _____ is(are) available for germination.
    a. water
    b. heat
    c. light
    d. nutrients

    Answer: C

35. Plants that are etiolated have not received enough
    a. water.
    b. light.
    c. heat.
    d. nutrients.

    Answer: B

36. Phytochrome may be involved in phototropism since the P_r form
    a. increases water uptake.
    b. increases stem elongation.
    c. inhibits stem elongation.
    d. increases auxin migrations.

    Answer: B

37. Senescence in plants is a(n)
    a. cessation of growth.
    b. shutting down of metabolic activities.
    c. wearing out of plant organs and tissues.
    d. energy-requiring process with new metabolic processors.

    Answer: D

38. The colorful hues of autumn leaves are due to
    a. flavin.
    b. carotenoid.
    c. altered chlorophyll.
    d. phytochrome.

    Answer: B

39. Plants prevent tissue freezing by synthesizing
    a. lipids.
    b. sugars.
    c. carotenoids.
    d. photochrome.

    Answer: B

40. Circadian rhythms in plants appear to be controlled by
    a. the amount of light.
    b. temperature.
    c. condition of the soil.
    d. genes.

    Answer: D

Chapter 32

1. Biotechnology started with
   a. immunization.
   b. wine making.
   c. pasteurizing milk.
   d. gene splicing.

   Answer: B

2. Gene manipulation in plants has been limited because
   a. it is expensive.
   b. the technology is difficult.
   c. the fear of introducing unknown genes into the environment.
   d. there is no practical application.

   Answer: C

3. Plant biotechnologists are still trying to
   a. insert a gene in a plant gamete.
   b. isolate and clone desired plant genes.
   c. grow a mature plant from a manipulated cell.
   d. determine which combinations of genes will give desired characteristics.

   Answer: D

4. Biotechnology offers an advantage over traditional plant breeding because it
   a. assures crop quality from season to season.
   b. is a cheaper means of getting new traits into plants.
   c. allows for new combinations of genes.
   d. increases genetic variation.

   Answer: A

5. Plants grown from somatic cells will be
   a. clones.
   b. close genetically but not identical.
   c. mutants.
   d. very different genetically.

   Answer: A

6. Protoplasts are fusions of cells from
   a. the same tissue of an organism.
   b. different tissues of an organism.
   c. two different organisms of the same species.
   d. two different species.

   Answer: D

7. To get two plant cells to fuse the _____ must be removed.
   a. cell membranes
   b. nucleus
   c. cell walls
   d. chloroplasts

   Answer: C

8. A somatic hybrid will have
   a. all the genetic material of both parents.
   b. one-half the genetic material of each parent.
   c. hybrid vigor.
   d. new genetic material found in neither plant.

   Answer: A

9. Monocot protoplasts differ from dicot protoplasts in that they
   a. have greater hybrid vigor.
   b. can only be produced with embryonic tissue.
   c. can propagate easily.
   d. cannot propagate easily.

   Answer: B

10. Somatic embryos differ from hybrid embryos because they have
    a. different genetic material than the parent.
    b. only one parent.
    c. been formed out of embryonic tissue.
    d. resulted from the fusion of two different species.

    Answer: B

11. Biotechnologists can select for mutants from calli by
    a. transplanting those that look different.
    b. adding mutagenic agents.
    c. altering nutrients and hormones.
    d. interbreeding individuals.

    Answer: C

12. The advantage of artificial seeds to farmers is
    a. they do not mildew.
    b. a longer shelf life.
    c. they contain more than one embryo.
    d. they are packages with fertilizers.

    Answer: C

13. Clonal propagation is used most often to increase
    a. genetic variation in a species.
    b. growth rate of slow-growing plants.
    c. seeds for a grower's production.
    d. the number of mutations per generation.

    Answer: B

14. A plant derived from somaclonal variation
    a. will be identical to the parent plant.
    b. will be more genetically different than an offspring produced from sexual production.
    c. is a result of somatic mutation.
    d. is a clone of the parent.

    Answer: C

15. Plant breeders believe somaclonal variation leads to increased variability in the offspring because of
    a. increased expression of mutations.
    b. increased mutation rates.
    c. decreased competition among variants.
    d. increased viability of variants.

    Answer: A

16. The benefit to plant breeders of gametoclonal variation is that they can produce
    a. new varieties through mutations.
    b. homozygous "pure" strains.
    c. clones of the parent plant.
    d. haploid offspring.

    Answer: B

17. Gametoclonal variants differ from somaclonal variants in that they are
    a. haploid.
    b. diploid.
    c. polyploid.
    d. aneuploid.

    Answer: A

18. Mutant selection is accomplished by
    a. placing mutagenic agents on plants.
    b. putting an agent on plants and recovering those that are resistant.
    c. crossing mutants.
    d. selecting mutants and finding out what they are resistant to.

    Answer: B

19. Mitochondria and chloroplasts are good candidates for transfer from cell to cell because they
    a. are easily isolated.
    b. have their own genes.
    c. can be cultivated easily.
    d. can be found in all cells.

    Answer: B

20. A cybrid is one cell with
    a. nuclei from two different cells.
    b. organelles from two different cells.
    c. nuclei from one cell and organelles from another.
    d. chloroplasts from one cell and mitochondria from another.

    Answer: C

21. Mitochondria are introduced into a cell by covering them with
    a. fat.
    b. protein.
    c. nucleic acid.
    d. carbohydrates.

    Answer: A

22. Recombinant organisms have reclined _____ from a donor.
    a. organelles
    b. nuclei
    c. cells
    d. genes

    Answer: D

23. In transgenic technology, foreign DNA is introduced into
    a. gametes.
    b. fertilized eggs.
    c. the embryo.
    d. the adult.

    Answer: B

24. To "stitch" two different species of DNA together, they are first
    a. melted at high temperature.
    b. cut with restriction enzymes.
    c. melted with linker DNA.
    d. mixed in a weak acid.

    Answer: B

25. Foreign DNA is introduced into cells by
    a. wrapping it in fat.
    b. wrapping it in sugar.
    c. passing it through the protein gates.
    d. another nucleic acid.

    Answer: D

26. Genes transferred to plants that offer resistance to disease often come from
    a. other plants.
    b. bacteria.
    c. blue-green algae.
    d. fungi.

    Answer: B

27. Unlike dicots, foreign DNA cannot be introduced into monocot cells through
    a. plasmids.
    b. liposomes.
    c. microscopic needles.
    d. gene guns.

    Answer: A

28. The "bullets" used in gene guns are DNA-coated
    a. protein.
    b. carbohydrates.
    c. minerals.
    d. fats.

    Answer: C

29. Transplant breeders determine the success of transgenic work by
    a. measuring replication of the inserted DNA.
    b. replication of the engineered cell in the tissue culture.
    c. expression of the trait in the mature plant.
    d. expression of the trait in the next generation.

    Answer: D

30. The biotechnology that is being used to produce herbicide-resistant crops is
    a. artificial seeds.
    b. clonal propagation.
    c. gametoclonal variation.
    d. mutant selection.

    Answer: D

31. The best method designed to measure the distance a newly engineered organism will travel is to
    a. make it radioactive.
    b. give it a gene that shows a color change when cultured.
    c. give it a plasmid that makes it stop multiplying in culture.
    d. take repeated cultures at known distances from the site of release.

    Answer: B

32. A "suicide plasmid" causes engineered organisms to die after
    a. three replications.
    b. the substance it is involved with, such as an insecticide, is no longer present.
    c. a nutrient needed for replication is depleted in its environment.
    d. any plant other than the one tested has to die if the engineered organism should infect it.

    Answer: B

33. Pests evolve in response to pesticides in their environment usually
    a. by one mutation at a time.
    b. by many mutations at a time.
    c. over a short period of time.
    d. over many generations.

    Answer: A

34. The biggest fear for the ecologist is that use of chemical-resistant engineered organisms will lead to
    a. decreased varieties of crop plants.
    b. noxious weeds that become chemically resistant.
    c. increased chemical use in the environment.
    d. pest species dying out completely.

    Answer: C

Chapter 33

1. The dodo bird and the Calvaria tree were said to have coevolved because
   a. they shared the same habitat.
   b. they were dependent on each other for survival.
   c. they first appeared in the same era.
   d. their fossils were found in the same strata.

   Answer: B

2. The extinction of many species on Mauritius Island, including the dodo bird, was due to
   a. volcanoes.
   b. hurricanes.
   c. human intervention.
   d. changing earth temperatures.

   Answer: C

3. Species that have limited habitation on new islands are
   a. birds.
   b. insects.
   c. plants.
   d. mammals.

   Answer: D

4. Often unique species are found on islands because
   a. they have unusual habitats.
   b. they are so isolated that inbreeding occurs.
   c. only rare organisms can get to the island.
   d. competition is so keen.

   Answer: A

5. Evolution is best defined by
   a. one species changing into another.
   b. a genetic change in a population.
   c. a genetic change in an individual.
   d. survival of the fittest.

   Answer: B

6. Two populations diverge into two different species when they
   a. look very different from each other.
   b. are geographically isolated.
   c. can no longer successfully interbreed.
   d. their behavior is distinctly different.

   Answer: C

7. Individuals who are adaptive
   a. live a longer life.
   b. reproduce more.
   c. alter their physiology to the environment.
   d. alter their behavior to adjust to the environment.

   Answer: B

8. The mechanism that allows for speciation is
   a. geographical isolation.
   b. behavioral isolation.
   c. gametic isolation.
   d. reproductive isolation.

   Answer: A

9. Microevolution includes
   a. speciation.
   b. extinction.
   c. single mutations.
   d. divergence into two species.

   Answer: C

10. Microevolution
    a. occurs when a new species arises.
    b. is only possible in organisms with short life spans.
    c. refers to a genetic change in the population.
    d. cannot be demonstrated in the laboratory.

    Answer: C

11. First to discuss the evolution of animals was
    a. Darwin.
    b. Aristotle.
    c. Empedocles.
    d. Bonnet.

    Answer: C

12. The "Great Chain of Being" that put all organisms in a single lineage from simple to complex was
    a. Empedocles.
    b. Aristotle.
    c. Bonnet.
    d. Lamarck.

    Answer: B

13. If Lamarck was correct in his prediction, you would expect
    a. that genetically fit individuals would reproduce more often than those who were less fit.
    b. the environment to select for the most fit individuals.
    c. people with genetic defects not to survive.
    d. characteristics like big muscles that resulted from exercise to appear in the next generation.

    Answer: D

14. Charles Bonnet envisioned that
    a. after catastrophes, new and more complex organisms appeared.
    b. only the strongest survived.
    c. acquired traits could be passed on.
    d. animals could adapt to their environment.

    Answer: A

15. The principle of superposition
    a. predicts that fossils are layered according to their age.
    b. states that lower rock layers are older than those above them.
    c. refers to the hierarchical order of organisms in the phylogenetic tree.
    d. states that the earth surface changes through tectonic plate movement.

    Answer: B

16. Both Neptunism and mosaic catastrophism explained changes in the earth's surface by
    a. tectonic plate movement.
    b. volcanoes.
    c. floods.
    d. earthquakes.

    Answer: C

17. Hutton's theory of uniformitarianism predicts that the earth's surface changes
    a. sporadically by catastrophic events.
    b. by tectonic plate movement.
    c. by gradual building up and breaking down of rock.
    d. constantly and over a relatively short period of time.

    Answer: C

18. The father of geology was
    a. William Smith.
    b. Charles Lyell.
    c. James Hutton.
    d. Charles Bonnet.

    Answer: C

19. The person who connected the position of fossils in rocks with their existence in time was
    a. William Smith.
    b. Charles Lyell.
    c. James Hutton.
    d. Charles Bonnet.

    Answer: A

20. The geological theories of Darwin's day
    a. were in conflict with what he had observed.
    b. supported biblical predictions.
    c. gave him the idea that organisms are under pressure to adapt.
    d. suggested that organisms had not changed at all.

    Answer: C

21. Darwin observed that similar-looking organisms are
    a. found in similar habitats.
    b. confined to specific geographical areas.
    c. capable of interbreeding.
    d. found at certain latitudes.

    Answer: A

22. Convergent evolution is when
    a. two species become one.
    b. two populations come together after being separated.
    c. different species evolve similar adaptations to similar environments.
    d. different species have homologous parts.

    Answer: C

23. Darwin observed that island dwellers
    a. were representative of the species found on the nearest mainland.
    b. were limited to those species that could swim or fly there.
    c. included many mammals and reptiles.
    d. were generally smaller than those found on the mainland.

    Answer: B

24. Darwin's study of finches revealed that species gradually change by
    a. pressure from their environment.
    b. chance.
    c. inbreeding of rare forms.
    d. predation.

    Answer: A

25. Darwin called the gradual change from the ancestral type
    a. evolution.
    b. convergence.
    c. descent with modification.
    d. variation.

    Answer: C

26. Malthus's "Essay on the Principle of Population" gave Darwin the idea that
    a. only the fit survive.
    b. variation occurs by mutation.
    c. species gradually change.
    d. more individuals are produced than can survive.

    Answer: D

27. How does the theory of natural selection relate to the theory of evolution?
    a. They are synonymous.
    b. Natural selection is one method by which evolution occurs.
    c. Natural selection relates to populations only.
    d. Evolution refers to speciation only.

    Answer: B

28. Adaptive radiation refers to
    a. a single line of descent from simple to more complex.
    b. divergence of several new types from a single ancestor.
    c. different adaptations of members of a population to a single problem or environmental pressure.
    d. two different species that become one.

    Answer: C

29. As is often the case in scientific research, new ideas occur to more than one individual at a time. Darwin shared the limelight on the theory of natural selection with
    a. Charles Lyell.
    b. Charles Bonnet.
    c. William Smith.
    d. Alfred Wallace.

    Answer: D

30. The survival of the fittest is measured by those individuals who
    a. can adapt to their environment.
    b. were already adapted to their environment.
    c. had the best genes.
    d. had reproductive success.

    Answer: D

31. Sexual selection refers to
    a. traits that boost reproductive success.
    b. male-male competition.
    c. female-male competition.
    d. selection for one sex over another in a population.

    Answer: A

32. Natural selection results in selection of species that
    a. are best adapted to their environment.
    b. are more perfect.
    c. are more complex.
    d. can acquire traits in response to their environment.

    Answer: A

33. Darwin thought that inherited variation was
    a. a result of changes in genes.
    b. due to blending of traits from each parent.
    c. acquired by an individual's somatic cells and then passed on to sex cells.
    d. acquired through want or need by an individual and then passed to the next generation.

    Answer: C

Chapter 34

1. Which of the following is considered a population?
    a. All the organisms found in a particular habitat.
    b. All the organisms found in a small forest.
    c. All the plants on earth.
    d. All the domestic cats found in a small city.

    Answer: D

2. Gene pool refers to all the genes found in
    a. an individual.
    b. a population.
    c. a species.
    d. living things.

    Answer: B

3. Gene frequency in a population cannot be changed by
    a. mutation.
    b. migration.
    c. natural selection.
    d. random mating.

    Answer: D

4. The Hardy-Weinberg equation measures changes in
    a. mutation rates in individuals.
    b. gene frequencies in individuals.
    c. mutation rates in populations.
    d. gene frequencies in a population.

    Answer: D

5. Populations can only reach equilibrium if _____ occurs.
   a. nonrandom mating
   b. natural selection
   c. migration
   d. mutation

   Answer: A

6. If Hardy-Weinberg equilibrium is met, then
   a. the number of homozygous genotypes is equal to the number of heterozygotes.
   b. the number of homozygous dominants is equal to the number of homozygous recessives.
   c. evolution is not occurring.
   d. the majority of genotypes are heterozygous.

   Answer: C

7. If the frequency of the trait for crooked little finger increased in a large population it would likely be due to
   a. nonrandom mating.
   b. natural selection.
   c. mutation.
   d. migration.

   Answer: A

8. Populations isolated on islands are more subject to
   a. mutation.
   b. founder effect.
   c. nonrandom mating.
   d. natural selection.

   Answer: B

9. Genetic drift is a change in gene frequency due to
   a. mutation.
   b. migration.
   c. chance.
   d. adaptation.

   Answer: C

10. When a population bottleneck occurs, the
    a. kinds of genes in the population increases.
    b. frequency of genotypes reaches equilibrium.
    c. new population has a restricted gene pool.
    d. new genes enter the population through migration.

    Answer: C

11. A change in a single base in the DNA of a gene
    a. will cause a change in the protein.
    b. may have no effect on the protein product.
    c. will be passed on to the next generation.
    d. will result in a new gene.

    Answer: B

12. The spontaneous mutation rate for a single human gene is about one in
    a. 1,000.
    b. 10,000.
    c. 100,000.
    d. 1,000,000.

    Answer: C

13. Genetic load refers to the number of
    a. recessive mutations in a population.
    b. deleterious alleles in a population.
    c. dominant mutations in a population.
    d. mutations of any kind in a population.

    Answer: B

14. Most mutations in a population are
    a. recessive.
    b. lethal.
    c. dominant.
    d. homozygous.

    Answer: A

15. Offspring from first-cousin marriages have a high frequency of
    a. homozygous dominant alleles.
    b. mutations.
    c. homozygous recessive alleles.
    d. DNA damage.

    Answer: C

16. In industrial melanism, dark peppered varieties of moths became more prevalent over white due to
    a. genetic drift.
    b. migration.
    c. the founder effect.
    d. natural selection.

    Answer: D

17. It is believed that as human populations migrated north from the equator a direct benefit was conferred on those with lighter skin. They were able to absorb more UV light, which converts cholesterol to vitamin D. Eventually, most northern populations became light-skinned. This is an example of
    a. genetic drift.
    b. bottleneck.
    c. natural selection.
    d. the founder effect.

    Answer: C

18. To be average in a population is often a benefit. This is called
    a. directional selection.
    b. stabilizing selection.
    c. disruptive selection.
    d. balanced polymorphism.

    Answer: B

19. Imagine a population of birds that migrate into a new area with two types of seeds available: a tiny seed and pine nuts. Those birds with beaks that allow them to pick up the tiny seeds eat mostly those. Others have a beak shape that allows them to pick up the pine nuts with ease. Other varieties cannot pick up either very well. Over time two species evolve with very different beak shapes and different diets. This is an example of
    a. directional selection.
    b. stabilizing selection.
    c. disruptive selection.
    d. balance polymorphism.

    Answer: C

20. Cystic fibrosis is a genetic disease found among northern Europeans. It is conceivable that the disease has remained in the population despite its devastating effects due to an advantage conferred to the heterozygote. If so, this would be an example of
    a. directional selection.
    b. stabilizing selection.
    c. disruptive selection.
    d. balance polymorphism.

    Answer: A

21. In areas where sickle cell disease is prevalent, the protozoan that causes malaria is best adapted to human populations that are
    a. homozygous normal.
    b. heterozygous.
    c. homozygous sickle cell.
    d. sickle cell carriers.

    Answer: A

22. What percent of children of carriers of sickle cell disease are likely to survive without medication to reproduce in areas where malaria is a problem?
    a. 0%
    b. 25%
    c. 50%
    d. 100%

    Answer: C

23. Individuals with the sickle cell allele survive malaria because
    a. their red blood cells are uninhabitable.
    b. they have high white blood counts.
    c. their spleens are abnormal.
    d. they run high fevers that prevent reproduction of the parasite.

    Answer: A

24. For speciation to occur, two populations must
    a. look different.
    b. have a significant number of genetic differences.
    c. be unable to breed.
    d. have a significant number of phenotypic differences.

    Answer: C

25. The first event in speciation is
    a. geographical isolation.
    b. behavior isolation.
    c. nonrandom mating.
    d. reproductive isolation.

    Answer: A

26. Populations that are separated geographically are referred to as
    a. patriarchal.
    b. allopatric.
    c. sympatric.
    d. biogeographical.

    Answer: B

27. Darwin attributed lack of transitional forms in the fossil record to
    a. the possibility that they did not exist.
    b. lack of preservation of hard body parts.
    c. poor record keeping.
    d. limited geological diggings.

    Answer: B

28. The theory of punctuated equilibrium proposed that
    a. macroevolution does not occur.
    b. speciation can occur rapidly under some conditions.
    c. mutation rates are higher than first imagined.
    d. selection pressure is much greater than predicted.

    Answer: B

29. Two closely related groups of organisms found in the same geographical area that are unable to reproduce are called
    a. patriarchal.
    b. allopatric.
    c. sympatric.
    d. biogeography.

    Answer: C

30. Ecological isolation occurs when members of two closely related groups fail to mate because they do not share
    a. the same timing of reproductive effort.
    b. the same habitats.
    c. compatible mating behavior.
    d. compatible reproductive anatomy.

    Answer: B

31. Two closely related species of the spring wildflower baby faces have temporal isolation. This means that they
    a. bloom at different times.
    b. live in different places.
    c. have incompatible reproductive organs.
    d. have different means of pollen dispersal.

    Answer: A

32. Organisms from different species that mate but produce infertile offspring probably
    a. are monoploids.
    b. have different chromosome numbers.
    c. have incompatible genes.
    d. put together recessive lethal alleles.

    Answer: B

33. Instantaneous isolation results from
    a. monoploidy.
    b. polyploidy.
    c. fusion of gametes from two different species.
    d. self-pollination.

    Answer: B

34. Allopolyploidy results from
   a. self-pollination of a polyploid.
   b. cross-pollination between two polyploids of the same species.
   c. fusion of gametes from two different species.
   d. self-fertilization by a tetraploid.

   Answer: C

35. Polyploidy is rarely seen in humans because it most often results in
   a. sterility.
   b. lethal combinations of genes.
   c. birth defects.
   d. reduced vigor.

   Answer: A

36. Extinction occurs when a population
   a. has become too inbred.
   b. can no longer compete in its habitat.
   c. cannot adapt to a new environment.
   d. has biological failure.

   Answer: C

37. Evidence for the impact theory comes from all but
   a. cracked quartz crystals.
   b. lack of fossils.
   c. iridium deposits.
   d. depressions on the earth surface.

   Answer: D

38. Plate tectonic theories propose that mass extinction occurred due to
   a. changing atmosphere.
   b. frequent volcanoes.
   c. changing habitats.
   d. flooding.

   Answer: C

39. Early mass extinctions on earth were probably due to the sudden presence of
   a. methane gas.
   b. ammonia.
   c. oxygen.
   d. carbon dioxide.

   Answer: C

40. The greatest cause of mass extinction in history during the Permian period probably was due to
   a. formation of the continent Gondwana over the South Pole.
   b. formation of a single gigantic continent known as Pangeae.
   c. break up of large continents into smaller ones.
   d. changing temperatures.

   Answer: B

41. The first mass extinction attributed to human causes resulted in loss of species of
   a. rare plants.
   b. small herbivores.
   c. large game animals.
   d. other primates.

   Answer: C

Chapter 35

1. Paleontologists use clues from both past and present species to determine
   a. specie habitat.
   b. phylogenies.
   c. analogies.
   d. homologies.

   Answer: B

2. In the study of evolution, evidence replaces _____ in the scientific method.
   a. observations
   b. the hypothesis
   c. the experiment
   d. the interpretation

   Answer: C

3. The rarest of fossils to be found are
   a. shells.
   b. plant structure.
   c. soft-bodied animals.
   d. bones.

   Answer: C

4. Complete fossilization of organisms required environments that were
   a. low in temperature.
   b. rich in nitrogen.
   c. arid.
   d. poor in oxygen.

   Answer: D

5. Whole organisms have been preserved in all of the media but
   a. ice.
   b. tree resin.
   c. tar.
   d. clay.

   Answer: D

6. The problem with relative dating is
   a. the fossil record is incomplete.
   b. it does not take carbon into account.
   c. rock strata forms at different rates.
   d. interference by radioactive atoms in the strata.

   Answer: C

7. A substance that has a half-life of 30 years will decay to 1/8 of its original mass by
   a. 15 years.
   b. 30 years.
   c. 60 years.
   d. 90 years.

   Answer: D

8. The radiometric date of a substance is determined by
   a. measuring the level of radioactivity.
   b. comparing relative amounts of radioactive isotopes of the same substance.
   c. comparing relative amounts of radioactive substances to nonradioactive substances.
   d. measuring the number of isotopes present in the substance.

   Answer: B

9. The problem with the radiometric method of dating is that
   a. rocks without fossils also emit isotopes.
   b. it is dangerous.
   c. it is not reliable.
   d. it is not very accurate.

   Answer: A

10. Fossils that cannot be measured by radiometric dating are those that are
    a. older than 40,000 years.
    b. younger than 300,000 years.
    c. younger than 40,000 years.
    d. between 40,000 and 300,000 years.

    Answer: D

11. Amino acid racemization measures
    a. changes in radioactivity.
    b. mutation rates.
    c. differences in isomers.
    d. differences in amino acid sequences.

    Answer: C

12. The problem with electron-spin resonance and thermoluminescence is that they are effected by
    a. radioactive interference.
    b. temperature.
    c. moisture.
    d. rock strata differences.

    Answer: A

13. Both electron-spin resonance and thermoluminescence measure differences in
    a. chemical composition.
    b. tiny holes in crystals.
    c. radioactivity.
    d. X-ray crystallography.

    Answer: B

14. Which of the following pairs are considered analogous?
    a. wings in bats and birds
    b. wings in bees and birds
    c. tails in monkeys and lizards
    d. beaks of birds and lips of mammals

    Answer: B

15. Which organ might be considered to be vestigial?
    a. toenails in humans
    b. molars in humans
    c. horses' tails
    d. tails on polliwogs

    Answer: A

16. Similarities in mammal embryos such as gill slits and a rudimentary tail probably
    a. are due to chance.
    b. are analogous structures.
    c. represent shared early genes in development.
    d. reflect shared structures in the adult form.

    Answer: C

17. Direct evidence of molecular evolution today comes from preserved _____ from extinct species.
    a. chromosomes
    b. proteins
    c. DNA
    d. fossils

    Answer: C

18. Comparing chromosome banding patterns of different species reveals all but
    a. inversions.
    b. deletions.
    c. additions.
    d. substitutions.

    Answer: D

19. By studying chromosome patterns of primates, the order of relatedness appears to be
    a. humans, chimpanzees, gorillas, orangutans.
    b. humans, gorillas, chimpanzees, orangutans.
    c. humans, chimpanzees, orangutans, gorillas.
    d. humans, orangutans, chimpanzees, gorillas.

    Answer: A

20. Evidence for common ancestry for all life on earth comes from the fact that all species
    a. have common proteins.
    b. use the same genetic code.
    c. have all 20 amino acids.
    d. have like chromosome structures.

    Answer: B

21. All eukaryotes have in common a sequence of 20 amino acids found in
    a. insulin.
    b. amylase.
    c. cytochrome C.
    d. adrenaline.

    Answer: C

22. DNA hybridization method involves
   a. DNA sequencing of related species.
   b. amino acid sequencing of related species.
   c. mixing single-stranded DNA from two different species.
   d. studying DNA from hybrid individuals.

   Answer: C

23. If humans and chimpanzees differ by approximately 2% of their DNA then it is estimated that they diverged as two species
   a. 1 billion years ago.
   b. 2 billion years ago.
   c. 1 million years ago.
   d. 2 million years ago.

   Answer: D

24. The best molecular clock for evolution is the rate of
   a. mutation in nuclear DNA.
   b. mutation in mitochondrial DNA.
   c. change in amino acid sequences.
   d. change in homologous structures.

   Answer: B

25. Evolutionary tree diagrams indicate all but
   a. the relative time when species diverged.
   b. how closely species are related.
   c. what their common ancestors were.
   d. why species diverged.

   Answer: D

26. The smallest division in the geological time scale is the
   a. epoch.
   b. era.
   c. period.
   d. year.

   Answer: A

27. The longest time of life's history was spent during the
   a. Precambrian era.
   b. Cambrian era.
   c. Mesozoic era.
   d. Cenozoic era.

   Answer: A

28. The most advanced organisms found in the Precambrian era were the
   a. cyanobacteria.
   b. bacteria.
   c. jellyfish.
   d. clams.

   Answer: C

29. Ancestors of all animal phyla debuted in the Cambrian era about
   a. 4 billion years ago.
   b. 1 billion years ago.
   c. 600 million years ago.
   d. 100 million years ago.

   Answer: C

30. The first vertebrate to appear during the Ordovician period was the
   a. bony fish.
   b. jawless fish.
   c. amphibian.
   d. reptile.

   Answer: B

31. The first organisms to move onto land during the Silurian period were related to modern-day
   a. plants.
   b. algae.
   c. fungi.
   d. animals.

   Answer: A

32. The Devonian period was called the "Age of
   a. Plants."
   b. Amphibians."
   c. Fishes."
   d. Dinosaurs."

   Answer: C

33. Lobe-finned fish gave rise to
   a. amphibians.
   b. bony fishes.
   c. porpoises.
   d. reptiles.

   Answer: A

34. The Carboniferous period was the "Age of
   a. Plants."
   b. Amphibians."
   c. Fishes."
   d. Dinosaurs."

   Answer: B

35. The amniote egg was an important evolutionary advancement because it
   a. better protected the embryo from predation.
   b. conserved the amount of energy required to produce an egg.
   c. allowed species to move their entire life cycle onto land.
   d. increased the number of eggs that could be produced by an organism.

   Answer: C

36. The Triassic period was the "Age of
   a. Dinosaurs."
   b. Amphibians."
   c. Reptiles."
   d. Mammals."

   Answer: C

37. The reptile that gave rise to mammals was the
   a. thecodonts.
   b. therapsids.
   c. ichthyosaurs.
   d. stegosaur.

   Answer: B

38. During the Cenozoic era the animals that were most prevalent were the
    a. monotremes.
    b. placental mammals.
    c. dinosaurs.
    d. hoofed animals.

    Answer: A

39. The placenta allowed for further advancement of mammals because
    a. the young were protected for a longer period of time.
    b. more young could be produced at one time.
    c. it enhanced migration.
    d. less energy was expended on each offspring.

    Answer: A

40. The marsupials of South America became extinct because
    a. geological changes resulted in altered habitats.
    b. they were hunted by humans.
    c. they could not compete with placental mammals.
    d. they converged into other species.

    Answer: C

41. The order of descent from *Ramapithecus* to modern humans was
    a. *H. habilis, H. erectus, Australopithecines, Neanderthals, Cro-Magnons.*
    b. *Australopithecines, H. habilis, H. erectus, Neanderthals, Cro-Magnons.*
    c. *Neanderthals, Australopithecines, H. habilis, Cro-Magnons.*
    d. *Neanderthals, Australopithecines, Cro-Magnons, H. habilis, H. erectus.*

    Answer: B

Chapter 36

1. Behavior in animals appears to be
   a. mostly inherited.
   b. mostly by learning.
   c. inherited but fine-tuned by the environment.
   d. instinct.

   Answer: C

2. A closed behavior program is best exemplified by
   a. singing in birds.
   b. talking in humans.
   c. escape behavior in ground squirrels.
   d. tool use in chimpanzees.

   Answer: C

3. Which of these organisms has the most open behavior programs?
   a. insects
   b. ground squirrels
   c. sparrows
   d. monkeys

   Answer: D

4. Physical evidence that learning has occurred comes from increased
   a. brain volume.
   b. weight of the brain.
   c. number of neurons.
   d. number of connections between neurons.

   Answer: C

5. Ethology examines behavior in terms of
   a. natural selection.
   b. physiology.
   c. learning.
   d. psychology.

   Answer: A

6. A fixed action pattern (FAP)
   a. can be modified by the environment.
   b. varies from individual to individual.
   c. is genetic.
   d. is not adaptive.

   Answer: C

7. A fixed action pattern releaser
   a. must be a very specific object.
   b. may vary from individual to individual.
   c. may be one aspect of a stimulus such as color.
   d. may not initiate the behavior.

   Answer: C

8. Neoteny refers to
   a. birth of offspring.
   b. juvenile appearance.
   c. behavior related to birthing.
   d. juvenile behavior.

   Answer: B

9. Pheromones are particularly important for identification of
   a. offspring.
   b. territory.
   c. mates.
   d. prey.

   Answer: C

10. A supernormal releaser occurs when an
    a. individual secretes abnormally high amounts of pheromones.
    b. individual responds to minimum amounts of releaser.
    c. exaggerated model causes a stronger response.
    d. individual responds to very small amounts of pheromones.

    Answer: C

11. A complex innate behavior requires
    a. a supernormal releaser.
    b. that many external stimuli be presented simultaneously.
    c. that many internal stimuli occur simultaneously.
    d. stimuli of any kind in a sequence.

    Answer: D

12. Behaviors that are adaptive
    a. are passed on by chance.
    b. will be more costly than beneficial.
    c. are shaped by natural selection.
    d. may not be passed on genetically.

    Answer: C

13. Ethologists determine if a behavior is genetic by
    a. breeding individuals with different behaviors.
    b. isolating genes.
    c. removing genes.
    d. causing mutations.

    Answer: A

14. Learning is defined as a
    a. new behavior that has not been seen before.
    b. change in a behavior as a result of experience.
    c. response to a stimulus.
    d. behavior that is repeated under different stimuli.

    Answer: B

15. If habituation has occurred,
    a. a learned behavior was elicited when the stimulus was presented.
    b. a learned behavior was elicited without the stimulus being presented.
    c. a harmless stimulus is ignored.
    d. all stimuli are ignored.

    Answer: C

16. Which of these behaviors is most likely a result of classical conditioning?
    a. A dog hides when it hears bathwater running and its name called.
    b. A cat arches its back when meeting a strange dog.
    c. A bird goes to sleep when its cage is covered.
    d. A pigeon pushes a lever to get food to drop at its feet.

    Answer: A

17. Extinction in classical conditioning occurs when the
    a. conditioned stimulus elicits the response only.
    b. unconditioned stimulus elicits the response.
    c. conditioned stimulus does not elicit the response.
    d. unconditioned stimulus does not elicit the response.

    Answer: C

18. Operant conditioning requires that
    a. a sequence of stimuli be given in a particular order.
    b. a sequence of behaviors be elicited in a particular order.
    c. the behavior be elicited without a stimulus.
    d. positive or negative reinforcement be given.

    Answer: D

19. Biofeedback is an example of
    a. classical conditioning.
    b. operant conditioning.
    c. habituation.
    d. imprinting.

    Answer: B

20. Imprinting requires that
    a. a stimulus occurs during a critical time period.
    b. an event is remembered by an individual.
    c. a stimulus be presented to elicit the behavior.
    d. behaviors of other members of the species serve as stimuli.

    Answer: A

21. A type of learning that does not require an obvious reward or punishment is
    a. classical conditioning.
    b. operant conditioning.
    c. insight learning.
    d. latent learning.

    Answer: D

22. A child who gets the idea to get a cookie after watching his mother is an example of
    a. imprinting.
    b. latent learning.
    c. insight learning.
    d. trial and error.

    Answer: C

23. Biological clocks require
    a. changes in light.
    b. changes in temperature.
    c. no environmental cues.
    d. changes in relative humidity.

    Answer: C

24. Which of these events is circadian?
    a. eating when hungry
    b. waking up at the same time everyday
    c. mating season of elk
    d. going to bed the same time every night

    Answer: B

25. Seasonal Affective Disorder (SAD) appears to upset the biological clock by
    a. overeating.
    b. altering light sensitivity.
    c. appetite loss.
    d. altering temperature sensitivity.

    Answer: B

26. The _____ is not considered a part of the biological clock.
    a. pituitary gland
    b. retina
    c. pineal gland
    d. suprachiasmatic nuclei (SCN)

    Answer: A

27. A person who stops breathing hundreds of times during the night has the condition
    a. narcolepsy.
    b. insomnia.
    c. dyslexia.
    d. sleep apnea.

    Answer: D

28. The hormone thought to be involved in biological timing is
    a. melatonin.
    b. thyroxin.
    c. adrenaline.
    d. insulin.

    Answer: A

29. Circalunadian refers to a period length of about
    a. 1 month.
    b. 25 hours.
    c. 48 hours.
    d. 1 year.

    Answer: B

30. Which of these animals is not known for migrating?
    a. insects
    b. fish
    c. crustaceans
    d. reptiles

    Answer: D

31. Natural compasses for birds may be all of these except the
    a. stars.
    b. sun.
    c. magnetic pull of the earth.
    d. wind.

    Answer: D

Chapter 37

1. Which of these traits would eliminate a group as being eusocial?
    a. Each generation dies at the end of a season, leaving eggs that hatch the next spring.
    b. Information about available food is shared.
    c. Individuals have a specific job to do.
    d. Juvenile members of the group work as "helpers at the nest."

    Answer: A

2. A social insect's job is determined by all of these factors except
    a. pheromones.
    b. temperature.
    c. nutrition.
    d. competition.

    Answer: D

3. Which of these statements regarding eusocial organisms is false?
    a. Insect societies are at least 200 million years old.
    b. Social behavior is continually evolving.
    c. The best-studied organisms are the nonhuman primates.
    d. Specialization of tasks results in maximizing efficiency of the colony.

    Answer: C

4. The worker in a honeybee hive
    a. fertilizes the eggs.
    b. lays the eggs for the next generation.
    c. takes care of the eggs.
    d. receives royal jelly.

    Answer: C

5. The haploid member of the hive is the
    a. queen.
    b. worker.
    c. drone.
    d. guard.

    Answer: C

6. Tent caterpillars cannot be classified as eusocial because they
    a. share mates.
    b. cooperate.
    c. communicate.
    d. have no specialization of labor.

    Answer: D

7. The defense behavior in which half of the group splits off and goes in one direction while the other half goes in another direction is called
    a. school effect.
    b. dilution effect.
    c. fountain effect.
    d. confusion effect.

    Answer: D

8. Which of these group behaviors applies to the ostrich hen when she allows other hens to deposit their eggs in her nest?
    a. school effect
    b. dilution effect
    c. fountain effect
    d. confusion effect

    Answer: A

9. Species that exhibit mobbing behavior are most likely to be
   a. in a highly competitive environment with others of the same species.
   b. in a high competitive environment with other species.
   c. prey species.
   d. predator species.

   Answer: C

10. The use of sights, sounds, and scents to alert other courting individuals is mainly for
    a. decreasing competition for mates.
    b. competing for mates.
    c. ensuring mating among members of the same species.
    d. synchronizing breeding.

    Answer: D

11. When members of a species breed at the same time, they gain the benefit of
    a. increased genetic variation.
    b. minimizing the number of offspring eaten by predators.
    c. decreased energy output on mating.
    d. increased mate selection.

    Answer: B

12. A behavior that does not increase with socialization is
    a. learning.
    b. foraging efficiency.
    c. competition for mates.
    d. parent investment.

    Answer: D

13. The greatest disadvantage of group living for a population is increased
    a. predation.
    b. competition for mates.
    c. parasitism.
    d. competition for resources.

    Answer: C

14. Aggression among members of a species is usually caused by limited
    a. nesting sites.
    b. mates.
    c. food.
    d. water.

    Answer: C

15. The means of long-distance communication most effective under water is
    a. chemical.
    b. visual.
    c. tactile.
    d. auditory.

    Answer: D

16. The means of short-distance communication most effective under water is
    a. chemical.
    b. visual.
    c. tactile.
    d. auditory.

    Answer: C

17. Elephants communicate with each other mostly through signals that are
    a. chemical.
    b. visual.
    c. tactile.
    d. auditory.

    Answer: D

18. The means of communication among social animals that reinforces social ties is
    a. chemical.
    b. visual.
    c. tactile.
    d. auditory.

    Answer: C

19. The round dance of honeybees communicates
    a. the type of food.
    b. the direction of food.
    c. that food is nearby.
    d. the distance of food.

    Answer: C

20. The waggle dance of honeybees is used to communicate all but
    a. the direction of food.
    b. the distance of the food.
    c. the sweetness of food.
    d. that food is far away.

    Answer: D

21. Inclusive fitness refers to like genes passed on by
    a. all relatives.
    b. siblings.
    c. offspring.
    d. nonrelatives.

    Answer: A

22. Siblings share approximately _____ of their genes.
    a. 25%
    b. 50%
    c. 75%
    d. 100%

    Answer: B

23. The purpose of territoriality may be to
    a. remove unfit members from the population.
    b. reduce aggression for limited resources.
    c. increase foraging efficiency.
    d. increase mating.

    Answer: B

24. Animals usually use all but _____ to defend territories.
    a. urine
    b. threat posture
    c. scent glands
    d. attack

    Answer: D

25. Appeasement behavior includes
    a. looking larger.
    b. exposure.
    c. running away.
    d. attacking.

    Answer: B

26. Dominance hierarchies are set up to
    a. ensure breeding among most fit members.
    b. reduce aggression.
    c. keep unfit members from breeding.
    d. ensure fair distribution.

    Answer: B

27. Courtship rituals are mainly for the purpose of
    a. increasing competition for mates.
    b. decreasing competition for mates.
    c. overcoming aggression long enough to mate.
    d. identifying members of species.

    Answer: C

28. When one male mates with only one female it is referred to as
    a. monogamy.
    b. polygamy.
    c. polygyny.
    d. polyandry.

    Answer: A

29. Members of a species who are sexually dimorphic usually
    a. share mates.
    b. have only one mate.
    c. compete for mates.
    d. are not mammals.

    Answer: C

30. The type of mating behavior practiced by most primates is
    a. monogamy.
    b. polygamy.
    c. polygyny.
    d. polyandry.

    Answer: B

Chapter 38

1. Mice grown under crowded conditions of scarce food and water show
    a. an increased number of offspring per female.
    b. increased interaction between individuals.
    c. premature aging.
    d. decreased mating.

    Answer: D

2. The growth rate of the human population has hastened in recent years due to
    a. increased fertility rates of modern women.
    b. decreased death rates due to advances in medicine.
    c. decreased death rates due to less war.
    d. increased longevity.

    Answer: B

3. Human population growth rate accelerated starting in approximately
    a. A.D. 500.
    b. A.D. 1250.
    c. A.D. 1850.
    d. A.D. 1950.

    Answer: C

4. There are presently over _____ billion people in the world.
    a. 5
    b. 6
    c. 7
    d. 8

    Answer: B

5. On a local scale, a population will most likely decrease by
    a. death by disease.
    b. death by disaster.
    c. death by starvation.
    d. emigration.

    Answer: D

6. Wildlife censuses are conducted by all of these methods except
    a. attempting to count all individuals.
    b. marking a few and comparing the percent captured to a mixed marked and unmarked population.
    c. marking a few and counting how many are recaptured.
    d. counting animal droppings.

    Answer: C

7. The increase in the United States population during 1981 was mostly due to
   a. increased birthrate.
   b. decreased emigration.
   c. increased immigration.
   d. decreased death rate.

   Answer: C

8. Population A is larger than population B. They have the same fertility rate and the same number of reproducing females. What can be said of their rate of growth?
   a. Population B will grow faster than Population A until it catches up.
   b. Population B will grow faster than population A and actually exceed it.
   c. Population A will grow faster than population B.
   d. They will grow at the same rate.

   Answer: C

9. Which of these is not a factor when discussing the exponential growth rate of the global human population?
   a. disease
   b. food
   c. space
   d. migration

   Answer: D

10. A population has a birthrate of 100/2,000 and a death rate of 60/2,000. What is its doubling time?
    a. 20 years
    b. 35 years
    c. 40 years
    d. 70 years

    Answer: B

11. A total fertility rate of 3.2 means
    a. an average number of 3.2 children are born to each woman of reproductive age.
    b. an average number of 3.2 children are born per adult of reproductive age.
    c. a 3.2% increase in the population per year.
    d. an average number of 3.2 children are born per 10 women per year.

    Answer: A

12. Infant mortality refers to the average number of deaths of infants per
    a. family.
    b. women of reproductive age.
    c. 100 births per year.
    d. 1,000 births per year.

    Answer: D

13. The rate of natural increase does not take into account
    a. migration rate.
    b. death rate.
    c. density independent factors.
    d. density dependent factors.

    Answer: A

14. Which of these parameters best indicates the age distribution of the population?
    a. crude birth and death rate
    b. rate of natural increase
    c. population doubling time
    d. total fertility rate

    Answer: A

15. For the human population to remain constant, total fertility rate must fall between
    a. 1.5 and 2.0.
    b. 2.1 and 2.5.
    c. 2.6 and 3.0.
    d. 3.0 and 3.5.

    Answer: B

16. The biotic potential for a human couple is _____ children.
    a. 15
    b. 25
    c. 35
    d. 45

    Answer: D

17. Population A and B have the same number of women of reproductive age, they are the same size, and they have the same fertility rate. Ten years later population A exceeds population B. What can be said about population A?
    a. It has an older population.
    b. It has a higher percent of women who bear children when they are very young.
    c. It has a higher infant mortality rate.
    d. It has a higher emigration rate.

    Answer: B

18. Mexico's predicted doubling time far exceeds that of the United States because
    a. their fertility rate is lower.
    b. their infant mortality rate is higher.
    c. the proportion of the population of reproductive age is higher.
    d. the number of people far exceeds that of the United States.

    Answer: C

19. The population growth rates of the United States and the Soviet Union are expected to slow down in the near future because the
    a. fertility rate is dropping.
    b. proportion of individuals in the population over 65 is increasing.
    c. amount of natural resources is declining.
    d. infant mortality rate is increasing.

    Answer: B

20. The biotic potential of a population is never met because of
    a. behavioral limitations.
    b. environmental resistance.
    c. physiological constraints.
    d. genetic limitations.

    Answer: B

21. Carrying capacity refers to the
    a. biotic and abiotic factors of an environment.
    b. maximum number of individuals an environment can support.
    c. constant rate of a population.
    d. exponential rate of growth of a population.

    Answer: B

22. Population crashes seen after a species colonizes a new area with abundant food are usually due to
    a. high predation rates.
    b. decreased fertility rates.
    c. starvation.
    d. limited space.

    Answer: C

23. The baby boom of the 1950s was mostly due to
    a. World War II.
    b. increased carrying capacity.
    c. low fertility rates in the depression years.
    d. decreased infant mortality rates.

    Answer: A

24. A population that exhibits a logistic growth curve has
    a. a higher birthrate than death rate.
    b. reached carrying capacity.
    c. a higher death rate than birthrate.
    d. no environmental resistance.

    Answer: B

25. Which of these is a density independent factor?
    a. earthquakes
    b. starvation
    c. migration
    d. aggression

    Answer: A

26. Scramble competition in a population usually results in a slow growth rate; this is due to
    a. aggression between individuals that may result in reduced viability for the entire group.
    b. increased emigration rates.
    c. distribution of scarce resources such that the entire population starves.
    d. increased immigration rates.

    Answer: C

27. Contest competition usually does not result in the elimination of
    a. an individual within the population.
    b. a population when only one species is involved.
    c. a population when two or more species are involved.
    d. species when one species is involved.

    Answer: B

28. The principle of competitive exclusion predicts that
    a. only one species will survive in an area of overlap.
    b. two species adapt by eliminating those individuals that cannot compete.
    c. both species will survive but growth rate will be reduced.
    d. each species will occupy a different niche within the habitat.

    Answer: A

29. A habitat includes the
    a. available food source.
    b. temperature.
    c. lighting.
    d. physical site.

    Answer: D

30. A niche includes the following factors except
    a. behavior.
    b. metabolic rate.
    c. physical site.
    d. rate of growth.

    Answer: C

31. A fundamental niche refers to
    a. any place that a species can survive.
    b. the minimal needs of a species.
    c. the actual place where a species survives due to competition.
    d. each of the subdivisions of a habitat.

    Answer: A

32. Predator-prey interactions result in
    a. reduced fertility of prey if too many predators exist.
    b. removal of the weakest from both populations.
    c. increased fertility of predators if the numbers of prey are low.
    d. increased fertility of prey if the numbers of predators are high.

    Answer: B

33. Coyotes either hunt in packs or are solitary hunters. One would predict that they hunt alone when
    a. the amount of prey is abundant.
    b. the only prey available is very small.
    c. the territories are very large.
    d. there is a close genetic relationship among members of the pack.

    Answer: B

34. Predicting human population growth is difficult because
    a. an accurate census is impossible.
    b. human behavior can alter fertility rates unpredictably.
    c. we can permanently alter the carrying capacity.
    d. the effect of density dependent factors is uncertain.

    Answer: D

35. The phrase coined by John Calhoun, "point of no return," refers to
   a. a population crash that results in extinction of a species.
   b. the maximal population size that results in deaths related to overcrowding.
   c. the destruction of habitats that cannot be revitalized.
   d. the time when a population reaches optimal density.

   Answer: B

36. The United Nations predicts that human populations will peak at
   a. 6.5 billion.
   b. 7.5 billion.
   c. 10.5 billion.
   d. 12.5 billion.

   Answer: C

Chapter 39

1. Ecology is the study of relationships between
   a. organisms and their environment.
   b. members of a species.
   c. two different species.
   d. plants and animals.

   Answer: A

2. Ecosystems require the input of
   a. carbon dioxide.
   b. energy.
   c. water.
   d. oxygen.

   Answer: B

3. An ecological community refers to
   a. all the species living in a given area.
   b. a physical site and the organisms living there.
   c. members of a species living in an area.
   d. the interactions among organisms.

   Answer: A

4. Individual ecosystems are mostly categorized into larger ecological units called biomes according to
   a. habitats.
   b. latitudes.
   c. species composition.
   d. temperature.

   Answer: C

5. The biosphere refers to
   a. that part of the planet where there is life.
   b. the earth and all of the organisms that inhabit it.
   c. all living organisms on earth.
   d. the physical earth excluding life forms.

   Answer: A

6. An owl eats a snake, which swallowed a grasshopper that munched on grass. This is an example of a(n)
   a. ecosystem.
   b. food chain.
   c. food web.
   d. trophic level.

   Answer: B

7. Gross primary production refers to the energy
   a. converted from one trophic level to the next.
   b. stored as starch at the end of photosynthesis.
   c. converted from sunlight to chemical energy in photosynthesis.
   d. stored in chemical bonds in all living things.

   Answer: C

8. The net primary production is the energy used for
   a. metabolism.
   b. movement.
   c. growth and reproduction.
   d. maintaining homeostasis.

   Answer: C

9. Most of the energy stored in chemical bonds is found in
   a. producers.
   b. primary consumers.
   c. secondary consumers.
   d. decomposers.

   Answer: A

10. Most of the energy that is passed from one trophic level to the next is in the form of
   a. light.
   b. chemical bonds.
   c. potential energy.
   d. heat.

   Answer: D

11. Of the 1,500 calories that you consumed today, how many could be used for growth and development?
   a. 1.5
   b. 15
   c. 150
   d. 1,500

   Answer: C

12. Chemosynthetic organisms utilize _____ energy.
   a. light
   b. ultraviolet
   c. geothermal
   d. chemical

   Answer: C

13. Lions are considered to be
    a. producers.
    b. primary consumers.
    c. secondary consumers.
    d. decomposers.

    Answer: C

14. Complex food webs are different from food chains in that they
    a. include organisms from each trophic level.
    b. include organisms that function at more than one trophic level.
    c. include organisms that eat different kinds of organisms from the same trophic level.
    d. do not include decomposers.

    Answer: B

15. Stable isotope tracing allows biologists to determine the
    a. trophic level of the organism.
    b. energy utilized at each trophic level.
    c. relative biomass of each trophic level.
    d. energy loss at each trophic level.

    Answer: B

16. In an energy pyramid which trophic level represents the base?
    a. producers
    b. primary consumers
    c. secondary consumers
    d. decomposers

    Answer: A

17. A pyramid of numbers that has a very narrow base most likely represents a population of
    a. grasses.
    b. flowering plants.
    c. shrubs.
    d. trees.

    Answer: D

18. Biomass refers to the
    a. total wet weight of organisms found in an area.
    b. total dry weight of organisms found in an area.
    c. number of organisms in an area times a constant N = 50.
    d. amount of energy in kilocalories found in organisms in a specific area.

    Answer: B

19. In an inverted biomass pyramid, the amount calculated for producers is less than for primary consumers because
    a. there are more consumers than producers in number.
    b. of the fast rates of growth and consumption of producers.
    c. there is limited sunlight available for producers.
    d. there are limited nutrients available for producers.

    Answer: B

20. Biomagnification refers to the increase
    a. of certain chemicals as they are passed up a food chain.
    b. of energy as it is passed down food pyramids.
    c. in biomass from the apex to the base of the food pyramid.
    d. in carbon as it is passed up a food chain.

    Answer: A

21. DDT levels increase with each trophic level because it is
    a. similar to a natural chemical in organisms.
    b. fat soluble.
    c. nondigestible.
    d. unstable.

    Answer: B

22. Infants of pregnant women who ate mercury-contaminated fish were born
    a. dead.
    b. retarded.
    c. blind.
    d. deaf.

    Answer: B

23. Mercury poisoning from smoking marijuana may result in
    a. death.
    b. paranoia.
    c. blindness.
    d. deafness.

    Answer: B

24. Decomposition of organisms by microbes results in the recycling of
    a. heat.
    b. chemical energy.
    c. nutrients.
    d. ATP.

    Answer: B

25. Carbon enters the carbon cycle in photosynthesis in the form of
    a. glucose.
    b. CO_2.
    c. limestone.
    d. methane gas.

    Answer: B

26. Nitrogen is required for the production of
    a. glucose.
    b. fats.
    c. carbohydrates.
    d. proteins.

    Answer: D

27. Nitrifying bacteria extract nitrogen from
    a. the air.
    b. ammonia.
    c. nitrites.
    d. nitrates.

    Answer: B

28. Phosphorus is an important component of
    a. carbohydrates.
    b. fats.
    c. nucleic acids.
    d. proteins.

    Answer: C

29. The earth's supply of phosphorous comes from
    a. air.
    b. water.
    c. rocks.
    d. soil.

    Answer: C

30. Ecological succession refers to the
    a. pathway of energy through a community.
    b. gradual change in a community.
    c. depletion of nutrients in a community.
    d. recycling of nutrients in a community.

    Answer: B

31. Climax communities
    a. are made up of pioneer species.
    b. remain fairly constant.
    c. are found in disturbed areas.
    d. are more diverse than other communities.

    Answer: B

32. Pioneer species
    a. are highly competitive with other species.
    b. reproduce at a faster rate than other species.
    c. break rock into soil.
    d. can tolerate any kind of environment.

    Answer: C

33. Secondary succession occurs
    a. when pioneer species die off.
    b. after present species are destroyed by natural disaster.
    c. slower than primary succession.
    d. after the climax species is established.

    Answer: B

34. Terrestrial biomes are determined by
    a. temperature.
    b. altitude and latitude.
    c. soil type.
    d. humidity.

    Answer: B

35. Tropical rain forests receive annual rainfall of
    a. 50-100 centimeters.
    b. 100-200 centimeters.
    c. 200-400 centimeters.
    d. 400-600 centimeters.

    Answer: C

36. The biome highest in diversity is the
    a. rain forest.
    b. temperate coniferous forest.
    c. temperate deciduous forest.
    d. tundra.

    Answer: A

37. The taiga refers to the cold, snowy mountains, which provide an environment for mostly
    a. deciduous trees.
    b. tundra.
    c. grasses.
    d. woody shrubs.

    Answer: D

38. Freshwater biomes are distinguished by
    a. temperature.
    b. turbidity.
    c. salinity.
    d. water flow.

    Answer: C

39. Most of the phytoplankton and zooplankton are found in the
    a. littoral zone.
    b. limnetic zone.
    c. profundal zone.
    d. thermocline.

    Answer: B

40. Light never reaches the _____ zone.
    a. neritic
    b. benthic
    c. abyssal
    d. pelagic

    Answer: C

Chapter 40

1. Most of the people affected by the nuclear disaster at Chernobyl
   a. were killed.
   b. got radiation sickness.
   c. got cancer.
   d. got thyroid disease.

   Answer: B

2. The most obvious effect of radiation on farm animals in the vicinity of Chernobyl was
   a. germ mutation.
   b. somatic mutation.
   c. poisoning.
   d. death.

   Answer: A

3. Landfills may be hazardous to the environment because of
   a. infectious microorganisms.
   b. depletion of nutrients.
   c. toxic chemicals.
   d. fires.

   Answer: C

4. Lichens, which provide the base of food webs, are dying in the Alaska's tundra because of the _____ pollutant.
   a. fluorocarbons
   b. sulfur
   c. carbon monoxide
   d. nitrous oxide

   Answer: B

5. Depletion of the ozone layers over the poles is due to high levels of
   a. fluorocarbons.
   b. sulfur.
   c. carbon monoxide.
   d. nitrogen oxide.

   Answer: A

6. The greatest contributor to acid rain is
   a. heavy metal smelters.
   b. gasoline engines.
   c. diesel fuel engines.
   d. coal-burning factories.

   Answer: D

7. The most devastating effect of acid rain on lakes is the
   a. extinction of all life.
   b. increased rates of eutrophication.
   c. decrease in diversity.
   d. extinction of decomposers.

   Answer: C

8. A natural deterrent to the effects of acid rain on lakes is
   a. calcium carbonate in the earth.
   b. heavy metals in the earth.
   c. aquatic organisms that can alter the pH.
   d. forests.

   Answer: A

9. The greenhouse effect is mostly due to high levels of
   a. sulfur oxide.
   b. carbon monoxide.
   c. carbon dioxide.
   d. nitrogen oxide.

   Answer: C

10. Global warming may affect animals because
    a. of increased radiation.
    b. of changes in feeding habits.
    c. there will be fewer producers.
    d. there will be more microorganisms.

    Answer: B

11. Evidence of global warming has been questioned, since
    a. it is difficult to measure slight differences in worldwide temperatures.
    b. there is too much variance in local conditions.
    c. there is natural variation in temperature that tends to be cyclic.
    d. only a few locales have been measured for a long period of time.

    Answer: C

12. A natural condition that may offset the greenhouse effect so global warming does not occur is the absorption of heat by
    a. plants.
    b. animals.
    c. rocks.
    d. oceans.

    Answer: D

13. Which of these is not caused by exposure to ultraviolet light?
    a. skin cancer
    b. death of cells
    c. birth defects
    d. sunstroke

    Answer: D

14. Products that greatly contribute to the depletion of the ozone are made out of
    a. plastic.
    b. petroleum.
    c. foam.
    d. polyurethane.

    Answer: C

15. The destruction of rain forests is due to
    a. the harvesting of wood for lumber.
    b. burning of wood for fuel.
    c. the harvesting of wood for paper.
    d. farming practices.

    Answer: D

16. The loss of rain forests has resulted in all of the following except
    a. loss of nutrients in the soil.
    b. loss of diversity of organisms.
    c. increased air pollution.
    d. increased gross primary productivity.

    Answer: D

17. The spreading of the Sahel Desert has been accelerated by
    a. changes in the ozone layer.
    b. air pollution, which has altered weather patterns.
    c. farming practices.
    d. acid rain.

    Answer: C

18. The biome most affected by farming and industry in the United States is the
    a. rain forest.
    b. temperate coniferous forest.
    c. temperate deciduous forest.
    d. tundra.

    Answer: C

19. A sign that a lake is beginning eutrophication is the
    a. presence of algal blooms.
    b. absence of aquatic plant life.
    c. absence of aquatic animal life.
    d. changes in plant life around the lake.

    Answer: A

20. Eutrophication can be expected to increase with all but
    a. air pollution.
    b. pollution from sewer systems and fertilizer run-off.
    c. mining.
    d. the greenhouse effect.

    Answer: A

21. Natural protection against eutrophication caused by humans comes from
    a. natural sodium bicarbonate in the soil.
    b. wetlands ringing a lake that block sediments.
    c. increased herbivore populations.
    d. low pH of the water.

    Answer: B

22. Ecotones refer to
    a. various strata within an ecosystem.
    b. habitats that serve as bridges between ecosystems.
    c. a collection of animals within an ecosystem.
    d. two interdependent species within an ecosystem.

    Answer: B

23. Oxygen depletion in an aquatic environment is caused by excess
    a. phytoplankton.
    b. microorganisms.
    c. zooplankton.
    d. fish.

    Answer: B

24. The organisms first affected by oxygen depletion in estuaries are
    a. phytoplankton.
    b. microorganisms.
    c. zooplankton.
    d. fish.

    Answer: A

25. All of the following offer natural protection against chemical pollution of oceans except
    a. the dilution by the large volume of water.
    b. the alteration by salt, which results in a heavy sediment that sinks.
    c. microorganisms that can break down certain chemicals.
    d. natural buffers.

    Answer: D

26. Paralytic shellfish poisoning is caused by a species of
    a. algae.
    b. phytoplankton.
    c. zooplankton.
    d. microorganisms.

    Answer: A

27. One of the most devastating human products on marine life is
    a. the rubber tire.
    b. plastic.
    c. foam rubber.
    d. Styrofoam.

    Answer: B

28. The mechanism most likely to lead to the greatest consequences to humans from a nuclear war is
    a. the blast wave.
    b. nuclear fallout.
    c. mass starvation.
    d. the thermal pulse.

    Answer: C

29. Organisms prevail through local and sometimes drastic changes in the environment by
    a. adaptation.
    b. isolation.
    c. genetic selection.
    d. rapid reproduction.

    Answer: C

30. What is the most critical factor for the maintenance of life as we know it?
    a.   Lower the reproductive rate of humans.
    b.   Increase food production for human consumption.
    c.   Prevent the collapse of intricate worldwide food webs.
    d.   Increase the number of nonhuman species.

Answer: C

Lewis Life - Bio Sci II Videodisc Directory

This section is a list of topics from the text with related photos, drawings, charts, diagrams and short movies from the Bio Sci II videodisc by Videodiscovery. These selections can be used for lecture support, individual or group study.

The number in the left-hand column is the figure from the text to which the following videodisc events are associated. The common and scientific names are listed followed by the five-digit frame number. (This is the starting frame number for movies or multiple-frame sequences.)

If you use the standard player remote control, you can enter this number and search for the corresponding frame. Press the Play button after the frame is found if the event is identified as a motion sequence or use the Step Forward / Step Reverse buttons for still-frame sequences.

The barcode is for use with LaserBarcode readers and will command a frame search when scanned. The bottom of each right-hand page contains codes to step reverse, play and step forward for use with motion and multi-frame sequences.

The LaserBarcodes can be reproduced on most copiers and cut and pasted for your customized presentation.

Refer to your videodisc player and LaserBarcode reader user's manuals for more information.

1.1

Corn
Zea mays

20318

Growing from city sewer, showing the myriad of places that plants can grow.

1.1b

Forest fire

18278

Burned approximately 25 years ago.

18969

Crown fire in a forest.

18970

Ground fire.

1.3

Scientific method

7 frame sequence

12981

A-G show steps in the development of scientific principles.

2.1

Diatom, Asterionella
Asterionella sp.

20576

A colonial, planktonic diatom.

Diatom, Dinobryon
Dinobryon sp.

13975

Each cell secretes a lorica in a typical colony like this one.

Diatom; attached to red alga
Licmophora sp.

20583

A group of cells attached to a red alga by a mucilaginous pad.

Diatom; filamentous
Tabellaria sp.

20565

(dark field illumination).

Diatom; freshwater, sigmoid-shaped
Pleurosigma sp.

20550
A sigmoid-shaped fresh water diatom.

Diatom; freshwater
Cymbella sp.

20572

Diatom; freshwater planktonic

Osterionella sp.

20623

Diatom; marine

Chaetoceros sp.

20584

A chair forming planktonic marine diatom.

Licmophora sp.

20563

A marine diatom growing on the surface of a filamentous brown alga.

Diatom; pennate

20648

Diatom

13963

Individuals adhere to debris by their sticky posterior ends, producing clusters of individuals.

20622

Predominantly diatoms.

Coscinodiscus sp.

20575

Large diatom showing plastids filled with carotenoids and xanthrophylls, and frustule pattern.

Cymbella sp.

20574

Showing frustules and mucilaginous stalks.

Diatom, round

Arachnodiscus

20646

A representative frustule of the marine, centric diatoms.

2.2

Earth

19251

From space the earth looks very complex, having oceans, continents and clouds.

Earth, from outer space

19141

Continent of Africa visible.

19142

North coast of South America visible.

19143

2.3

Root, legume; xs

20237

Nodules of nitrogen-fixing bacteria in legume root.

Reading 2.1

Lichen

17346

17347

17459

17460

17682

17805

17893

Lichen growing on a branch of a dead tree.

17908

Lichens and moss on rocks.

18066

18237

18420

18456

18646

Lichens among the hardiest of organisms, quickly colonize bare rock surfaces in the Arctic. Three species on this substrate.

18647

Competition in lichens straightforward, as one species overgrows another.

Cladonia sp.

17259

Onia sp.

17458

Fruiting bodies on the lichen cladina sp.

Lichen, British soldier

Cladonia sp.

17258

Lichen, SEM

20246

SEM, algal and fungal components visible.

Lichen

17254

Lichen on building.

17255

17256

17257

2.10a

Amoeba, arcella

14039

A side view of an indented test. The cell is being touched by pseudopod of Amoeba proteus.

2.10b

Paramecium moving

Paramecium caudatum

9.0 second play sequence.

31663- 31933

Rowing along with thousands of cilia, the paramecium is able to produce coordinated movement.

2.10c

Euglena moving

Euglena gracilis

7.0 second play sequence.

31934- 32143

The flagella can be seen thrashing around these euglena which feel trapped under a cover glass on a microscope slide. They are also exhibiting classic 'euglenoid' movement by changing the shapes of their bodies.

2.11a

Alga, green; fresh water

Chlorococcum sp.

20542

Chlorophyll-producing green algae.

2.11b

Kelp, bladder (brown algae)

Nereocystis leutkeana

20250

Showing holdfast, bulbous float & blades. Western U.S.

Alga, red

17074

Red algae growing on snow.

2.11c

Dinoflagellate, dinophysis

Dinophysis sp.

20585

An armored marine dinoflagellate.

2.13, 2.14

Slime mold

20223

It moves slowly along the forest floor using dead organic matter as its food. In this way, slime molds are like fungi, which also feed on dead organic matter.

Mushroom, Cathathelasma

Cathathelasma ventricosa

16994

2.15

Fungi, puffball

Scleroderma citrinum

17021

2.16

Sponge, elephant ear

14601

An elephant ear sponge. From the Philippines.

2.17a

Sponge anatomy; diagram

30349

Generalized sponge anatomy.

30350

Showing water flow through sponge.

30351
Collar cell anatomy.

2.17b

Hydra anatomy, diagram

30354
Polyp form of cnidarian showing stinging cells.

2.17c

Octopus, giant Pacific

Octopus dofleini

14567
The giant Pacific octopus. Here showing a mottled color pattern. Compare with next two images.

2.17f

Worm, annelid

Sabellastarte magnifica

14504
Diameter of tentacle crown was 5 cm. Sabellidae.

Earthworm; crawling

Lumbricus terrestrius

6.6 second play sequence.

36188- 36387
The thrusting and pulling movements seen in the earthworm are very effective for undergound travel where they have to make their own tunnel through a solid substrate.

2.17g

Grasshopper

Tropidacris cristata

18716
One of the world's largest grasshoppers, with a wingspread of nearly 25 cm; adults and larvae feed on plants rich in secondary chemicals and are aposematically colored.

2.17i

Sea urchin

14585
Tropical sea urchin with iridescent blue test. From the Philippines.

2.17j

Hippopotamus

Hippopotamus amphibius

16788

16856
16906
Aquatic during day, leaves water at night to graze on terrestrial vegetation. Widespread in Africa.

3.1

Booby, Blue-footed

Sula nebouxii

16358
Pair in courtship display; male on left recognizable by smaller pupil.

3.4

Gorilla kid/baby taunt

6.2 second play sequence.

49769- 49956
Look familiar? Little kid gorilla is in a mischievous mood and moves to pick on baby gorilla, precipitating the parental reaction.

3.4a

Butterfly, viceroy

Limenitis archippus

14867
Mating.

3.4b

Butterfly, monarch, adult

15065
Newly emerged adult. After wings have dried (7).

3.7

Electron shells

12917
Diagram showing electron shells in hydrogen, carbon and magnesium.

3.10

Ionic bonding

3 frame sequence

12918
Diagrams showing ionic bonds in A) NaCl B) MgCl and C) CaCl$_2$.

3.11

pH scale

12925

Diagram shows relationship between pH and the log of hydrogen ion concentration.

3.12

Covalent bonding

3 frame sequence

12913

Diagram showing covalent bonding in A. chlorine, B. nitrogen, C. water.

3.14

Water molecule

12921

Hydrogen bonds are weak attractions between hydrogen and other atoms.

3.15d

Cellulose, microfibrils

20653

From the primary cell walls of the endosperm of persimmon, Diospyros sp. Viewed through crossed polarizers bright pattern indicates cellulose microfibrils are highly oriented.

3.16

Glycogen

12938

Diagram

Starch

12979

Diagram

3.17, 3.18

Fatty acids

2 frame sequence

12940

Diagram

3.19

Amino acid; general

2 frame sequence

12926

Diagram

3.20

Protein structure

12935

Diagram showing primary, secondary and tertiary structure of protein.

Amino acid; protein structure

6 frame sequence

12905

3.23

Enzymatic reaction

12994

Enzymes encourage reactions by bringing reactants together at their active sites.

3.24

DNA structure

12972

Diagram showing structure overview from helix to complementary strands.

3.26

Origin of life

4 frame sequence

30312

Gases existing in primitive atmosphere, some washed into the oceans by rainfall.

3.36

Fern, resurrection; dry
Polypodium polypodioides

20331

This plant can withstand tremendous drying-out.

20332

Under wet conditions the dry plant rehydrates and presents its normal looking fern anatomy.

4.1b

Cell culture, nerve

21779

SEM X2500.

4.1c

Elodea; chloroplasts
Elodea

20155

Shows numerous chloroplasts inside an Elodea leaf cell. x1000

4.2

Virus, influenza

13916

An RNA virus that has a membraneous envelope outside of its protein capsid.

Virus, TMV

13918

Tobacco mosaic virus is an RNA virus that causes mosaic disease in tobacco and other plants.

4.7

Prokaryotic, photosynthetic; diagram

22617

4.9

X-ray crystallography; diagram

30167

Diagram showing how X-ray crystallography is carried out. See frames 30070 and 30071 for actual photos of DNA x-ray diffraction patterns.

4.10

Cell, organelles; animation

60.0 second play sequence.

11094- 12893

This 3D computer animation flies through the cell membrane, past organelles into the nucleus, from there out through a nuclear pore, through the rough ER, smooth ER, golgi, mitochondrion, lysosomes and out of the cell. Narration.

Table 4.4

Mitochondrion

21854

x121,500.

Golgi complex

22325

Endoplasmic reticulum, rough

22324

Endoplasmic reticulum, smooth

22613

Diagram showing ER which lacks ribosomes.

Table 4.5

Epithelium, columnar

21577

Med. mag. (Trichrome stain) Some columnar epithelium in this section of intestine has been cut at a plane suitable to show the shape of the cells and relation to underlying C.T. The tubular gland on the middle left offers the best view of the epithelium.

4.14

Skin; cross section, model

22443

Skin cross section model shows microscopic structure of skin and its three layers: epidermis, dermis, subcutaneous layer with fatty (adipose) tissue.

4.15a

Connective tissue, rectal

21576

Med. mag. (H+E stain) Section of rectal-anal junction of monkey showing dense irregular connective tissue and rectal mucous glands. Note the blue fibers are densely packed and randomly distributed.

4.15c

Blood cells

21801

Leukcocyte, surrounded by erythrocytes, platelets, TEM, x17,000.

Cartilage, elastic

21594

Med. mag. (trichrome stain) The chondrocytes within this section of young elastic cartilage are rounded and have little matrix surrounding them. Each chondrocyte is located in a cavity (lacuna) in the matrix.

4.15d

Bone, dense; Haversian systems

21605

Lo mag. (india ink stain) A transverse section of long bone showing cylindrical Haversian systems cut in cross section. The osteocytes and their matrix have been deposited in layers (lamellae) that appear like growth rings on a tree.

5.2

Lymphocytes, T and B cells; diagram

4 frame sequence

22478

C-D show process of T cell activating B cell and then B cell giving rise to clones of plasma cells and memory cells.

5.3

Blood typing

21807

Type A

5.8

Osmosis; diagram

22643

5.10

Paramecium; contractile vacuole

Paramecium

22322

Vacuole is enlarged with fluid.

22323

Vacuole has discharged fluid from contractile vacuole.

5.12

Diffusion, passive; diagram

22620

Molecules pass freely through the membrane.

Active transport; diagram

4 frame sequence

22626

Proteins actively move molecules through membranes with the use of energy.

Facilitated transport; diagram

5 frame sequence

22621

Molecules are helped through the membrane with certain proteins.

5.13

Endocytosis; diagram

4 frame sequence

22635

A particle is engulfed by a cell.

5.14

Exocytosis; diagram

4 frame sequence

22639

Diagram showing a particle being secreted by a cell.

5.15, 5.16, 5.17

Microtubules & Microfilaments; diagram

22614

ATP

6.1

ATP

4 frame sequence

12988

A. Complete diagram, B-D progressive reveal of breaking of phosphate bond.

6.3

Feedback of hormones

4 frame sequence

22564

A. Diagram, B. TSH feedback, C. thyroxin feedback to pituitary, D. thyroxin feedback to hypothalamus.

6.7

Chlorophyll & UV light

20309

Chlorophyll, the green plant photosynthetic pigment, absorbs ultraviolet (UV) light strongly, providing the driving mechanism for photosynthesis. Here, green chlorophyll fluoresces red in the visible spectrum under UV light, showing UV light absorption.

6.8

Chloroplast anatomy

22612

Diagram of the fine structures, stroma and thylakoid membranes within a chloroplast.

Chloroplast, EM

20154

Fine structure of the chloroplast showing stacked grana and stroma. x10,000

Chloroplasts, codium; in mollusk

20610

Codium chloroplasts are endosymbiotic in gut of sacoglossan mollusc Elysia viridis.

Chloroplasts, EM

20248

Shows the stroma of the chloroplast, with the stacks of grana composed of flattened membranous sacs (thylakoids). EM x90,000

6.9

Photosynthesis

4 frame sequence

13000

A. Shows combined diagram, B. shows non-cyclic photophosphorylation in the thylakoid membrane. C. shows Calvin Benson cycle within the stroma, D. summary repeated.

6.12

Glycolysis

1.4 second play sequence.

12981- 13022
Diagram

9 frame sequence

13010

A. Shows overview, B-I shows steps of glycolysis in progressive reveal.

1.4 second play sequence.

13028- 13069

A series of progressive reveal graphics showing entire pathway of glycolysis.

6.13

Fermentation

13007

Diagram showing chemical pathways in which sugar is broken down in the absence of oxygen.

6.15

Mitochondrion

22615

Showing cristae, matrix and membranes of a mitochondrion.

6.17

Krebs cycle

7 frame sequence

13021

A. Shows overview of Krebs cycle, B-G show progressive reveal of step within cycle.

6.20

Cellular respiration

2 frame sequence

13019

A. Block diagram showing pyruvate as a metabolic intermediate, B. showing continuation of process inside mitochondrion.

6.22

Photosynthesis/respiration

13006

Diagram comparing the reactions and structures of photosynthesis and aerobic cellular respiration.

7.2

Cell cycle

30097

Diagram showing relative lengths of phases of mitosis and phases of interphase.

7.4

Cytokinesis; animal cell

30000

Animal cell, pinching inward of cell membrane divide the cytoplasm forming two cells.

7.4a

Mitosis, animal; early interphase

30084

Diagram showing animal cells in early interphase.

Mitosis, animal; late interphase

30079

Diagram of animal cell in late interphase.

7.4b

Mitosis, animal; prophase

30080

Diagram of animal cell in prophase.

7.4c

Mitosis, animal; metaphase

30081

Diagram of animal cell in metaphase.

7.4d

Mitosis, animal; anaphase

30082

Diagram of animal cell in anaphase.

7.4e

Mitosis, animal; telophase

30083

Diagram of animal cell in telophase.

7.5

Mitosis, disrupted; ls

20218

X-ray disrupted mitosis in onion root tip cell. Broken pieces of chromosomes with no centromere can't attach to spindle fibers.

7.11

Cancer cells

21754

Human cancer cells showing ruffling edges. SEM x2000 mag.

7.11

Cervical carcinoma cells

21869

In culture, x6.

7.12

Cancer development; diagram

2 frame sequence

22650

Diagrams suggest a genetic basis for cancer.

7.14

Cancer summary

4 frame sequence

30194

A. proto-oncogene exists in a normal cell, B. the environment and viruses affect it, C. the gene expresses itself as a cancerous cell, D. and the immune cell tries to fight off the cancerous cell.

Chromosome mutations; translocation

30141

In translocation entire sections of genes come to lie on different chromosomes.

8.1

Reproductive system, male

22570

diagram

8.2

Reproductive system, female

22575

Diagram showing cross section of female reproductive anatomy.

Step <||	Play >	Step ||>

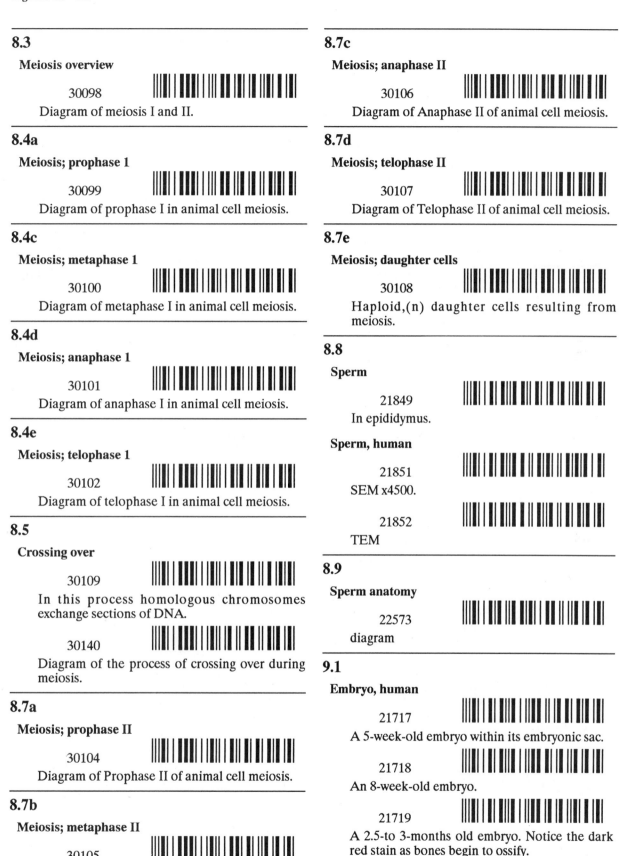

8.3

Meiosis overview

30098

Diagram of meiosis I and II.

8.4a

Meiosis; prophase 1

30099

Diagram of prophase I in animal cell meiosis.

8.4c

Meiosis; metaphase 1

30100

Diagram of metaphase I in animal cell meiosis.

8.4d

Meiosis; anaphase 1

30101

Diagram of anaphase I in animal cell meiosis.

8.4e

Meiosis; telophase 1

30102

Diagram of telophase I in animal cell meiosis.

8.5

Crossing over

30109

In this process homologous chromosomes exchange sections of DNA.

30140

Diagram of the process of crossing over during meiosis.

8.7a

Meiosis; prophase II

30104

Diagram of Prophase II of animal cell meiosis.

8.7b

Meiosis; metaphase II

30105

Diagram of Metaphase II of animal cell meiosis.

8.7c

Meiosis; anaphase II

30106

Diagram of Anaphase II of animal cell meiosis.

8.7d

Meiosis; telophase II

30107

Diagram of Telophase II of animal cell meiosis.

8.7e

Meiosis; daughter cells

30108

Haploid,(n) daughter cells resulting from meiosis.

8.8

Sperm

21849

In epididymus.

Sperm, human

21851

SEM x4500.

21852

TEM

8.9

Sperm anatomy

22573

diagram

9.1

Embryo, human

21717

A 5-week-old embryo within its embryonic sac.

21718

An 8-week-old embryo.

21719

A 2.5-to 3-months old embryo. Notice the dark red stain as bones begin to ossify.

21720

A 10-week-old embryo. Notice the dark red stain as bones begin to ossify.

21721
An 18-week-old embryo.

21722
A vestigial tail taken from a human embryo. This indicates that we must still have the genes for this structure but they are usually not expressed.

7 frame sequence

22452
Shows embryo with placenta, inside uterus.

9.2b

Sperm, model

22334
Sperm model shows its parts: head with acrosome and nucleus, mid-piece, tail. Egg models (to the right) show process of mitosis in egg after penetration of sperm into egg.

9.2c

Sperm, human; on Egg

21855
SEM x3000.

9.4

Blastulation; models

22335
Models show blastulation from 2-cell stage, through morula stage, to the hollow blastula.

9.7

Placenta and egg implantation; model

22339

22340
Model shows development of chorionic villi and their penetration into maternal tissues to form the placenta.

9.9

Germ layers, frog

9 frame sequence

2113
The three tissue layers are color coded, blue is ectoderm, red is mesoderm, yellow is endoderm.

9.11

Placenta and umbilical cord, human

21877

Placenta; model

22341
Placenta and embryo; umbilical cord has external covering removed to show umbilical vein and arteries.

9.16

Fetus and umbilical cord

3 frame sequence

22609
A-E show stages of early development of fetus from implantation to formation of the umbilical cord.

10.2

Sperm, human

21851
SEM x4500.

21852
TEM

11.2

Genetics, pea phenotypes

30024
Pea phenotypes with which Mendel experimented.

11.3, 11.4, 11.7

Genetics, monohybrid cross

30028

11.5, 11.10

Genetics; Punnett square

30030

11.7

Genetics; recombination

30023
Mendelian ratio 3:1.

Step <||	Play >	Step ||>

Table 11.3

Genetics; multiple alleles, rabbit

5 frame sequence

30051

Multiple alleles determine coat color in rabbits. The gene for full-color is dominant over the gene for chinchilla and the gene for white is recessive to both the genes for full-color and chinchilla.

11.13

Genetics; incomplete dominance, chicken

3 frame sequence

30048

Chicken color inheritance demonstrates dominant and recessive traits.

30049

30050

Genetics; incomplete dominance, flower

2 frame sequence

30046

12.2

Chromosome, sex

30137

Diagram showing the way in which the X and Y chromosomes determine the sex of offspring.

12.3

X-linked inheritance

2 frame sequence

30138

A. Punnett square for sex-linked color blindness gene, B. phenotypes resulting from various genotypes involving color blindness.

X-linked pedigree chart

30155

Recessive X-linked genetic disorders such as color blindness.

13.1

DNA, x-ray diffraction patterns

30071

These patterns provided the evidence of the helical form of DNA for Watson and Crick when they were determining the molecular structure of DNA.

13.1a

DNA, bacteriophage

29984
TEM X110,000

Bacteriophage, T4 virus; diagram

13928

13929

13.1b

DNA, xenopus; TEM

Xenopus sp.

30061

TEM

13.1b, 13.7, 13.8

Chromosome; bacterial

30076

The circular bacterial chromosome begins division into individual strands during replication.

13.2

DNA structure

12972

Diagram showing structure overview from helix to complementary strands.

13.3

Nucleotides

4 frame sequence

12968

Diagram comparing purine with pyrimidine bases.

13.4

DNA, base pairing

15 frame sequence

13168

13.5

DNA, supercoiled

29986
TEM X110,000

Chromosome; and DNA

30173

Diagram shows the intensely coiled and compacted structure of DNA within the chromosome.

13.6

DNA replication

30171

DNA replication is semi-conservative, so that replicant helixes are composed of one old strand and one new complementary strand.

13.7, 13.8

DNA; replication animation

2.8 second play sequence.

13084- 13167

13.10

DNA; transcription

30182

Diagram, showing how the DNA unzips and one complementary strand of mRNA is synthesized on one of the DNA strands.

13.11, 13.17

Ribosomes; protein synthesis

3 frame sequence

13202

These are ribosomes on mRNA, polysomes.

13.14

Central dogma of biology

30181

DNA codes for messenger RNA, whose code is read 3 bases at a time by the transfer RNA's to sequence the amino acids in forming a protein strand.

13.16

Protein synthesis

5 frame sequence

13206

Protein synthesis animation

15.0 second play sequence.

13458- 13907

From the nucleus, a strand of messenger RNA passes into the cytoplasm. During protein synthesis the ribosome moves along the mRNA strand. Transfer RNA brings amino acids to the ribosome. Bonding occurs, and the protein chain grows.

14.1

Down's Syndrome; child

29990

Note wide spaced eyes due to broad saddle nose, low set dysplastic ears, epicanthalskin folds and slanted palpebral fissures of eyes.

Down's Syndrome; karyotype

29989

Trisomy 21, Male Karyotype.

14.2

Chromosome; karyotyping

30067

These human chromosomes are photographed; then cut out; matched by size, banding pattern, and centromere location and pasted in matched pairs so that any abnormalities can be easily seen.

30068

Matched pairs of human chromosomes making up the karyotype, matched by size, banding pattern, and centromere location and arranged by size. Shows 22 pairs of autosomes plus both the female and male sex chromosome combinations. Normal pattern.

30069

Female human karyotype.

14.2d

Chromosome mutations; inversion

30143

Here a section of chromosome becomes flipped upside down.

Chromosome mutations; deletion

30142

Here a segment is lost entirely.

Step < ||	**Play >**	**Step || >**

14.2e

Chromosome mutations; duplication

30144

Here a section of chromosome is repeated.

15.1

Pedigree chart

2 frame sequence

30152

A. pedigree for autosomal recessive genetic disorders, B. autosomal dominant genetic disorders.

16.2

Neuron anatomy, interneuron

22506

diagram

Neuron anatomy, motor

22505

diagram

Neuron anatomy, sensory

22504

Diagram showing parts of sensory neuron.

16.6

Nerve fiber

21778

Myelinated and non-myelinated axons, TEM, x47,000.

Nerves, Myelinated

21784

16.8

Synapse anatomy

6 frame sequence

22507

A. Full diagram, B. cell body, C. close-up, D. synapse, E. close-up, F. synaptic vesicles.

17.2

Spinal cord, human

21775
Thoracic.

17.3

Nervous system, autonomic

11 frame sequence

22514

A. Diagram of all organs which are innervated, B-K detail each organ's attachment to the parasympathetic and sympathetic N.S.

17.4

Reflex arc

22513

diagram

17.5

Brain, human; model

22290

Frontal view of both hemispheres of a right-brained person showing the central sulcus separating the right and left portions of the prefrontal, and premotor areas. Position of brain stem and cerebellum are visible.

17.7, 17.8, 17.9

Brain, human; model

22283

Right hemisphere, functional areas, (right-brained person): purple-striped sensory cortex with anatomical areas; red-left=hearing input,right=interpret; orange-motor cortex; Right prefrontal for creativity, spatial perception, intuition, art & music.

18.3

Taste buds

21616

Low mag. (trichrome stain) Section through circumvallate papillae of rat tongue showing stratified squamous epithelium (purple tissue on right) and underlying connective tissue.

21617

Hi mag. (gold toned) Three barrel-shaped taste buds project up through the stratified squamous epithelium. The sensory cells within each taste bud are very long and have tapered apical ends.

18.7

Nerve, optic

21740

Cross section of the eye showing the place where the optic nerve attaches to the retina. This place has no rods and cones which results in a blind spot.

Eye; retina

21741

Cross section of the retina.

18.8

Eye, human, model

22305

Eyeball with outside sclera removed to show choroid layer. On choroid surface, diagram of microscopic section of retina: rod and cone layer, bipolar cell layer, ganglionic cell layer, optic nerve fiber layer.

18.11

Eye, human

21737

Large pupil (dilated) as it would appear in the dark.

21738

Small pupil constricted as it appears in bright light.

21739

Cross section of the eye.

Ear, human

21743

Horizontal section through entire external, middle and inner ear. 1.3 mag.

18.13

Ear, human; cochlear canal

21745

Section of ear showing the Cochlear canal, the auditory organ. x27 mag.

18.15

Ear, human; cochlea

21746

Section of ear showing the Cochlea. x4 mag.

18.19

Ear, human, model

22311

Semicircular canals on the left, the cochlea in the middle with auditory nerve.

19.1

Hormone crystals, progesterone

21863

Progesterone, x125.

19.2

Endocrine glands

3 frame sequence

22556

A. unlabeled, B. labeled, C. quiz letters

19.4

Hormones, protein

22559

Method of action of protein hormones is explained.

19.5

Hormones, steroid

22560

Method of action of steroid hormones is explained.

19.7

Pituitary gland, human

21838

x64.

19.8

Hypothalamic-pituitary hormones

22574

Shows the way in which the anterior pituitary controls sperm production.

2 frame sequence

22584

A. overview, B. FSH and LH effect on developing follicle.

19.8, 19.11

Pituitary, action of anterior

22563

diagram

Pituitary, action of posterior

22561

diagram

19.13

Thyroid gland

21672

Med. mag (H+E stain) Thyroid gland of mammal. Note the cells secrete a precursor to the hormone into the follicular cavity before it is released into the circulatory system. Like the above endocrine glands you can see many small blood vessels.

Thyroid & parathyroid glands

21837

19.15

Adrenal gland

21834

Entire gland, x1.6.

19.17

Pancreas

22568

Diagram shows gross anatomy of the kidneys, pancreas and gallbladder.

Pancreas

21647

Section of pancreas showing acinar tissue and islets of Langerhans.

19.19

Ovarian cycle

7 frame sequence

22576

A. Overview, B. three phases, C. developing follicle, D. mature follicle, E. ovulation, F. early corpus luteum, G. regressive corpus luteum.

Reading 20.1

Cat; skeleton

22219

20.5

Bone, dense

21607

Low mag. (india ink stain) Longitudinal section of dense bone demonstrates the haversian canal (dark line on left) and many small elongated lacunae where osteocytes were located.

20.7, 20.8

Bone; development

21724

Sagital section of developing finger bone of a fetus. Notice the area of elongation.

20.11

Skeleton, human

22267

Front view of whole articulated skeleton.

20.12

Skull, human

22263

Top view, showing sutures.

20.12

Skull, human

22262

Calvaria sectioned, showing the inside of the skull.

20.14

Vertebral column, human

22265

Close-up shows intervertebral discs and articulation points.

20.16

Joint; finger

21608

Low mag. (trichrome stain) Longitudinal section through a finger joint of a monkey. Note one bone (left side of screen) has an epiphyseal plate. Also note articulating cartilage (hyaline) does not have a perichondrium.

21.1a

Muscle, smooth

21587

Hi mag. (trichrome stain) In longitudinal aspect smooth muscle has long spindle-shaped cells with elongated nuclei.

21.1b

Muscle, cardiac

21583

Hi mag. (trichrome stain) Cardiac muscle has both striations and branched fibers.

21.1c

Muscle, skeletal

21578

Med. mag. (H+E stain) Section through the tongue of a rat showing striated skeletal muscle in both transverse and longitudinal section.

21.9

Skeletal muscle contraction physiology

24779

21.11

Muscle experiment; gastrocnemius

99.5 second play sequence.

24780- 27764

A frog muscle is stimulated and the contractions measured on a chart recorder. Effects of stimulation frequency, strength and loading are studied. Narrative.

22.2

Blood circulation

10.0 second play sequence.

27768- 28068

Blood circulation in a frog tadpole tail. Narration

Circulation, amphibian; diagram

22460

In amphibians,blood returns to three-chambered heart going into right atrium and then ventricle. Blood has returned from gills to left atrium into ventricle where oxygenated and deoxygenated blood mix. Ventricle sends partially oxygenated blood to tissue.

Circulation, birds & mammals; diagram

22461

Blood returns from body to four-chambered heart,going into right atrium and then into right ventricle,which pumps blood under pressure to the lungs. Blood returns to left atrium and into left ventricle, which sends it under pressure to body tissues.

Circulation, fish; diagram

22459

In fish,deoxygenated blood returns from the body to the atrium and then the ventricle of the two-chambered heart. Ventricle sends blood under pressure to the gills, where it is oxygenated. With greatly reduced pressure, blood goes directly to body tissue

Circulation, mammalian; diagram

22462

Diagram of circulatory system shows: heart; right & left lung capillaries; pulmonary artery & vein;head & arm capillaries;anterior & posterior vena cava; gut,liver,kidney,lower trunk and leg capillaries;hepatic & hepatic portal veins;lymph vessel & node.

22.3

Blood cells, red & white

21796

SEM

22.6

Blood cells, white

21802

WBC ingesting sperm in vagina, phagocytosis, x14,000.

Blood cells, basophilic leukocyte

21708

Hi mag. (Wright's stain) The basophilic leukocytes have large bright purple granules that fill the cytoplasm and obscure the nucleus.

Blood cells, lymphocytic leukocyte

21710

Hi mag. (Wright's stain) The lymphocytes are about the same size as a RBC and the thin, crescent-shaped cytoplasm contains no granules.

Figures 22.7 - 22.24

Blood cells, white (monocytes)

21709

Hi mag.(Wright's stain) Monocytes are about three times the size of RBCs. They have large indented nuclei with pale staining cytoplasm. These cells are migratory in the circulating blood and differentiate into various phagocytes upon entering the tissue.

Blood cells, white (neutrophil)

21706

Hi mag. (Wright's stain) Neutrophils have very small neutral staining granules (not distinctly blue or pink) in the cytoplasm.

Blood cells, plasma cell leukocyte

21711

Hi mag. (Wright's stain) A plasma cell is one type of lymphocyte. It has been activated for rapid antibody synthesis. The cytoplasm is a larger portion of the cell volume and appears granular due to the increased volume of rough endoplasmic reticulum.

Blood cells, white (eosinophil)

21707

Hi mag. (Wright's stain) The eosinophilic leukocyte has large bright red granules that fill the cytoplasm and partly obscure the nucleus.

22.7

21794

Red blood cells & platelets in capillary TEM, size comparison.

22.8

Blood clot fibrin

21798

Fibers SEM x2600.

22.11

Capillary

21793

Red blood cells in capillary TEM, x40,000.

22.12

Capillary Exchange; diagram

3 frame sequence

22470

A. Diagram of the parts of the capillary network. B. shows direction of flow of nutrients and waste products. C. Osmotic pressure.

22.13

Veins

21792

Red blood cells in veins, x130.

22.15

Blood pressure and velocity; diagram

22467

The thick line shows the total cross sectional area of the vessels which reaches a maximum in the capillaries. The thin lines show blood velocity and pulse which vary throughout the system.

22.17

Heart, human; model

22314

Whole heart viewed from left ventricle The large blue vessel is the pulmonary artery near the right and left branch.

22.19

Heart muscle contraction physiology

22654

22.20

Heart, beating

10.4 second play sequence.

28370- 28683

A movie shows the operation of the heart valves and sequence of contraction and the nervous control of heart contraction.

22.21

Lymphatic system; diagram

22468

Diagram showing major aspects of lymphatic system:right shoulder vein,left lymph duct, left shoulder vein, thoracic duct, small intestine, lymph nodes. Also is included is diagram of villi enlarged to show location of lacteals.

22.24

Lymph node

21804

L.S.,x5.

23.2

Respiratory system, insect; diagram

22488

Diagram shows anatomy of insect respiratory system: air sacs,spiracles,tracheae and tracheole.

Respiration, aquatic insect; diagram

22356

Diagram of pathway of oxygen and carbon dioxide in an aquatic insect. Oxygen enters spiracle and trachea from an air bubble and diffuses into tissues. Carbon dioxide enters trachea and diffuses into air bubble and then into the water.

23.2d

Respiratory system, arthropod; diagram

22351

Cross section of an arthropod (diagram) through a spiracle to show respiratory organs (spiracle, respiratory siphon, trachea) and diffusion of oxygen into tissues and diffusion of carbon dioxide out through the spiracle.

Reading 23.2

Lung, human

21772

x.s. small, Bronchus.

23.7

Respiratory system, human; diagram

22492

Diagram shows anatomy of respiratory system, with magnification of lung to show alveolus,artery,vein and capillary network.

Trachea Lining

21828

Ciliated pseudostratified columnar epithelium of the trachea. SEM x1200.

23.8

Bronchiole; SEM

21773

23.10

Lung; alveoli; SEM

21774

23.12

Inspiration & expiration, human

3 frame sequence

22489

Diagram shows lungs,rib cage,and diaphragm during inspiration and then expiration.

23.13

Respiratory system, human; diagram

22493

A.Complete system, B. systemic circulation, C. Lung circulation, D. Gas exchange in capillaries, E. Gas exchange in lung detail.

24.2

Centipede; eating a frog

Scolopendra sp.

15163

Eating a frog. From Thailand.

Centipede; feeding on mouse

Scolopendra sp.

15164

Feeding on a mouse. From Thailand.

Snake; eating deer mouse

31293

Copperhead eating.

24.5a

Esophagus

21829

x.s. wall.

24.6

Tooth; developing

21728

Longitudinal section of a developing tooth. x16 mag.

24.7

Stomach

21637

Section through the stomach of amphiuma.

24.8

Gastric glands

Fundic Glands

21623

Med. mag. (H+E stain) Fundic glands of the stomach mucosa showing chief (blue) and parietal (bright pink) cells.

Stomach; fundus

21622

Med. mag. (H+E stain) Section through the fundic region of the stomach of a monkey showing the surface epithelium, gastric pits and fundic glands. This tissue includes only the mucosal layer of the last frame.

24.15

Digestive glands

21628

Hi mag. (trichrome stain) Small intestine of a monkey showing crypts of Lieberkuhn, Paneth cells and enteroendocrine cells.

24.16

Intestinal lining; diagram

22486

Diagram shows the villi, underlying connective tissue, muscle, solitary lymph node.

Intestine, small; villi close-up

22449

This close-up view of the intestinal lining shows the structure of each villus, with columnar epithelial cells on the outside and lacteals (lymph vessels) and capillary network of blood vessels on the inside.

Villi

21813

In small intestine.

24.18

Intestine, large

21629

Low mag. (trichrome stain) Cross section of the colon of monkey demonstrates three tissue divisions (mucosa, submucosa, & muscularis externa). Note the thickened layer of muscularis externa in the intestinal wall.

24.20

Liver, human

21816

Normal, x16.

24.21

Gallstones

21824

Gall bladder wall

21823

x18.

Reading 25.3

Breast feeding

21864

26.1

Eland

Taurotragus oryx

16843

Reading 26.1

Iguana, marine

Amblyrhynchus cristatus

16130

Only marine lizard, unusual species of Galapagos Islands. Dives beneath water to feed on marine algae.

26.8

Excretory organs, human; diagram

22498

Diagram showing all organs involved in excretion.

22499

Schematic of excretion process.

26.9

Kidney, mammal

21821

Long sect.

26.10

Kidney; Bowman's capsule

21661

Hi mag. (trichrome stain) Section of kidney showing detail of renal capsule surrounded by proximal and distal tubules of the nephrons. Note opening from Bowman's space into proximal tubule.

26.11

Kidney; glomeruli

21822

With Podocytes, SEM x10,000.

Kidney; glomerulus

21660

Med. mag. (trichrome stain) Section of human kidney demonstrating renal corpuscles. Note tuft of glomerular capillaries surrounded by the capsule of bowman and the glomerular space.

27.1

Blood cells, white

21802

WBC ingesting sperm in vagina, phagocytosis, x14,000.

27.5

Blood cells, plasma cell leukocyte

21711

Hi mag. (Wright's stain) A plasma cell is one type of lymphocyte. It has been activated for rapid antibody synthesis. The cytoplasm is a larger portion of the cell volume and appears granular due to the increased volume of rough endoplasmic reticulum.

27.9

Thymus gland

21833

C.S., x128.

27.18

Antibodies, monoclonal

21818

Spleen removed from 'nude' mouse with antibodies.

28.1a

Bananas

18744

Bananas (large, planted) and heliconias (in same family-Musaceae, but smaller, native) growing in second growth at forest edge.

28.2a

Coffee

Coffea arabica

18731

Coffee in fruit; characteristically grown at lower-montane elevations in tropical belt; native to Africa.

28.4

Wheat; grains

20387

Harvested grains.

28.7

Corn

20395

Shelled corn at feed plant.

28.10

Rice; workers

Oryza sativa

20530

Workers planting rice in the tropics.

28.12

Sunflower, Kansas

Helianthus annuus

17167

17168

17169

Reading 29.1

Wood, annual ring; xs

20173

Large circles are vessels which grow larger in the Spring when plentiful water is available but attain less size in summer. After summer, growth stops until the following Spring when the availability of water again produces large vessels. x.s.

29.3

Stomata

20171

Dicot leaf epidermis, showing stomates with guard cells.

29.4

Xylem; ts

Fraxinus americana

20164

Xylem with vessels, and fibers in long section and ray fibers in cross section.

29.5

Phloem; sieve plate

Curcurbita sp.

20158

Phloem cells in cross section showing sieve plates. Evident is the difference in size between sieve tube members and companion cells.

29.6

Stem, buttercup; xs

Ranunculus

20192

A section of the herbaceous dicot, Ranunculus, which has a hollow stem. From the top down: epidermis, collenchyma, red, heavy-walled sclerenchyma, phloem, xylem, with cortex on either side of the vascular bundle. No pith.

Stem, corn; xs

20188

Large, thick-walled cells are xylem vessels, smaller thick-walled cells near them are also xylem; the thin-walled, cells in the center are phloem, the 'boxes' are companion cells. Smaller, red walled cells at the upper right are sclerenchyma.

29.7

Plant tendril

17107

29.8

Leaf, grass blade; SEM

20217

Typical monocot vascular bundle, but encircled by bundle sheath cells (with stored photosynthate) of a C4 plant. SEM

Leaves, squirrel corn

Dicentra canadensis

20265

Dicot with deeply cut leaves.

29.9

Leaf, variegated; aluminum plant

Pilea sp.

20269

Dicot variegated leaves.

29.10

Leaf, privet; xs

20208

Shows typical leaf structure, with upper epidermis, palisade mesophyll cells, spongy mesophyll with air spaces and stomates in the lower epidermis. A small vein is seen to the left. x90

29.11

Venus fly trap

Dionaea muscipula

20368

Small trigger hairs visible on inside of trap surface trigger trap to close when touched by insects.

Venus fly trap; shutting

9.0 second play sequence.

40838- 41107

The trigger hairs on this specialized plant leaf pass a signal to close the trap when an insect (usually a fly) walks across. The leaf also has specialized digestive glands to break down the prey into absorbable nutrients for the plant.

29.12a

Roots, Violet

20300

Fibrous roots of a violet.

29.12b

Roots, dandelion

20304

Dandelion tap roots.

29.13

Root tip; model

20737

Long section of the root cap, apical meristem, and growing zones.

29.14

Root hairs

Ranunculus

20181

Shows numerous hairs growing out from root epidermis. x15

29.15

Root, corn; xs

20244

Corn root shows typical monocot root structure: central pith, xylem with large metaxylem elements, phloem, pericycle, endodermis, cortex, exodermis, epidermis.

29.16

Root; starch storage; xs

Ranunculus

20249

Young root stained to show starch granules in the parenchyma cells of the cortex. Tetrarch arrangement of the xylem. x90

29.17

Wood; ts

Tilia glabra

20234

Tangential section shows long section view of xylem elements, cross section view of ray parenchyma. This picture is sideways to the axis of tree growth.

30.1

Wasp, Scoliid, orchid

15621

The flower, Ophrys speculum, attracts scoliid orchid wasps for pseudocopulation.

30.2

Moss life cycle

7 frame sequence

20746

The sporophyte generation of moss grows on top of the gametophyte.

30.3

Gametophytes, fern; wm

20233

Shows gametophyte structure and shape.

30.4

Flower anatomy

14 frame sequence

20782

diagrams showing flower parts, with and without labels.

30.5

Flower fertilization

11 frame sequence

20833

Diagrams showing the growth of the pollen tube and transport of sperm nuclei to polar nuclei and egg cell.

Flower, female gamete

16 frame sequence

20809

diagrams showing development of flower female gametes.

Flower, male gamete

6 frame sequence

20825

diagrams showing development of flower male gametes.

30.6

Hummingbird, Anna's

Calypte anna

16710

Female feeding in garden on cultivated flowers; note gradual maturation of flowers on stalk.

20044

The fragrant flowers of this plant are frequently visited by both bumblebees & hummingbirds. They are scented for bees (birds can't smell) and red for birds (bees can't see red). Eastern USA.

30.8a

Seedling, bean

Phaseolus vulgaris

20315

Dicot seed germinating, with two cotyledons visible.

30.8b

Seeds; germination movie

23.5 second play sequence.

20865- 21570

A time lapse movie showing the germination of a corn seed.

30.10

Seeds; dispersal

17782

Mammalian dispersal of seeds. Seeds entrapped in hairy coat.

Seeds; mule-grab, grapple plant

Harpagophytum sp.

20361

A seed dispersed by grazing animals — it clings to their feet.

Seeds; Jack-go-to-bed-at-noon

Tragopogon porrifolius

20356

Wind-dispersed seeds.

Seeds; limoncocha pods & seeds

20354

Winged seeds for dispersal.

Seeds; hitchhiking

20360

Hitchhiking seeds that are dispersed by mammals.

30.10d

Dandelion

Taraxacum officinale

17109

Seed head.

30.11

Seeds; germinating

20727

These seeds have imbibed water and swollen and just begun to sprout. The little white root tip (hypocotyl) is visible on some.

30.12

Pine, white, female cone

Pinus strobus

20095

A female cone of the white pine is shown here. It is fertilized by pollen from a male cone which is borne by the wind. Eastern USA.

Gymnosperm; life cycle

8 frame sequence

20767

diagram

Pine; megagametophyte, ls

Pinus sp.

20722

Female gametophyte in the ovule of a pine. From the outside in (central right): integument, megasporangium, female gametophyte (pale blue) which contains the archegonium (usually 2, but here one is disintegrated), which contains the egg nucleus.

31.1

Bee, digger

Eucera sp.

15619

The digger bee, Eucera sp., pseudocopulating with the flower Ophrys scolopax (2). From Tunisia.

Sundew

20375

Damselfly trapped by round-leaved sundew.

31.6

Phototropism

3 frame sequence

20860

A. shows growing seedling, B. light is applied from one side and plant without shoot tip doesn't respond, C. excised shoot tips placed on agar block provide growth hormones which in turn cause differential growth in unexposed shoot.

31.7

Geotropism

20325

Trees are growing straight upwards, opposite the direction of gravity, even though the ground surface is sloped. Shoots show negative geotropism.

20326

Even though this tree had to grow sideways to get out from under the shade of the surrounding trees, it is growing upwards, opposite the direction of gravity. Shoots have negative geotropism.

32.2

Mustard

Brassica campestris

20028

Mustard plants contain a variety of highly toxic chemicals in their leaves that protect them from being eaten by both invertebrate (insect) and vertebrate (mammal) herbivores. Europe.

33.3

Tortoise, Galapagos

Geochelone elephantopus

16112

Largest tortoises, subject of evolutionary studies since Darwin's time. Galapagos islands.

33.4

Grand Canyon

30215

Sedimentary layers.

33.9

Sheep, bighorn

Ovis canadensis

16917

Copulation about to occur; note great difference in head adornment.

33.9a

Bowerbird, satin

Ptilonorhynchus violaceus

16743

Bower of the male, constructed of vegetation and filled with conspicuous objects to attract female for mating. Australia.

33.9b

Bird of paradise, emperor

Paradisaea gulielmi

16524

Communal ('lek') displays by male birds of paradise among most spectacular of bird behaviors. This species restricted to Huon Peninsula, New Guinea; calls varied and musical.

34.4

Cheetah

19113

Threatened species; Masai Mara G.R. Kenya.

34.11

Humans hunting; Diorama

30296

Early humans hunting.

35.1

Condor, California

Gymnogyps californianus

16410

Largest wing span of any North American bird; formerly restricted to California, now extinct in wild, with small population in zoos.

35.2

Fossil; insect

30265

Dinosaur bones

30268

Bones being excavated.

Dinosaur tracks

30288

Tracks found in Tuba City Az.

Reading 35.2

Dinosaur, duckbill; juvenile

30269

Dinosaur; eggs

30286

30287

35.9

Evolutionary tree; primate

30308

Showing DNA differences over past 40 million years in primates.

35.16

Dinosaur

Allosaurus

30281

Jurassic period.

Heterodontosaurus

30276

A small ornithopod dinosaur.

36.1

Squirrel, Columbian ground

Spermophilus columbianus

16757

Ground squirrels retire to burrows at night, for hibernation, and when danger threatens.

Reading 36.1

Mockingbird, Northern

Mimus polyglottos

16731

Male able to imitate songs of almost any bird in area; biological significance not understood.

36.11

Owl, long-eared, head

Asio otus

16489

Close-up, showing large, night-adapted eyes, raptorial beak, and head tufts ('ears') effective in camouflage.

37.1

Ant, fire

Solenopsis sp.

15553

Fire ants and fly larva.

Reading 37.2

Midshipman, plainfin

Porichthys notatus

15721

15938

Bee, carpenter; nest

Ceratina sp.

15564

Nest of carpenter bee in a twig, showing adult, pollen storage area and larva.

37.3

Caterpillar, eastern tent

Malacosoma americana

14859

Eastern tent caterpillar.

37.5

Ostrich

Struthio camelus

16341

Females and immatures. Resident widely in Africa.

37.10

Lobster, northern

Homarus americanus

14812

37.11

Elephant, African; with calf

Loxodonta africana

19128

Samburu G.R., Kenya.

37.12

Capuchin

Cebus nigrivittatus

16875

From South America.

37.13

Bee dance

4 frame sequence

31387

A. dance when food is between sun and hive, B. when hive is between food and sun, C-D. when food is at some angle between hive and sun.

10.0 second play sequence.

52120- 52419

The very precise pattern of movement and beating of the wings describes to the interested on-looking bees the exact location of the food source.

37.15

Bee eater, blue-breasted

Merops variegatus

16505

Tall grasslands or marshes of central Africa, feeding primarily on bees.

37.18

Gorillas

31267

Mt. gorillas ascending trail; Volcano NP, Rwanda.

38.2

Population, human

19205

Diagram of population growth from 4OOO B.C. until present.

Reading 38.1

16653

Flocking birds in flight are evenly spaced to facilitate flock coordination.

38.6

Mt. St. Helens and Spirit Lake

18962

What is left of Spirit Lake after the devastating eruption of Mt. St.Helens in May, 1980.

38.10

Cheetah

16938

Carrying young Grant's gazelle. This fastest land animal can run down any prey in straight line but misses many artful dodgers.

39.1

Food web, in forest ecosystem

19225

Diagram of food web in forest ecosystem.

Food web, in pond ecosystem

19226

Diagram of food web in pond ecosystem.

39.4

Moose; browsing

Alces americana

14.2 second play sequence.

43753- 44180

Moose will reach up and grab branches of saplings and selectively nibble all the tender shoots off. This provides food for the hungry herbivore but also devastates the tree.

39.7

Energy dissipation

12992

Energy from the sun is passed from the producers to level after level of consumers with waste at each step.

39.8

Biological magnification

19229

Diagram showing how certain materials can become more concentrated within a food chain.

39.9

Carbon cycle

3 frame sequence

19233

A. Complete diagram, B. carbon fixation, C. oxidation of carbon compounds.

Biogeochemical cycle

19232

Some chemicals pass from reservoirs in and out of the biological system.

39.10

Nitrogen cycle

5 frame sequence

19236

A. Background, B. nitrogen fixation, C. nitrification, D. denitrification, E. aquatic nitrogen fixation.

39.11

Phosphorous cycle

19241

Phosphorous would not cycle were it not for birds, migratory fish and geology.

39.12

Succession, secondary

19219

Diagram showing process of plant succession from grass, to shrub, to mature forest.

39.13

Succession, secondary; bog

18997

This bog is slowly filling in, and specialized pioneer species are invading the edge formed between the bog and the forest.

39.14

Forest, alpine

18149

Forest, deciduous and coniferous

17733

Mixed deciduous and coniferous forest in the fall.

Grassland and oak woodland

17968

Forest, tropical rain

18732

Palm seedlings common in understory of many Neotropical rain forests.

Polar pack ice

18841

Pack ice in Bismarck Strait breaking up in spring (late December).

Biomes, temperature & rainfall effects

19267

Chart shows distribution of biomes according to interaction of temperature and rainfall.

39.15

Lake temperature profiles

19209

Diagrams show the characteristic temperatures of a lake in each season.

Reading 39.1 fig. 2

Fire; recovery, Yellowstone Park

18985

Notice the green, growing vegetation along the stream.

40.5

Acid rain, scientific monitoring

19017

Scientist is monitoring acid rain on the shore of affected lake.

Lake; acid rain- affected

19016

It's still scenic, but nearly sterile from low ph.

40.6

Greenhouse effect

2 frame sequence

19248

A. Diagram, B. rise in global temperatures with increased production of carbon dioxide.

40.7

Forest, tropical rain; clearing

18736

Much of the tropical rainforest areas have been cleared by small peasant farmers, seeking an area in which to grow food. Unfortunately, once the forest has been cleared, the thin soil layer only supports crops for a few years.